RETINAL COMPUTATION

RETINAL COMPUTATION

GREGORY WILLIAM SCHWARTZ

Departments of Ophthalmology and Physiology, Feinberg School of Medicine, Northwestern University, Chicago, IL, United States

Department of Neurobiology, Weinberg College of Arts and Sciences, Northwestern University, Chicago, IL, United States

Northwestern Interdepartmental Neuroscience Program (NUIN), Northwestern University, Chicago, IL, United States

ELSEVIER

ACADEMIC PRESS

An imprint of Elsevier

Academic Press is an imprint of Elsevier
125 London Wall, London EC2Y 5AS, United Kingdom
525 B Street, Suite 1650, San Diego, CA 92101, United States
50 Hampshire Street, 5th Floor, Cambridge, MA 02139, United States
The Boulevard, Langford Lane, Kidlington, Oxford OX5 1GB, United Kingdom

Library of Congress Cataloging-in-Publication Data
A catalog record for this book is available from the Library of Congress

British Library Cataloguing-in-Publication Data
A catalogue record for this book is available from the British Library

ISBN 978-0-12-819896-4

For information on all Academic Press publications
visit our website at https://www.elsevier.com/books-and-journals

Publisher: Nikki Levy
Acquisitions Editor: Joslyn Chaiprasert-Paguio
Editorial Project Manager: Hilary Carr
Production Project Manager: Punithavathy Govindaradjane
Cover Image Artist: Isabel Romero Calvo

Typeset by SPi Global, India

Contents

Contributors

Gautam B. Awatramani
Biology Department, University of Victoria, Victoria, BC, Canada

Devon Greer
Northwestern Interdepartmental Neuroscience Program (NUIN), Northwestern University, Chicago, IL, United States

Jared Levine
Laboratory of Neurogenetics of Language, Rockefeller University, New York, NY, United States

Benjamin L. Murphy-Baum
Biology Department, University of Victoria, Victoria, BC, Canada

David Swygart
Northwestern Interdepartmental Neuroscience Program (NUIN), Northwestern University, Chicago, IL, United States

Introduction

When I told a friend of mine that I was writing a book on retinal computation, his sincere reply was, "I didn't know that the retina does math!" My friend is a successful and intelligent engineer, but not an academic scientist. His reply got me thinking about how I can explain succinctly that the retina does, in fact, "do math," and why I find *how* and *why* it does that math important enough to write a book about it.

Some of you, like my friend, are approaching the subject of retinal computation with little prior knowledge. Others were making foundational discoveries before I was born, let alone when I entered the field less than 20 years ago. But I hope that everyone who reads this book will learn something new, as I do every day studying the retina, and share some of my fascination and wonder at its amazing ability to make us feel as if we have a high-resolution video camera with a pipeline right to conscious perception when what we really have is a bunch of noisy neurons doing math.

At my public PhD defense, in front of an audience mostly of non-neuroscientists, I tried to sum it up by saying that, "The retina is not a video camera; it is the world's most advanced image-processing machine." But that is not quite right. The camera on my iPhone has more pixels, the photomultiplier on my microscope has far superior temporal resolution, and algorithms used by security agencies around the world can recognize millions of faces, far exceeding the abilities of any human. What makes the retina powerful is not easily quantified as the *smallest*, the *fastest*, or the *most*.

The word "flexibility" comes closer to capturing the right concept. Flexibility in the brain is well documented. It is the property that allows us to learn, adapt, and grow over the course of weeks, months, and years. Synaptic connectivity changes every time we learn a new skill, and entire regions of the brain can be rewired and repurposed after injury. But in the retina, flexibility means something different. Synaptic plasticity and rewiring are minimal. The retina's circuits must represent all the relevant information in the visual world—brightness, contrast, color, motion, texture, threat, and more—all simultaneously in parallel. The same circuits must work in the moonlight and under the noon sun when there are a billion times as many photons. They must enable us to read street signs with small letters while at the same time judging the speed of nearby vehicles in the periphery.

The biological substrate for all that computation is highly constrained to be thin, transparent, and must transmit all its information through the optic nerve's bottleneck. Human-made imaging devices do not come close to that kind of computational flexibility despite being made of faster and more reliable electronic components than the retina's biological components. And Apple has not made an iPhone camera with an integrated image processor that is transparent and less than 200 μm thick, at least not yet!

Purpose and scope

This is not a textbook designed to summarize the enormous body of literature on the anatomy and physiology of the retina. Several other excellent books serve that purpose (Dowling, 1987; Kolb et al., 2011). The perspective of this book is that of computation. Each chapter covers a different computation and broadly asks four questions.

1. Why did the retina evolve to perform this computation?
2. How do the signals transmitted to the brain carry information about this particular feature of the visual world?
3. What mechanisms in retinal circuits implement the computation?
4. How are the signals from the retina for this computation used to drive behavior?

The degree to which the answers to each of these questions are known varies widely, so in some chapters, instead of providing answers, I will pose more specific versions of these questions as topics for future work.

For graduate courses

I assume that the reader has a basic understanding of cellular and systems neuroscience and retinal anatomy and physiology at the early graduate level. While I hope this book will appeal to readers at many different career stages, from advanced undergraduates to established investigators, and in disparate fields, from psychophysics to sensory biology outside vision, I wrote it with neuroscience and vision science graduate students in mind. The chapters are independent modules, and each chapter contains discussion questions from one to three landmark papers that could be assigned in a seminar-style course. Each chapter also contains a programming exercise in MATLAB® for students who want to develop an intuition for retinal computation by implementing and exploring simulations. The content could fill a full-semester graduate course, or individual chapters could be incorporated into a broader course on sensory computation or visual processing in the retina and the brain.

Model species

Retinal neurobiology is somewhat unique in sensory neuroscience in its use of many different model species. Prominent vertebrate species in the field over the years have included salamander, goldfish, mudpuppy, frog, zebrafish, chicken, mouse, rabbit, guinea pig, ground squirrel, ferret, tree shrew, macaque, and marmoset. While a proper comparative study including the visual ecology of each of these animals is of great importance (Baden et al., 2020) and fascinating to me personally, it is unfortunately beyond the scope of this book. There is also an entire field of study on invertebrate vision, including a detailed understanding of several retinal computations that I will not cover here—perhaps in a future edition! My own work has been on salamander, mouse, and nonhuman primate retina, and many of the examples will be from these species. In particular, mouse has become the model of choice for retinal neurobiology over the last decade because of the vast knowledge of cell types and genetic tools available. Fortunately, where they have been studied, most of the principles of retinal computation are well conserved among vertebrates. I will highlight some exceptions to that rule, like in color vision, where the visual niche of different animals has placed very different constraints on the computation.

Guided by the scientific community that welcomed me

I have thoroughly enjoyed my career so far as a retinal neuroscientist in large part because of the extraordinarily thoughtful and caring community of scientists that makes up our discipline. I have tried to thank many of them personally, but I want to thank them collectively here. At every stage in my journey, including the current one, I have benefited from the insights, wisdom, advice, collegiality, openness, and trust of many of the people who make up the foundation of modern retinal neuroscience. Table 1 lists my colleagues who formally reviewed chapters of this book, but the list of additional people with whom I discussed aspects of what I wrote would comprise the majority of the field. Part of the joy of these professional relationships, of course, is that there are many voices to tell me when I am wrong or what I

Table 1 Chapter reviewers.

Chapter	Individuals	University
1. Photon detection	Petri Ala-Laurila	Aalto University and University of Helsinki
2. Luminance adaptation	Felice Dunn	University of California San Francisco
3. Absolute luminance detection	Tiffany Schmidt	Northwestern University
4. Contrast sensitivity	Michael Freed	University of Pennsylvania
5. Contrast adaptation and sensitization	Michael Manookin	University of Washington
6. Contrast suppression	Ben Sivyer	Oregon Health & Science University
7. Texture sensitivity	Timothy Golllisch	University of Göttingen
10. Orientation selectivity	Stewart Bloomfield	SUNY College of Optometry
12. Object motion sensitivity	Steven Baccus	Stanford University
13. Motion anticipation	Michael J. Berry II	Princeton University
14. Threat detection	Thomas Münch	University of Tübingen
16. Color processing	Thomas Euler and Katrin Franke	University of Tübingen

missed. And our understanding of retinal computation is evolving so rapidly that I predict, and hope, that parts of this book will be out of date in the next decade. For both of these reasons, I encourage feedback from readers. I am already taking notes for suggestions for future editions.

Acknowledgments

I thank everyone on the Elsevier team who gave me this opportunity that turned into an extremely rewarding way to spend a largely "locked-down" 2020 and Isabel Romero Calvo for the beautiful chapter title and cover illustrations. I also thank my formal mentors in the field—Michael Berry and Fred Rieke—as well as the many informal mentors and colleagues who have welcomed me into retinal neuroscience and have created an intellectually exciting, open, and supportive community. My colleagues who were kind enough to review chapters are listed in Table 1, but many others contributed through conversations. Some of the chapters were cowritten with major contributions from my trainees who are credited on those chapters, and Chapter 11 (Direction selectivity) was primarily written by my colleagues Benjamin L. Murphy-Baum and Gautam Awatramani. These contributions were vital and much appreciated. I am grateful to the members of my lab who were endlessly patient with me in accommodating this time commitment and to my department chair, Nicholas Volpe, for his tremendous support and belief in me. Finally, and most importantly, I thank my family—Sarah, Nathan, and Erica—for their love, support, and encouragement throughout this journey.

References

Baden, T., Euler, T., and Berens, P. (2020). Understanding the retinal basis of vision across species. Nature Reviews Neuroscience *21*(1), 5–20.

Dowling, J.E. (1987). The Retina: An Approachable Part of the Brain (Cambridge: Harvard University Press). https://doi.org/10.1016/0092-8674(88)90378-9.

Kolb, H., Nelson, R., Fernandez, E., and Jones, B.W. (2011). Webvision—The Organization of the Retina and Visual System. https://webvision.med.utah.edu/.

PART 1

Luminance

In the dimmest conditions, the visual world is no longer continuous but is instead made up of discrete, sparse photons. Nocturnal hunters like owls rely on their keen vision in these extreme conditions. Remarkably, retinal circuits are optimized to detect sparse arrivals of single photons amidst the dark and mysterious background.

CHAPTER 1

Photon detection

Gregory William Schwartz[a,b,c]
[a]Departments of Ophthalmology and Physiology, Feinberg School of Medicine, Northwestern University, Chicago, IL, United States
[b]Department of Neurobiology, Weinberg College of Arts and Sciences, Northwestern University, Chicago, IL, United States
[c]Northwestern Interdepartmental Neuroscience Program (NUIN), Northwestern University, Chicago, IL, United States

Photon detection can be considered the most elementary retinal computation, but as we will see, achieving exquisite sensitivity to dim lights using the hardware of biology is no simple task; it requires specializations from molecules to circuits to behavior. Amazingly, our ability to detect light approaches the physical limits set by its quantization into discrete photons. This is one of the most extraordinary examples in biology of evolution's power to optimize computation. The adaptive advantage of maximizing photon detection is clear; an animal that can detect the fewest photons gains information about the world that remains invisible to its predators, prey, and competitors.

As one of the best-studied retinal computations, much is known about the components of photon detection from the level of the molecules involved in phototransduction up to the neurons and synapses of the downstream circuit. Decades of behavioral studies, including some brilliant classical works, have made quantitative links between photons captured by the retina and perception. The long and accomplished history of the study of photon detection makes summarizing it in a single chapter difficult, and I will first point the reader to several reviews by some of the researchers who made seminal contributions to the field (Donner, 1992; Field et al., 2005; Kiani et al., 2020; Nelson, 2017; Rieke and Baylor, 1998b).

My goal is not to cover all the details of the steps between photon arrival and retinal ganglion cell (RGC) spikes, but rather to examine the signal and noise sources at several processing stages and discuss the neural mechanisms that enable the signal to be amplified and detected over the noise. I will bookend this discussion of neural mechanisms with accounts of the behavioral literature. Insights from behavioral studies in humans provided the analytical framework for studying photon detection decades before electrophysiological measurements in the retina caught up. Our contemporary understanding

Retinal Computation
https://doi.org/10.1016/B978-0-12-819896-4.00002-0

of photon detection in the retina has inspired a new set of behavioral questions that are increasingly accessible with the rapid advance of technologies in animal models.

How many photons does it take to create a percept?

This question has captivated people for over a century. Only two years after Einstein proposed that light is made up of quanta, the first psychophysics experiments claimed that humans could detect 50 quanta arriving at the cornea (von Kries and Eyster, 1907). But it was not until 1942 that Hecht, Schaler, and Pirenne made a breakthrough, demonstrating that as few as 5 photons absorbed by the retina can drive perception and that rods must be single-photon detectors (Hecht et al., 1942).

The experimental paradigm used by Hecht et al. was essentially the same that had been used in all previous works on the absolute sensitivity of vision, and it is about as simple an experiment as you will find in neuroscience. Observers (in this case, the three authors) sat in a dark room and watched brief, small, very dim flashes, and they reported on each trial whether or not they had seen a flash. What made this work revolutionary was not the data but rather the authors' insight into its analysis.

As in previous studies, Hecht et al. could use optical measurements to estimate the number of photons from each flash reaching the cornea, but they had no way to measure how many of those photons were ultimately absorbed by the rods. So, they took an approach to the problem based on signal detection theory. They supposed that the observer acted like a perfect threshold detector, responding that he had seen a flash when the number of counted photons exceeded a fixed threshold. The emission of photons from a light source is a Poisson process, so even with identical settings on their lightbulb, the number of photons reaching the observer, or his rods, should follow a Poisson distribution; the mean photon count should be equal to its variance. As the number of photons increases, the Poisson distribution becomes more and more Gaussian, so its shape varies very little. In the limit of small counts, however, the shape of the Poisson distribution changes substantially with each increment of its mean, and this fact proved essential.

Returning to the observer as a threshold detector, Hecht et al. proposed that a "seen" flash is one that emits n or more photons, where n is the threshold. For the photon detection task, the probability of seeing a flash is then equivalent to the Poisson cumulative distribution function (CDF). Before they could compare their frequency-of-seeing data to Poisson CDFs, however, Hecht et al. needed one more insight. Still lacking a measurement of the

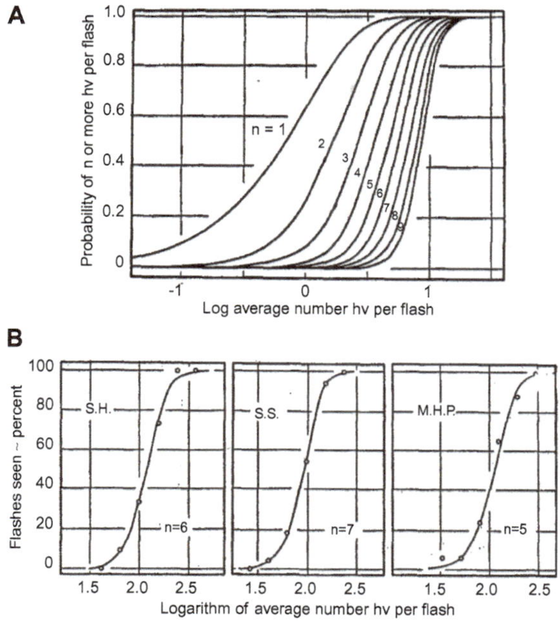

Fig. 1.1 The frequency-of-seeing experiment by Hecht, Schaler, and Pirenne. (**A**) Cumulative Poisson distributions plot the probability of absorbing "*n*" or more photons for $n = 1$–9 as a function of the mean number of photons per flash plotted with a logarithmic x-axis. (**B**) Psychometric curves for the frequency-of-seeing experiment for the three subjects (authors) in the study plotted as in (**A**). The best fit curve from (**A**) is noted for each subject. *(From Hecht et al. (1942).)*

fraction of photons incident on the cornea that was absorbed by the photoreceptors, they could not plot their data on an absolute light intensity axis. Instead, they used a logarithmic x-axis in which changes in this unknown factor became translations along x without distorting the shape of the curve (Fig. 1.1A). In the end, the act of estimating photon detection thresholds was reduced to finding which Poisson CDF with a logarithmic x scale best matched their data. This single-parameter fit yielded $n = 6$, 7, and 5 photons for the three observers (Fig. 1.1B). From anatomical studies of the human retina, the authors estimated that the flashed spots in their experiment covered an area of ~500 rods. Therefore, they concluded that humans can perceive as few as 5 photons among a pool of 500 rods and that rods themselves must be single-photon detectors.

Increment threshold and dark light

The experiment by Hecht, Schlaer, and Pirenne was performed in absolute darkness, but how does the problem of photon detection change in the

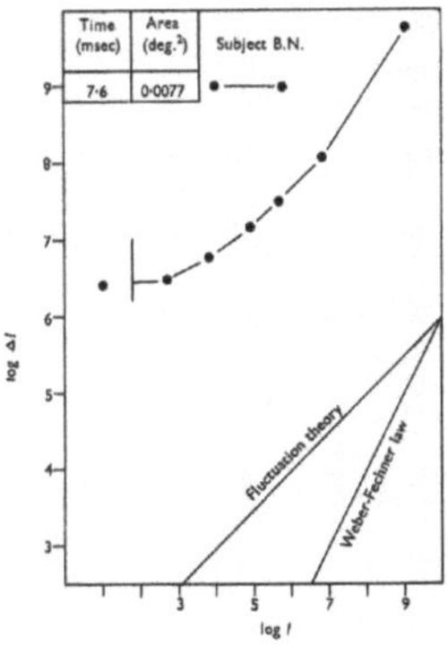

Fig. 1.2 The threshold increment detection curve in human perception. Minimum detectable threshold in luminance (y-axis) versus the mean background luminance (x-axis) both on logarithmic scale. *Labeled lines* in the *bottom right* quadrant show predictions from fluctuation theory (slope = 0.5) and from the Weber-Fechner law (slope = 1.0). *(Modified from Barlow (1957b).)*

presence of background light? In this situation, the subject must detect a flash containing ΔI photons on a background of I photons. Horace Barlow (1957b) performed such an experiment and, like Hecht, Schlaer, and Pirenne, he used the frameworks of signal detection theory and Poisson statistics to analyze the data. Barlow plotted the intensity of the dimmest detectable flash (ΔI) against the intensity of the background (I), both on logarithmic axes (Fig. 1.2). This resulting curve, called the "increment threshold" curve, has three distinct regions when the test spot is small and brief. I will discuss these regions from right to left, beginning with the highest luminance portion of the curve.

At background intensities greater than 7 log units in Barlow's study, the detection threshold rose proportionally with the background. This gives a slope of 1 on a log-log plot, and it is referred to as the Weber-Fechner region. In this luminance regime, the retina is not optimized for photon detection but instead implements luminance adaptation to support contrast constancy. We will examine this computation in the next chapter.

Next, in the middle region of the increment threshold curve lies a regime defined by "fluctuation theory," where the slope is 0.5. This regime has been interpreted as evidence of a fixed signal-to-noise ratio (SNR) threshold in photon detection by the following logic. In the increment detection task, an observer must detect $I + \Delta I$ photons in the spot location as greater than I photons in the background. Recall that SNR is the mean signal divided by the standard deviation, σ, of the noise, in this case, the standard deviation of the background photon count, σ_I. Because photon counts are Poisson, the mean light intensity, I, is always equal to its variance, σ_I^2. Therefore, as I increases, its standard deviation increases as $\sqrt{I}$. Thus, to maintain a fixed

SNR, the increment threshold, ΔI, must increase in proportion to the square root of the background intensity, I, leading to a slope of 0.5 on the log-log increment threshold detection curve.

Finally, the threshold detection curve has a flat region at the dimmest background intensities. In this regime, changing the background light intensity does not affect the detection threshold, and ΔI is the same as the absolute detection limit in darkness, as measured by Hecht, Schlaer, and Pirenne. Barlow argued that the light intensity at which the curve reaches this flat region is a measure of the system's intrinsic biological noise, since extrinsic noise from background light must be comparable to intrinsic noise to affect the threshold. He called this value the "intrinsic retinal noise" or, more poetically, the "dark light." Dark light is an equivalent light intensity, so its units are events per second (or events per rod per second converting spot area to number of rods), but it represents the total noise of the system, from phototransduction all the way to the motor output to initiate a response. Barlow (1957b) calculated a dark light of 1260 events per degree of visual angle per second, equal to ~0.01 events per rod per second. As we will see, this value has a striking correspondence to biological phenomena, even though the experiments were purely behavioral.

Signal and noise for sparse photon detection through the retina

The behavioral work of Hecht, Schaler, and Pirenne, Barlow, and others framed the question of photon detection in a new way as a retinal computation that must preserve sparse signals within the background of biological noise (Koenig and Hofer, 2011; Sakitt, 1972). We now know that the most sensitive RGCs collect from ~10,000 rods and that single-photon responses in only two or three of those rods within an ~50 ms integration period are sufficient to drive the percept of a flash (Ala-Laurila and Rieke, 2014). This remarkable sensitivity creates two constraints that have driven research on the mechanisms of the computation. First, the signal originating from a single photon absorbed by a single rhodopsin molecule must be amplified enormously to create a sufficiently large voltage change in a rod to alter neurotransmitter release. Eventually, through additional amplification, it must create one or more action potentials in an RGC. Second, the entire pathway from photon detection in rods to the motor command to initiate a

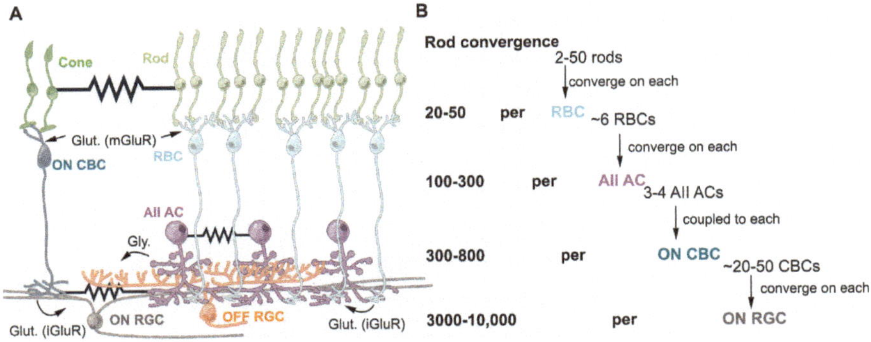

Fig. 1.3 The rod bipolar circuit of the mammalian retina and its convergence. (**A**) Schematic of the rod bipolar circuit in mammalian retina. Chemical synapses marked with arrows and neurotransmitters. Electrical synapses marked with resistors. Abbreviations: *CBC*, cone bipolar cell; *RBC*, rod bipolar cell; *AC*, amacrine cell; *RGC*, retinal ganglion cell; *Gly*, glycine; *Glut*, glutamate; *iGluR*, ionotropic glutamate receptor; *mGluR*, metabotropic glutamate receptor. (**B**) Rod convergence at different levels in the circuit. Note that there is also divergence, so the number of converging rods is not simply the product of convergence of each cell type.

response must be essentially noiseless, such that signals in the tiny proportion of responding neurons are not swamped by noise in the vast majority of cells that carry no signal.

The pathway in the retina for dim signals is well known (Fig. 1.3). Rods are the only photoreceptors sensitive in this regime, and they synapse onto rod bipolar cells (RBCs). Each RBC in the primate or mouse retina collects from ~20 rods. RBCs synapse onto AII amacrine cells, which form an electrically coupled network both with each other and with ON cone bipolar cells (CBCs) (Kolb and Nelson, 1983; Sterling et al., 1988; Tsukamoto et al., 2001). Thus, convergence at the level of the AII-ON CB network is not a straightforward anatomical measurement, but estimates in mice, cats, and primates range from 300 to 800 rods pooled at this level in the circuit (Grimes et al., 2018; Sterling et al., 1988; Tsukamoto et al., 2001). Finally, ON CBCs synapse onto ON RGCs. In the primate retina, ON parasol RGCs are among the most sensitive to dim flashes, and depending on eccentricity, they can collect from up to 10,000 rods (Ala-Laurila and Rieke, 2014; Grimes et al., 2018). There are also OFF circuits that operate in dim conditions converging onto OFF RGCs. Since behavioral measurements described later in this chapter have shown that the ON pathway is responsible for photon detection (Smeds et al., 2019), this circuit will be our focus here.

Amplification in rod phototransduction

Single-photon responses in rods were first measured using the suction pipette technique developed by Baylor, Lamb, and Yau (Baylor et al., 1979a,b) (Fig. 1.4). Not only did these recordings serve as experimental confirmation that rods, indeed, transduce single photons into an electrical response, but they also allowed researchers, for the first time, to quantify both signal and noise at the first stage in vision. Measurements of noise and signal reproducibility that came from these recordings are detailed in the following sections, but first, we will consider the amplification that converts the absorption of a single photon into a macroscopic current in the rod outer segment.

Phototransduction is arguably the best understood G-protein signaling cascade in biology. There are countless reviews on the subject, including

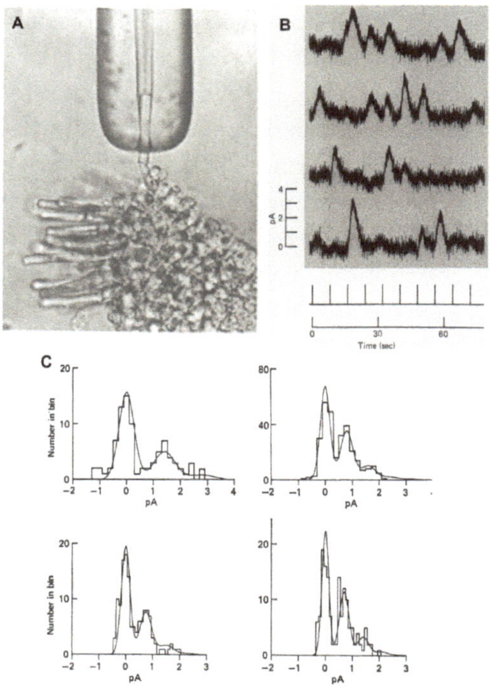

Fig. 1.4 Suction recordings of single-photon responses in rods. (**A**) Image of a rod inside a suction pipette. (**B**) Raw traces of currents from a suction recording showing single-photon events. Flash times are indicated below. (**C**) Distributions of the amplitudes of rod responses at increasing flash strengths *(top left to bottom right)*. Lines on histograms show Poisson fits. *(Modified from (A) Baylor et al. (1979a) and (B, C) Baylor et al. (1979b).)*

detailed equations that model every step in the process with great accuracy. I will cite only a single particularly comprehensive review (Pugh and Lamb, 1993). For our purposes, it will suffice to know the main steps in the pathway and the amplification at each step to appreciate (1) how a single-molecule event can be amplified into a macroscopic current in the rod outer segment and (2) the impact of noise at different steps in the process. Quantitative analysis of phototransduction has been performed in many species, and there are some interesting differences, particularly between mammals and nonmammals, but these are beyond the scope of this chapter. The numbers in the following paragraph are for mouse, and they are generally within a factor of two of measurements in nonhuman primates.

Phototransduction in rods is diagrammed in Fig. 1.5. A single absorbed photon causes the activation of a single molecule of rhodopsin, which is a G-protein-coupled receptor. During its active lifetime, this rhodopsin molecule activates ~12–16 G-protein molecules called transducin (Yue et al., 2019). This is the first amplification step. Transducins then activate the enzyme phosphodiesterase (PDE) in a step with little or no amplification. PDE, however, is a highly efficient enzyme, and each active PDE molecule hydrolyzes ~150 molecules of cyclic guanosine monophosphate (cGMP). This second amplification step brings the total number of cGMP molecules hydrolyzed per rhodopsin isomerization to ~2000 (Arshavsky and Burns, 2014; Yue et al., 2019). Finally, the reduction in the concentration of cGMP closes the cGMP-gated cation channels in the rod membrane, leading to a reduction in the steady inward "dark" current. Each channel that is closed prevents about 10^4 ions from entering the cytoplasm during the time of the

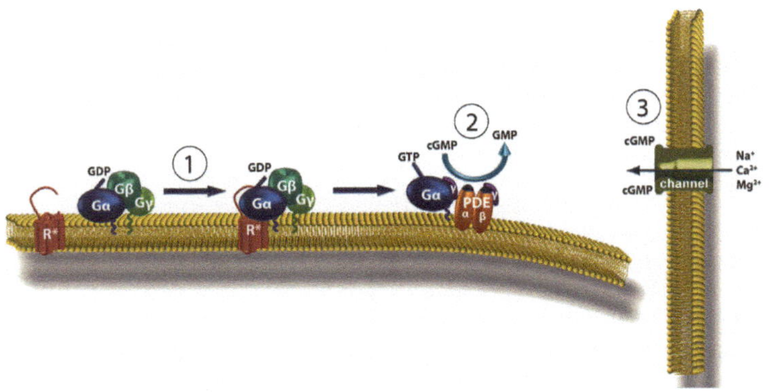

Fig. 1.5 Amplification steps in phototransduction. Diagram of the phototransduction cascade with the three key amplification steps labeled. *(Modified from Arshavsky and Burns (2014).)*

single-photon response. Thus, the total amplification is $\sim 10^7$ ions prevented from entering the rod per photon absorption. This results in the current of ~ 1 pA as measured by suction recordings (Fig. 1.4).

Continuous versus discrete noise in rods

Large amplification in the phototransduction cascade in rods means that noise is amplified along with single-photon responses. There are two distinct types of noise in rods, and they have different implications for downstream circuits. Continuous noise is the ongoing fluctuation of current in the rod outer segment, and it is caused by stochastic fluctuations in the activity of elements of the phototransduction pathway downstream of rhodopsin, chiefly the phosphodiesterase (Baylor et al., 1980, 1984; Rieke and Baylor, 1996). It is smaller in amplitude and higher in frequency than single-photon responses. As we will see in the following sections, these properties are important to how retinal circuitry reduces the impact of continuous noise while retaining single-photon responses.

Discrete noise is caused by spontaneous thermal isomerizations of rhodopsin. Thus, it is indistinguishable from single-photon responses at the level of rods. The spontaneous isomerization rate is ~ 0.004 events per rod per second (Field et al., 2019). Remarkably, this value is within a factor of 2–3 of the dark light calculated by Barlow (1957b) in his behavioral experiment. While this may seem like a small number, consider that among the 10,000 rods in the collecting area of a single RGC, it amounts to ~ 40 events per second. We will see in the following sections how the retinal circuitry deals with this barrage of noise events while still maintaining sensitivity to just a few real single-photon events.

One may wonder why rhodopsin isomerizes spontaneously at such a high rate. It would seem that evolution would have favored the most stable molecule possible to reduce noise. In fact, the rhodopsin molecule is remarkably stable; it undergoes a spontaneous thermal isomerization about once per 700 years (Field et al., 2005)! The reason the spontaneous isomerization rate of rods becomes a relevant noise source is because of the enormous number of rhodopsin molecules per rod: $\sim 10^8$ in humans (Calvert et al., 2001). Thus, the spontaneous thermal isomerization rate per rod is about once per 250 s. So, why did nature not make rhodopsin even more stable? Thermodynamics is the key constraint. The energy from a single photon must reliably cause a conformational change in the molecule. The lower this energy barrier, the more likely it is that random thermal fluctuations will push it over the threshold. The thermodynamic constraint is related to wavelength sensitivity; longer wavelength photons carry less energy, so the opsin molecules that transduce them must be less stable, leading to higher rates of discrete noise (Ala-Laurila et al., 2004; Barlow, 1957a; Luo et al., 2011). The relationship between wavelength and thermal isomerization rate has been proposed as the reason rhodopsin evolved to have a relatively short-wavelength peak sensitivity and why infrared visual pigments are not found in eyes, despite the presence of behaviorally relevant infrared signals in the environment (Barlow, 1957a).

Reproducibility of the single-photon response

Filtering in the retinal circuitry is essential for maintaining single-photon responses and discarding continuous noise. The effectiveness of such an event filter can be improved if the shape of the events is highly reproducible. As discussed earlier, however, phototransduction in rods involves several amplification steps of stochastic biochemical reactions. How can such a process produce the highly reproducible single-photon response currents measured in suction recordings (Fig. 1.4)?

Part of the answer involves the shutoff of rhodopsin (Doan, 2006; Field and Rieke, 2002a; Hamer et al., 2003; Rieke and Baylor, 1998a). As the first amplification step in phototransduction, variance in the active lifetime of rhodopsin gets multiplied in the subsequent steps, potentially causing large variance in the integrated current of the single-photon response. If rhodopsin shutoff were a Poisson process, the integrated single-photon responses would be exponentially distributed. Exponential distributions have a standard deviation equal to their mean, so the ratio of these quantities, called the coefficient of variation (CV), is 1. The CV of integrated current in single-photon responses measured from mouse rods is 0.34, far below the Poisson prediction (Doan et al., 2006). Doan et al. (2006) showed that the increased reproducibility (decreased CV) of the single-photon response is a result of a multistep shutoff process for rhodopsin that involves six phosphorylation events followed by the binding of a protein called arrestin. The authors used mutant mice retaining 0, 1, 2, or 5 of the phosphorylation sites, and they showed that single-photon responses in rods from each of these mouse lines had increased variability with an inverse relationship to the number of remaining sites. Single-photon responses in the mutant with no phosphorylation sites followed a single-step Poisson shutoff with a CV near 1 (Fig. 1.6).

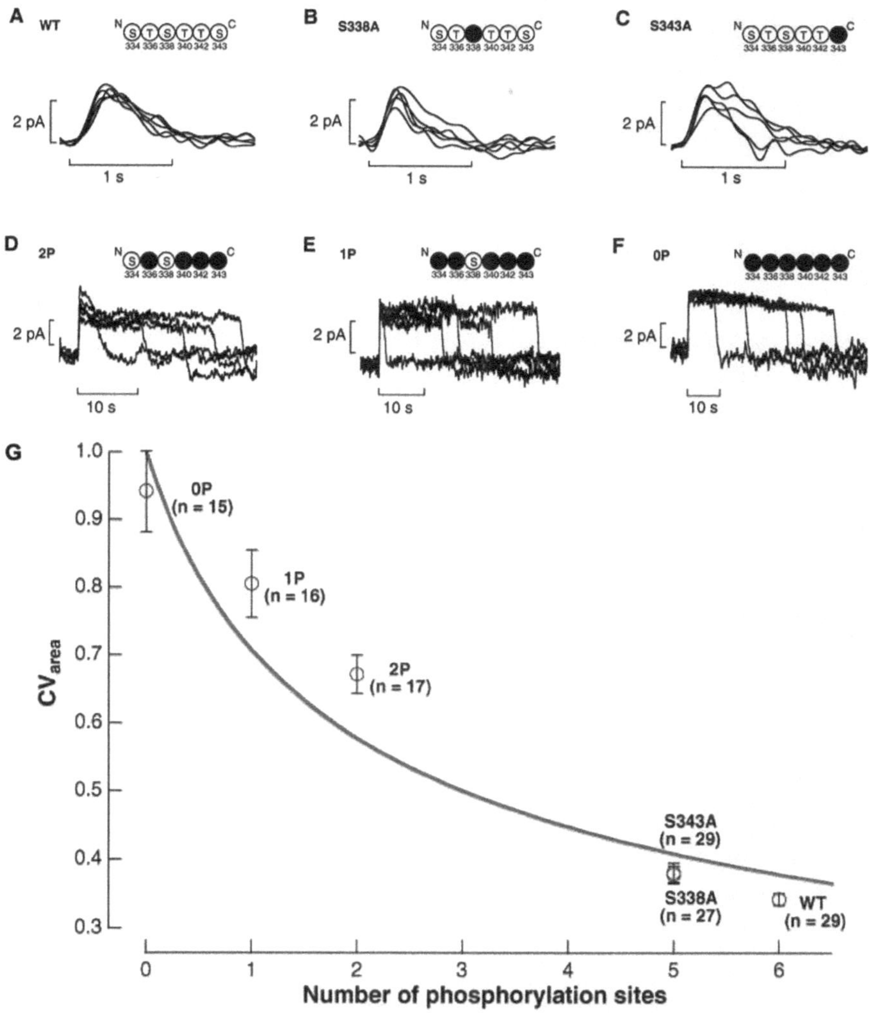

Fig. 1.6 Reproducibility of the single-photon response depends on six phosphorylation sites on rhodopsin. (**A**) *Top*, diagram of the portion of the rhodopsin protein that contains the six phosphorylation sites, either serine (S) or threonine (T) residues. *Bottom*, representative single-photon responses from the WT animals showing high reproducibility. (**B**) Same as (**A**) but for an animal in which the serine at position 338 has been mutated to an alanine (S338A). (**C–F**) Additional mutants with between one and all six sites mutated to alanine. (**G**) Coefficient of variation of single-photon response as a function of the number of remaining rhodopsin phosphorylation sites. Line is $1/\sqrt{7}$ for the seven steps of rhodopsin shutoff (six phosphorylation events plus arrestin binding). *(Modified from Doan et al. (2006).)*

Retinal mechanisms for noise reduction in sparse photon detection

To this point, I have described the signal and noise properties of rods in the single-photon detection regime, and I have set up the problem of reading out a very sparse signal in only a handful of the 10,000 rods in an RGC's collecting area. Now, we will look at the mechanisms within the retinal circuitry that solve the sparse signal detection problem.

Noise reduction at the rod output synapse

The first of these mechanisms is at the synapse between rods and RBCs, and it is tasked with removing continuous noise. Because such a small fraction of rods contains signal while they all contain noise, simple averaging is not viable for RBCs; the noise would swamp the signal. Instead, the synapse implements a thresholding nonlinearity, eliminating a large fraction of the continuous noise at the cost of losing some of the single-photon responses. Field and Rieke (2002b) identified this nonlinearity by measuring responses to flashes of increasing intensity in both mouse rods and RBCs. While average rod responses increased linearly with light intensity, RBCs displayed a supralinear regime for dim flashes (Fig. 1.7A–F). Sampath and Rieke later showed that the nonlinear threshold at the rod to RBC synapse is caused by saturation of glutamate receptors on the RBC in the dark. Small fluctuations in glutamate release from rods are not transmitted to the RBC because the receptors remain in saturation; larger pauses in glutamate release bring the receptors out of saturation and trigger a biochemical cascade in the RBC to amplify the response (Sampath and Rieke, 2004).

Proper positioning of this thresholding nonlinearity is critical to the computation. Too low a threshold would allow some of the continuous noise to pass into the RBC; too high a threshold would eliminate many of the single-photon responses. Field and Rieke (2002b) modeled the rod to RBC synapse with different placements of the threshold and computed the SNR for dim flashes. They found that the optimal placement of the threshold could improve SNR by a factor of 400 over a linear readout and that the measured nonlinearity was remarkably close to its optimal position (Fig. 1.7G–I). Surprisingly, since single-photon responses are so sparse in this regime, the nonlinearity actually removes a sizeable proportion of them—almost half—so that it can effectively eliminate most of the continuous noise.

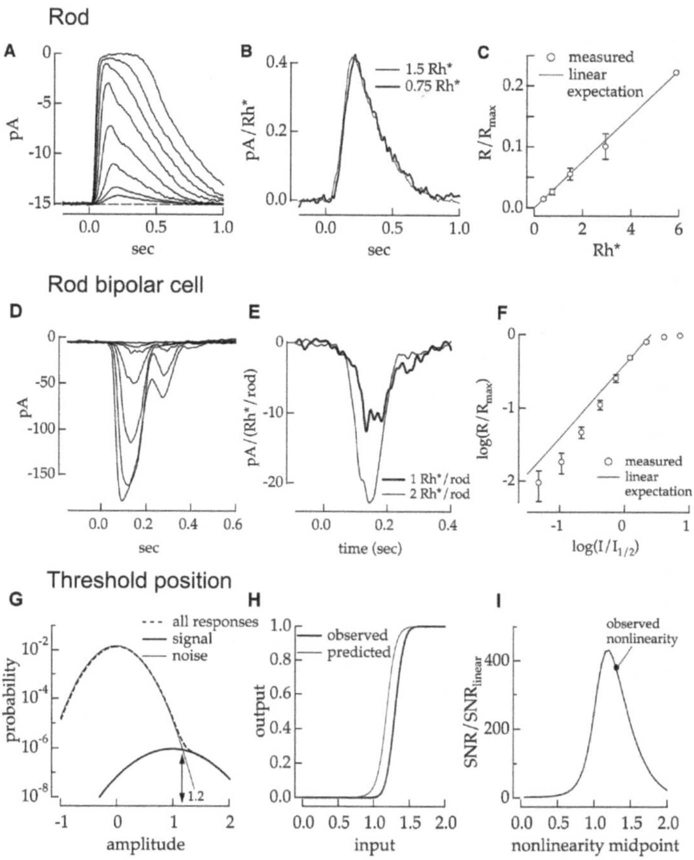

Fig. 1.7 Nonlinear thresholding at the rod-to-rod bipolar cell synapse. (**A**) Rod transduction currents for a family of flashes of increasing intensity. (**B**) Gain (in units of pA per isomerization) for responses at two flash intensities as indicated. (**C**) Rod response as a fraction of maximum versus flash intensity (points). Line indicates linear prediction (fixed gain). (**D–F**) Same as (**A–C**) but for the excitatory input currents measured in rod bipolar cells. Axes in (**F**) are logarithmic and show sublinear responses (below unity line) indicating a thresholding nonlinearity. (**G**) Signal and noise distributions of peak amplitudes of single-photon responses in rods. *Arrow* indicates the intersection point at 1.2 pA. (**H**) Sigmoidal nonlinearity at the rod-to-rod bipolar cell synapse *(thick line)* and the nonlinearity predicted to maximize SNR *(thin line)*. (**I**) SNR fold-improvement for single-photon detection from the linear model as a function of the midpoint of the nonlinearity. Point indicates observed nonlinearity midpoint. *(Modified from Field and Rieke (2002b).)*

An additional threshold in the inner retina

While the nonlinearity at the rod to RBC synapse largely eliminates continuous noise from rods, it cannot filter out spontaneous thermal isomerizations since they have the same amplitude and shape as single-photon responses. Recall that the spontaneous event rate within an ON parasol RGC collecting from 10,000 rods is ~40 events per second. Much of this noise must also be removed if there is any hope in perceiving flashes with just a few extra photons among this pool of rods. An additional nonlinear threshold is responsible for filtering out this noise source, and it occurs in the inner retina, where rod convergence is much higher (Fig. 1.3).

In one of the landmark papers for this chapter, Ala-Laurila and Rieke (2014) measured this nonlinearity and its effects on photon detection. The authors began with a simple observation that led to an apparent paradox. ON parasol RGCs in nonhuman primate retina are almost completely silent in absolute darkness. If each single-photon event in a rod eventually leads to a spike in the RGC, as suggested by previous work (Barlow et al., 1971; Mastronarde, 1983), then spontaneous events alone should make the RGC fire at nearly 40 Hz. Even accounting for half of the spontaneous events to be eliminated at the rod to RBC synapse, the rate should still be 20 Hz. The quiet state of ON parasol RGCs was not a reflection of a lack of sensitivity. On the contrary, cells responded to flashes producing as few as 3.6 photons on average across the entire pool of rods in the collecting area of the RGC (Fig. 1.8A). Similar to the nonlinearity in the outer retina, the spike data for the RGC had a slope > 1 on the log-log plot of response versus light intensity (Fig. 1.8C). The same experiment in OFF parasol RGCs revealed a linear relationship between spike response and light intensity (Fig. 1.8).

The fact that the ON pathway was nonlinear, but the OFF pathway was linear, suggested that the location of the nonlinearity was after these pathways diverge, that is, past the AII amacrine cell (Fig. 1.3). To isolate the location of the nonlinearity further, the authors recorded excitatory input currents to ON parasol RGCs. Excitatory current was also nonlinear as a function of light intensity, with a slope similar to that of the spike output (Fig. 1.8C). Thus, the synapse between the ON CB and the ON parasol RGC implements a thresholding nonlinearity that effectively eliminates spontaneous events while allowing light-driven events.

But how is such a computation possible if spontaneous and light-driven events have the same amplitude and shape? The answer lies in their timing.

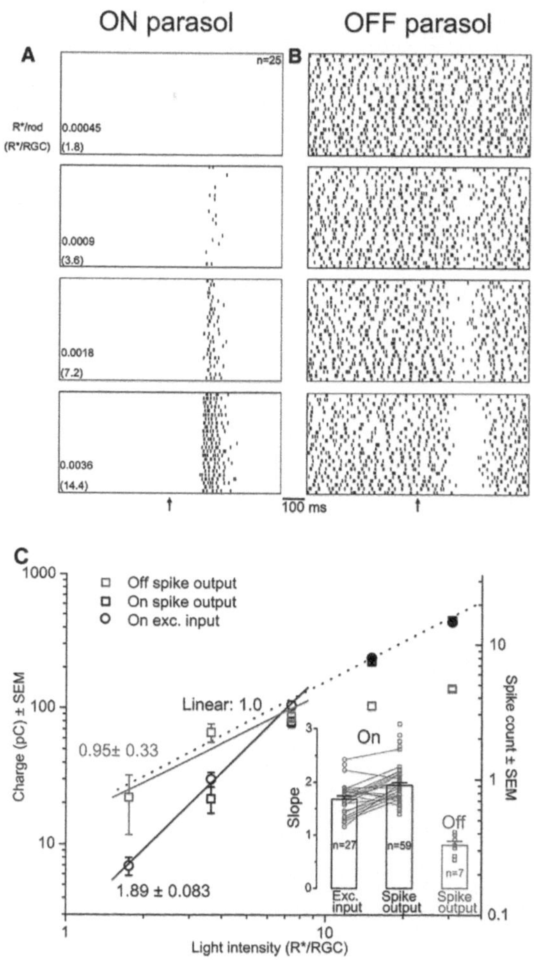

Fig. 1.8 A thresholding nonlinearity between ON cone bipolar cells and ON RGCs. (**A**) Spike rasters for an ON parasol RGC responding to very dim flashes at the luminance values indicated. (**B**) Same as (**A**) for an OFF parasol RGC. (**C**) Spike count *(right axis)* or excitatory input charge *(left axis)* as a function of flash intensity on a log-log scale for OFF parasol RGCs *(gray)* and ON parasol RGCs *(black)*. Slope of each relationship is indicated. Inset, log-log slope for each relationship for each recorded RGC. Lines connect points from the same cell. *(Modified from Ala-Laurila and Rieke (2014).)*

Spontaneous events occur at random times across the rod array, while events caused by a real light flash should be effectively simultaneous. Ala-Laurila and Rieke (2014) measured the timescale of the nonlinearity at the ON CB to ON parasol RGC synapse by presenting flashes at different time lags and recording excitatory currents in the RGC. The flashes interacted nonlinearly only if they were presented within ~50 ms (Fig. 1.9). Thus, the inner retinal nonlinearity is almost perfectly poised to eliminate individual spontaneous events occurring randomly at 20 Hz (mean interval = 50 ms) while transmitting real flashes of light that cause 2–3 simultaneous events across the pool of rods.

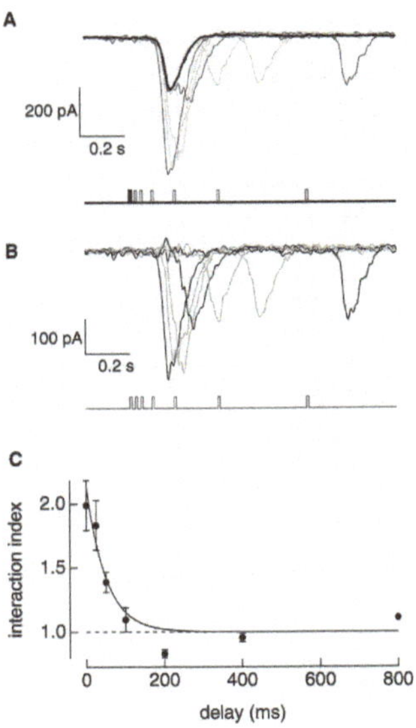

Fig. 1.9 Coincide detection in the inner retinal nonlinearity eliminates spontaneous rhodopsin isomerizations while allowing sparse signals to pass. (**A**) Excitatory input currents in an ON parasol RGC in response to a single dim flash *(thick line)* or pairs of flashes with the second flash at different intervals from the first *(thin lines)*. (**B**) Traces from (**A**) with initial flash response subtracted. (**C**) Interaction index quantifies the enhancement of the second flash response as a function of time after the first flash. Line is a single exponential fit. *(From Ala-Laurila and Rieke (2014).)*

The next steps in linking retinal physiology to behavior

The study of photon detection began with behavior, and as we have seen, quantitative links between the detection limits of humans and measurements of signal and noise in the nonhuman primate retina are compelling. Nonetheless, these links are correlations. More complete tests of our understanding of the computation must come from experimental manipulations of the retinal circuits with behavioral measurements from the same animals. Smeds et al. (2019) performed such a causal experiment in mice and demonstrated that the nonlinear ON pathway drives behavior in this task.

As in primates, mice have an ON RGC, called the ON alpha, and an OFF RGC, called the OFF sustained alpha, and these are the most sensitive RGCs for detecting dim flashes (Murphy and Rieke, 2006; Smeds et al., 2019). The upstream circuits for photon detection are also similar between mouse and primate with shared components (rods, RBCs, and AII amacrine cells) followed by separate ON and OFF cone bipolar cells that synapse onto the ON and OFF RGCs, respectively. Smeds et al. (2019) showed that also like in primate retina, the ON pathway has a nonlinearity at the CB to RGC synapse while the OFF pathway is linear. Critically, however, there is a suite of transgenic mice with specific modifications in the phototransduction pathway components, so researchers can alter the signal and/or noise properties in rods and measure how these manipulations play out in RGCs and in behavior.

In their study, Smeds et al. used a mouse line, called OPN1LW, in which a small fraction of the human L-cone opsin was coexpressed with rhodopsin in rods (Fu et al., 2008). While the fraction of L-opsin was only 0.4%, it dramatically affected the amplitude of single-photon responses, reducing them by 2.8-fold (Fig. 1.10A). In WT mice, the linear OFF pathway and the nonlinear ON pathway were similarly sensitive to dim flashes. In OPN1LW mice, due to the additional thresholding nonlinearity, the ON pathway was more strongly affected by the reduced single-photon response amplitude

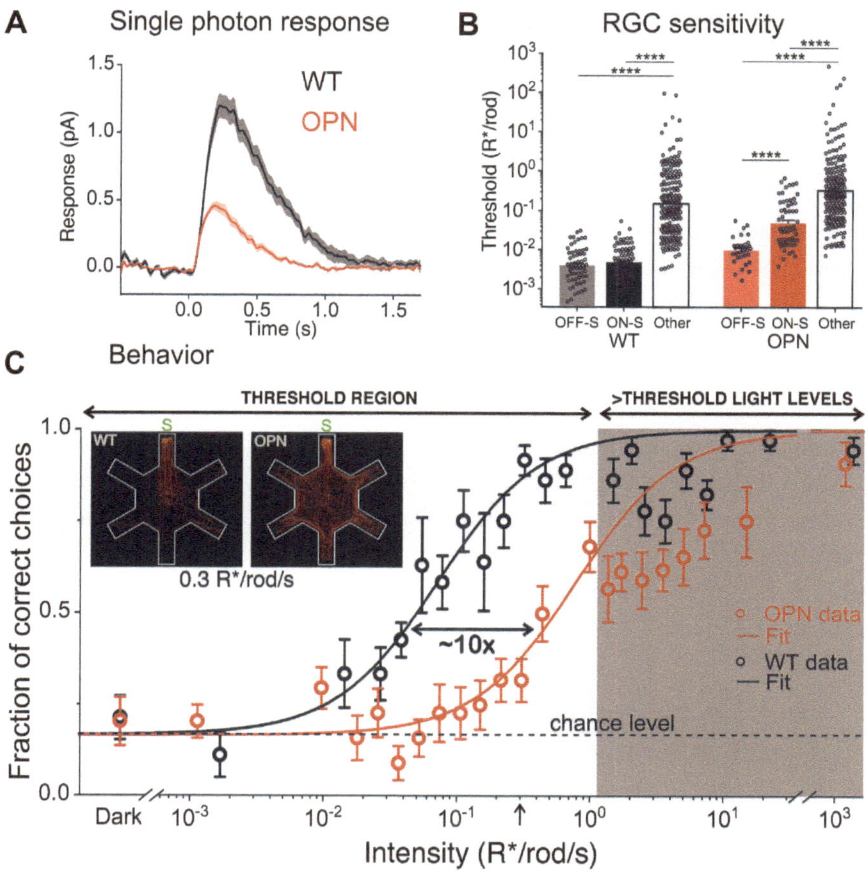

Fig. 1.10 Specific manipulations of retinal circuits show that photon detection behavior uses ON RGCs. (**A**) Single-photon responses from WT *(black)* and OPN1LW (OPN, *red*) mice. (**B**) Response thresholds of different RGC types in WT and OPN mice. Note logarithmic y-axis. (**C**) Behavioral measurements of dim flash sensitivity in WT *(black)* and OPN *(red)* mice. Data points are fraction of correct choices in a six-arm detection task as a function of the intensity of the target spot. Note logarithmic x-axis. Insets, examples of trajectories of WT *(left)* and OPN *(right)* animals looking for the target arm (marked S) at the same light intensity (0.3 R*/rod/s). *(From Smeds et al. (2019).)*

(10-fold lower sensitivity) than the OFF pathway (2.5-fold lower sensitivity), as measured in the spike outputs of ON and OFF RGCs, respectively (Fig. 1.10B). Having broken the ON:OFF symmetry of sensitivity, the authors could then perform a version of the dim flash detection task (in this case in a water maze with one of the six arms displaying a dim light) and test whether behavioral sensitivity was degraded in the OPN1LW mouse relative to WT and, critically, to a degree consistent with a readout of the ON RGCs (10-fold) or with the OFF RGCs (2.5-fold). The result was that OPN1LW mice were 10-fold less sensitive than WT mice in the photon detection task, demonstrating that this behavior relies on the nonlinear ON pathway even when the OFF pathway is more sensitive (Fig. 1.10C).

This study established a framework for future work linking vision in the quantal regime to animal behavior, and many questions remain. Is the ON pathway similarly dominant in the increment detection task? Does the OFF pathway drive perception of decrements in light? How are sparse signals integrated across multiple RGCs in space and time? Can signal detection theory explain behavioral limits on motion perception in the sparse photon regime?

Field et al. (2019) published a model of signal and noise in primate rods, including updated measurements and all the noise sources mentioned in this chapter. While this model is a satisfying culmination of decades of work on photon detection, the authors also point out that much remains to be done in connecting this understanding at the circuit level to the limits of behavior in dim illumination. In particular, they discuss the issues of temporal precision and the perception of motion. In later chapters, we will explore motion computations in higher luminance, but the sparse photon detection regime offers unique opportunities to connect detailed models of signal and noise in the retina to behavior.

Landmark paper: Field, G.D., and Rieke, F. (2002b). Nonlinear signal transfer from mouse rods to bipolar cells and implications for visual sensitivity. Neuron *34*(5), 773–785.

Discussion questions:

1. What is the key difference between the signals recorded in rod bipolar cells versus those in OFF bipolar cells, as presented in Fig. 3?
2. In Fig. 4, the authors compare the mean response squared to the variance in both bipolar cell types. Explain why the difference they observed was significant and why it supports the data from Fig. 3.
3. Explain in your own words why setting the threshold at a position that cuts out half the single-photon responses is the best way to optimize SNR at the rod to RBC synapse. Write an equation using Bayes Theorem to explain this result mathematically.

4. The optimality of the position of the nonlinear threshold appears to fall apart for higher light intensities (Fig. 6F). How do the authors interpret this result? Design an experiment to test whether the position of the nonlinearity could be dynamically adjusted.

Landmark paper: Ala-Laurila, P., and Rieke, F. (2014). Coincidence detection of single-photon responses in the inner retina at the sensitivity limit of vision. Current Biology: CB *24*(24), 2888–2898.

Discussion questions:

1. What is the firing rate of OFF parasol RGCs in darkness? What is the significance of this value?
2. What is the log-log slope of each of the three regions in Fig. 2E? What is the authors' explanation for these three regions?
3. What level of background light is required to suppress the nonlinearity in ON parasol RGCs? How does this relate to Barlow's (1957b) experiment and calculation of dark light?
4. Fig. 6E compares the SNR of the nonlinear model to that of a linear model. This is the same measure plotted in Fig. 6 of Field and Rieke (2002b), but the x-axis of the plot is different. Compare these results. What evidence is presented in Ala-Laurila and Rieke (2014) that the nonlinear threshold is placed near its optimal position?
5. The SNR ratio in Fig. 6E drops below 1 for very dim flashes. What does this mean? How does this relate to the difference between ON and OFF parasols shown in Fig. 1?
6. How did the authors isolate cone responses from rod responses? How did they use the cone responses to demonstrate that the nonlinearity is at the ON cone bipolar cell to ON RGC synapse?

Programming exercise

To access the programming exercise, please visit https://www.elsevier.com/books-and-journals/book-companion/9780128198964.

This exercise explores signal-to-noise ratio (SNR) for the detection of sparse photons by an array of rods.

Open the script "PhotonDetection.m." The "setup" section contains the initial parameters of the simulation that you will be editing.

1. Run the "setup" and the "Make rod signals" sections. Use "rodArrayPlotter.m" to visualize the signals across the rod array for a given trial. Test yourself on the sparse signal detection task by plotting the first 10 trials and writing down which ones contained a flash. Check your performance against the "flash_trials" variable. Adjust the values of "N_photons_mean" and

"rod_noise_sd" until the task becomes difficult. How close are these values to the real limits of photon detection in the retina?

2. Run the "One layer detection" section. This simulates a single optimized threshold following averaging of the whole rod array. How does this model perform? Explain why. What is the value of the optimal threshold that it found? What is the mean number of photons that were actually present on trials in which the model detected a flash? Explain your intuition for these values.
3. Now run the "Two layer detection" section. This simulates a nonlinear threshold at the rod to RBC synapse followed by another threshold in the RGC. How much better was the performance of this model? Explain why. Now the model finds two thresholds and again reports the mean number of photons present on detected trials. Explain your intuition for these values.
4. Make a new plotter that displays signals in the RBC array. Plot the rod array and RBC array on the same trials and explain the key differences in the pattern of activation.
5. Adjust the value of "RB_noise_sd." This is noise added at the level of the RBC. Thus, the noise is absent in the single-layer model but present in the two-layer model. What value of RBC noise is required to cancel the effect of the additional thresholding nonlinearity?
6. Now increase "single_photon_response_sd." This variable models the CV of the single-photon response. What value would model a rhodopsin molecule with three phosphorylation sites, instead of six? Use this value and explain how it impacts the simulation.
7. Now try the three-layer model. Does this give you an additional increase in performance? Why or why not?
8. There are several other parameters in "setup" to explore. Explore how performance of the two-layer and three-layer models depends on several of these parameters. Make a plot of what you discovered and explain it.

References

Ala-Laurila, P., and Rieke, F. (2014). Coincidence detection of single-photon responses in the inner retina at the sensitivity limit of vision. Current Biology: CB *24*(24), 2888–2898.

Ala-Laurila, P., Donner, K., and Koskelainen, A. (2004). Thermal activation and photo-activation of visual pigments. Biophysical Journal *86*(6), 3653–3662.

Arshavsky, V.Y., and Burns, M.E. (2014). Current understanding of signal amplification in phototransduction. Cellular Logistics *4*, e29390.

Barlow, H.B. (1957a). Purkinje shift and retinal noise. Nature *179*(4553), 255–256.

Barlow, H.B. (1957b). Increment thresholds at low intensities considered as signal/noise discriminations. The Journal of Physiology *136*(3), 469–488.

Barlow, H.B., Levick, W.R., and Yoon, M. (1971). Responses to single quanta of light in retinal ganglion cells of the cat. Vision Research *11*(Suppl. 3), 87–101.

Baylor, D.A., Lamb, T.D., and Yau, K.W. (1979a). The membrane current of single rod outer segments. The Journal of Physiology *288*, 588–611.
Baylor, D.A., Lamb, T.D., and Yau, K.W. (1979b). Responses of retinal rods to single photons. The Journal of Physiology *288*, 613–634.
Baylor, D.A., Matthews, G., and Yau, K.W. (1980). Two components of electrical dark noise in toad retinal rod outer segments. The Journal of Physiology *309*(1), 591–621.
Baylor, D.A., Nunn, B.J., and Schnapf, J.L. (1984). The photocurrent, noise and spectral sensitivity of rods of the monkey *Macaca fascicularis*. The Journal of Physiology *357*(1), 575–607.
Calvert, P.D., Govardovskii, V.I., Krasnoperova, N., Anderson, R.E., Lem, J., and Makino, C.L. (2001). Membrane protein diffusion sets the speed of rod phototransduction. Nature *411*(6833), 90–94.
Doan, T. (2006). Multiple phosphorylation sites confer reproducibility of the rod's single-photon responses. Science *313*(5786), 530–533.
Doan, T., Mendez, A., Detwiler, P.B., Chen, J., and Rieke, F. (2006). Multiple phosphorylation sites confer reproducibility of the rod's single-photon responses. Science *313* (5786), 530–533.
Donner, K. (1992). Noise and the absolute thresholds of cone and rod vision. Vision Research *32*(5), 853–866.
Field, G.D., and Rieke, F. (2002a). Mechanisms regulating variability of the single photon responses of mammalian rod photoreceptors. Neuron *35*(4), 733–747.
Field, G.D., and Rieke, F. (2002b). Nonlinear signal transfer from mouse rods to bipolar cells and implications for visual sensitivity. Neuron *34*(5), 773–785.
Field, G.D., Sampath, A.P., and Rieke, F. (2005). Retinal processing near absolute threshold: from behavior to mechanism. Annual Review of Physiology *67*(1), 491–514.
Field, G.D., Uzzell, V., Chichilnisky, E.J., and Rieke, F. (2019). Temporal resolution of single-photon responses in primate rod photoreceptors and limits imposed by cellular noise. Journal of Neurophysiology *121*(1), 255–268.
Fu, Y., Kefalov, V., Luo, D.G., Xue, T., and Yau, K.W. (2008). Quantal noise from human red cone pigment. Nature Neuroscience *11*(5), 565–571.
Grimes, W.N., Baudin, J., Azevedo, A.W., and Rieke, F. (2018). Range, routing and kinetics of rod signaling in primate retina. eLife 7, e38281.
Hamer, R.D., Nicholas, S.C., Tranchina, D., Liebman, P.A., and Lamb, T.D. (2003). Multiple steps of phosphorylation of activated rhodopsin can account for the reproducibility of vertebrate rod single-photon responses. Journal of General Physiology *122*(4), 419–444.
Hecht, S., Shlaer, S., and Pirenne, M.H. (1942). Energy, quanta, and vision. The Journal of General Physiology *25*(6), 819–840.
Kiani, R., Ala-Laurila, P., and Rieke, F. (2020). Seeing with a few photons: Bridging cellular and circuit mechanisms with perception. In The Senses: A Comprehensive Reference, B. Fritzsch, ed. (2nd ed., Vol. 1–7) (Academic Press).
Koenig, D., and Hofer, H. (2011). The absolute threshold of cone vision. Journal of Vision *11*(1), 21.
Kolb, H., and Nelson, R. (1983). Rod pathways in the retina of the cat. Vision Research *23*(4), 301–312.
Luo, D.G., Yue, W.W.S., Ala-Laurila, P., and Yau, K.W. (2011). Activation of visual pigments by light and heat. Science *332*(6035), 1307–1312.
Mastronarde, D.N. (1983). Correlated firing of cat retinal ganglion cells. II. Responses of X- and Y-cells to single quantal events. Journal of Neurophysiology *49*(2), 325–349.
Murphy, G.J., and Rieke, F. (2006). Network variability limits stimulus-evoked spike timing precision in retinal ganglion cells. Neuron *52*(3), 511–524.
Nelson, P. (2017). From Photon to Neuron: Light, Imaging, Vision (Princeton University Press). https://www.physics.upenn.edu/biophys/PtN/.

Pugh, E.N., and Lamb, T.D. (1993). Amplification and kinetics of the activation steps in phototransduction. Biochimica et Biophysica Acta-Bioenergetics *1141*(2–3), 111–149.
Rieke, F., and Baylor, D.A. (1996). Molecular origin of continuous dark noise in rod photoreceptors. Biophysical Journal *71*(5), 2553–2572.
Rieke, F., and Baylor, D.A. (1998a). Origin of reproducibility in the responses of retinal rods to single photons. Biophysical Journal *75*(4), 1836–1857.
Rieke, F., and Baylor, D.A. (1998b). Single-photon detection by rod cells of the retina. Reviews of Modern Physics *70*(3), 1027–1036.
Sakitt, B. (1972). Counting every quantum. The Journal of Physiology *223*(1), 131–150.
Sampath, A.P., and Rieke, F. (2004). Selective transmission of single photon responses by saturation at the rod-to-rod bipolar synapse. Neuron *41*(3), 431–443.
Smeds, L., Takeshita, D., Turunen, T., Tiihonen, J., Westö, J., Martyniuk, N., Seppänen, A., and Ala-Laurila, P. (2019). Paradoxical rules of spike train decoding revealed at the sensitivity limit of vision. Neuron *104*(3), 576–587.e11.
Sterling, P., Freed, M.A., and Smith, R.G. (1988). Architecture of rod and cone circuits to the on-beta ganglion cell. Journal of Neuroscience *8*(2), 623–642.
Tsukamoto, Y., Morigiwa, K., Ueda, M., and Sterling, P. (2001). Microcircuits for night vision in mouse retina. Journal of Neuroscience *21*(21), 8616–8623.
von Kries, J., and Eyster, J.A.E. (1907). Uber die zur Erregung des Sehorgans efforderlichen Energiemenzen. Z. Sinnesphysiol *41*, 394.
Yue, W.W.S., Silverman, D., Ren, X., Frederiksen, R., Sakai, K., Yamashita, T., Shichida, Y., Cornwall, M.C., Chen, J., and Yau, K.W. (2019). Elementary response triggered by transducin in retinal rods. Proceedings of the National Academy of Sciences of the United States of America *116*(11), 5144–5153.

Bright sunlight shines through a keyhole in a completely dark room. In real life this scene would include a dynamic range of light intensity that is impossible to reproduce on paper or a computer monitor. Fast retinal circuits for luminance adaptation allow us to see over a billion-fold range in light intensity.

CHAPTER 2

Luminance adaptation

Gregory William Schwartz[a,b,c] and Jared Levine[d]
[a]Departments of Ophthalmology and Physiology, Feinberg School of Medicine, Northwestern University, Chicago, IL, United States
[b]Department of Neurobiology, Weinberg College of Arts and Sciences, Northwestern University, Chicago, IL, United States
[c]Northwestern Interdepartmental Neuroscience Program (NUIN), Northwestern University, Chicago, IL, United States
[d]Laboratory of Neurogenetics of Language, Rockefeller University, New York, NY, United States

The rate of photons arriving at the retina varies over a billion-fold between a moonlit night and a sunny day, yet we can use vision in both conditions. While most of us are familiar with the slow timecourse of adaptation to these extremes from experiences like feeling blinded by the sun walking out of a movie theater, much of the luminance adaptation process takes place unconsciously. The luminance values in sequential saccades within the same visual scene can differ by 10^4 (Frazor and Geisler, 2006), yet we experience no lag in perception shifting gaze from a sunny reflection to a shadow. The goal of luminance adaptation is to adjust gain, defined as response amplitude per photon, dynamically so that visual neurons can respond at these vastly different light levels.

Luminance adaptation has been a prominent research focus in vision science for decades; its remarkable range and speed set interesting constraints on biological mechanisms. Most of the known mechanisms of luminance adaptation in the visual system occur in the retina. This makes sense since retinal ganglion cells (RGCs) must respond within their limited dynamic range (approximately two orders of magnitude) if there is any hope of recovering visual information downstream in the brain. The ability of luminance adaptation to cover nine orders of magnitude in photon flux strongly suggests that it involves multiple mechanisms. The "duplex theory" introduced over 150 years ago explains how rods and cones cover different luminance regimes because of their different sensitivities (reviewed by Ingram et al., 2016), but as we will see, there are a host of adaptation mechanisms within both rod and cone pathways. Luminance adaptation has been measured at each different level in retinal circuits, from phototransduction to spike generation in RGCs and at almost every synapse in between.

Retinal Computation
https://doi.org/10.1016/B978-0-12-819896-4.00003-2

Table 2.1 Multiple mechanisms of fast luminance adaptation in the retina.

Location	Threshold luminance (isomerizations/ photoreceptor/sec)	Photoreceptor convergence
AII AC or cone bipolar cell	0.08	~1000 rods
RBC to AII AC synapse	0.4	~20 rods
Rod phototransduction	~10	1 rod
Cone bipolar to RGC synapse	~100	~200 cones
Cone phototransduction	~1000	1 cone

From the perspective of computation, luminance adaptation involves several statistical questions about the estimation of luminance from a noisy set of samples and the optimal way to adjust gain dynamically. How does the system balance the need to adapt rapidly with the danger of overadapting to noise fluctuations around a fixed mean? What mechanistic differences are required to adapt to sparse photon arrivals and fluxes in the range of 10^7 photons per second? What is the appropriate spatial and temporal extent over which to adapt based on the statistics of the visual scene? These are the kinds of questions I will address in this chapter, inspired by an excellent review of luminance and contrast adaptation by Fred Rieke and Michael Rudd (2009). Table 2.1 summarizes the many known sites of fast luminance adaptation in the retina and the luminance range, and spatial convergence of each mechanism. The second part of this chapter will cover slower forms of luminance adaption, operating over seconds to hours, which comprise a wide variety of cellular and circuit mechanisms.

Rod vision: A gain control mechanism for each convergence point of two photons

In the previous chapter, we saw how the retina uses multiple filtering and thresholding steps to retain sensitivity to signals in which photons are extremely sparse, on the order of 1–2 per 1000 rods. Remarkably, for light levels only a bit higher, where photons are still quite sparse, luminance adaptation begins (Barlow, 1957). Just as the limits on photon detection are set by physical constraints in the quantization of photons and the thermodynamics of rhodopsin, luminance adaptation in the sparse photon regime has optimal sensitivity, given the architecture of the circuit. That is, at each point of convergence in the rod bipolar circuit (shown in Chapter 1, Fig. 1.3), a single photon reduces the gain of the next photon over a certain integration time.

A key principle, both in rod and cone vision, is that adaptation to lower luminance occurs further downstream in the retinal circuit (Donner et al., 1990). This is a result of the requirement of more spatial pooling to measure the luminance change. In the sparse photon regime, this limit is absolute in the sense that there are ranges of luminance adaptation in which most upstream response elements receive zero or one photon within their integration time, so adaptation is impossible. Rod bipolar cells (RBCs) get inputs from ~20 to 50 rods (Anastassov et al., 2019; Field and Rieke, 2002; Tsukamoto et al., 2001), so even if they integrated over a full second, they could not possibly adapt to signals dimmer than ~1/20 rhodopsin isomerizations (R*) per rod per second (or even higher because half the single-photon responses are eliminated at the first synapse (Field and Rieke, 2002)). Rods themselves cannot adapt within 1 s until the luminance reaches ~1 R*/rod/s. The statistical version of this logic applies to higher-luminance situations as well. Luminance changes are more reliable in a given time window when they are averaged over a larger pool of photoreceptors; the shorter the integration time, the larger the integration pool must be (Fig. 2.1).

What is the lowest luminance at which adaptation occurs in the retina? Fred Rieke and I studied this question in the primate retina by recording from ON parasol RGCs over a range of dim backgrounds (Schwartz and Rieke, 2013). We found that gain in both the excitatory inputs and spike outputs of these RGCs was reduced by 50% at a light level of 0.08 R*/rod/s (Fig. 2.2). We also measured the time constant of this adaptation mechanism

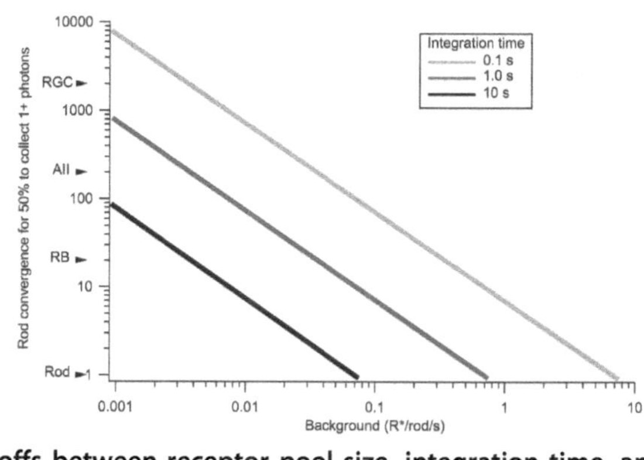

Fig. 2.1 Tradeoffs between receptor pool size, integration time, and luminance at different levels of convergence. Theoretical limits on luminance adaptation. The number of rods required for at least half of them to receive one or more photons is plotted versus background luminance. *Arrows* indicate rod convergence at three locations in the rod bipolar circuit. *Different shades* indicate different integration times. *(From Schwartz and Rieke (2013).)*

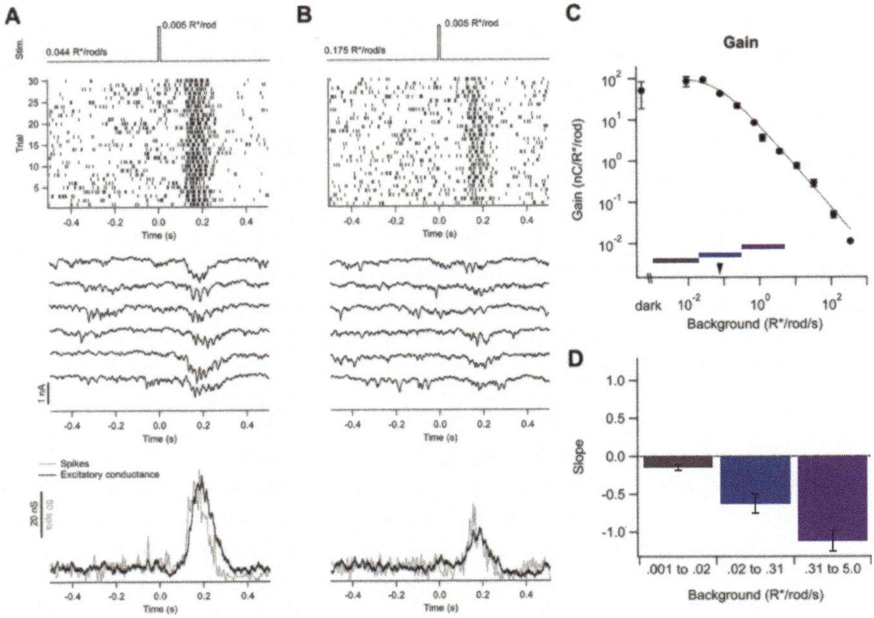

Fig. 2.2 Luminance adaptation in the primate retina near absolute threshold. (**A**) *Top,* spike rasters for a parasol RGC responding to the stimulus shown above. *Middle,* six trials of excitatory input currents recorded in the same stimulus conditions. *Bottom,* mean spike rate *(gray)* and excitatory conductance *(black)* across trials. (**B**) Recordings from the same cell as in (**A**) for the same flash intensity from a slightly brighter background. (**C**) Gain of excitatory conductance as a function of background intensity plotted on a log-log scale. Colored regions correspond to slope measurements below. (**D**) Slope of the graph in (**C**) at different background intensity regimes. *(Modified from Schwartz and Rieke (2013).)*

by presenting pairs of flashes at different time intervals and computed a value of 0.1 s. Thus, adaptation must rely on a pool of at least 200 rods, demonstrating that it occurs at a site downstream of RBCs. Adaptation in the rod bipolar pathway failed to transfer to cone-mediated flashes, further isolating the adaptation site to AII amacrine cells and/or their electrical synapses with ON cone bipolar cells. However, a precise cellular or synaptic mechanism for this most sensitive form of luminance adaptation has not been identified.

One synapse upstream lies the next known site of luminance adaptation in rod vision, at the synapse between RBCs and AII amacrine cells. The gain of this synapse is very high in darkness to support photon detection, but the transmission of a single-photon response reduces the gain of the next single-photon response within an integration time of ~200 ms (Dunn and Rieke, 2008) (Fig. 2.3). Synaptic depression at the ribbon synapses in RBCs was identified as the adaptation mechanism at this site. Even at this single-photon

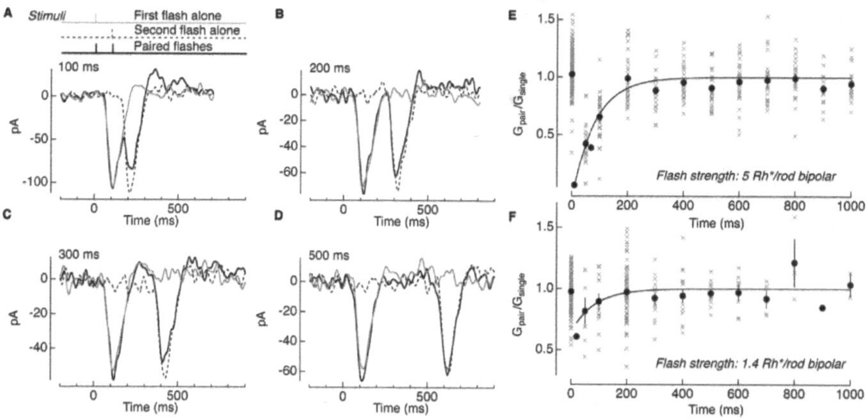

Fig. 2.3 Paired flash depression at AII ACs for dim flashes. (**A**) Excitatory currents measured in mouse AII ACs for single and paired dim flashes. *Gray trace* is first flash response, *black trace* is paired response, and *dotted trace* is second flash response alone. (**B–D**) Same as (**A**) for flashes at different intervals. (**E**) Ratio of second flash response when part of a pair to the response alone as a function of interval between flashes. (**F**) Same as (**E**) for a lower flash intensity. *(From Dunn and Rieke (2008).)*

sensitivity limit, the RBC-AII gain control mechanism requires higher luminance than the mechanism described earlier because of the RBC's smaller integration pool. The luminance at which this synapse halves its gain is ~0.4 R*/rod/s (Dunn and Rieke, 2008).

Finally, the phototransduction cascade in rods undergoes adaptation at light levels high enough for multiple photon absorptions to interact within the rod outer segment. The absorption of two photons in a rod within ~200 ms has a measurable effect on gain, and the luminance required to halve the rod gain is ~10 R*/rods (Dunn et al., 2006; Schneeweis and Schnapf, 1999) (Fig. 2.4).

Fig. 2.4 Adaptation in mouse rods. Gain of the photocurrent in mouse rods *(black)* and ganglion cells *(gray)* versus background light intensity expressed as a fraction of their respective values in darkness. *(Modified from Dunn et al. (2006).)*

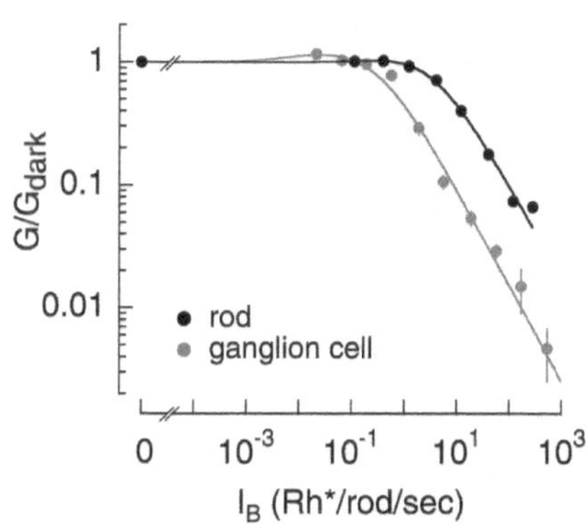

Cone vision: Separate mechanisms in the circuitry and in the cone

Photons are not sparse in cone vision, so adaptation is not limited by circuit elements collecting zero or one quantum. Nonetheless, the same principle applies; adaptation to lower light levels requires pooling over more cones, so it occurs further downstream within the retina. In the primate retina, adaptation from darkness to 1000 isomerizations (P*) per cone per second takes place at the synapse between cone bipolar cells and RGCs while adaptation from 1000 to 10,000 P*/cone/sec takes place in the cones themselves (Fig. 2.5) (Dunn et al., 2007). The division of adaptation in cone vision into receptor and postreceptor mechanisms has implications for human perception. Outside the midget pathway in the fovea and the specialized S-cone ON bipolar cells, cone bipolar cells collect from multiple cone types. Thus, postreceptor adaptation transfers across wavelengths, while receptor adaptation does not. At a moderate luminance, green light will adapt your subsequent perception of red light; at high luminance, adaptation to red and green is largely independent (Tuten et al., 2017).

Both the receptor and postreceptor forms of adaptation in the cone pathway follow Weber's law; gain decreases proportionally to the increase in background luminance (Dunn et al., 2007). The same functional relationship between luminance and gain was measured across most of the rod light regime (Donner et al., 1990; Schwartz and Rieke, 2013). Why then, do threshold versus intensity curves in perception, as measured by Barlow and others (see Chapter 1), show separate regions with a log-log slope of 0.5 (fluctuation theory) in the rod regime and 1.0 (Weber's law) in the cone regime? The answer lies in the different sources and behavior of rod and cone noise. Beyond absolute darkness, where spontaneous isomerizations are significant, rod noise is dominated by noise in the phototransduction pathway that adapts the same way as the signal. Cone noise is instead

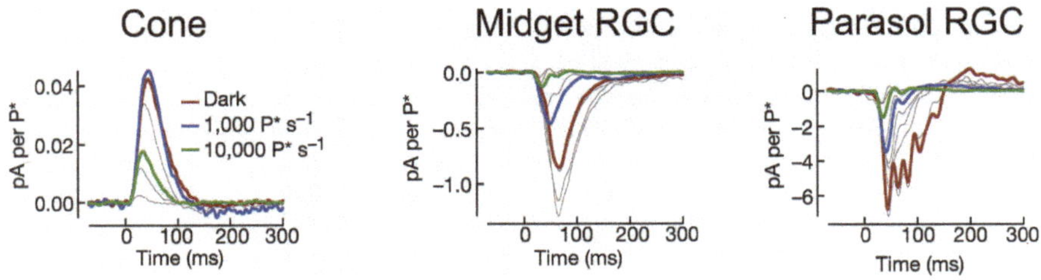

Fig. 2.5 Luminance adaptation in cones and RGCs of the primate retina. Light responses in cones, midget, and parasol RGCs at three different light intensities as indicated. *(Modified from Dunn et al. (2007).)*

dominated by fluctuations in the state of cGMP-gated cation channels (Fig. 2.6A–D) (Angueyra and Rieke, 2013). Unlike the noise in rods, this channel noise in cones is relatively unaffected by luminance. Thus, for rods, where signal and noise adapt similarly, the signal-to-noise ratio (SNR) follows fluctuation theory (see Chapter 1), while for cones, SNR changes according to Weber's law because it is controlled by the adapting signal divided by a relatively fixed level of noise (Fig. 2.6E and F) (Angueyra and Rieke, 2013).

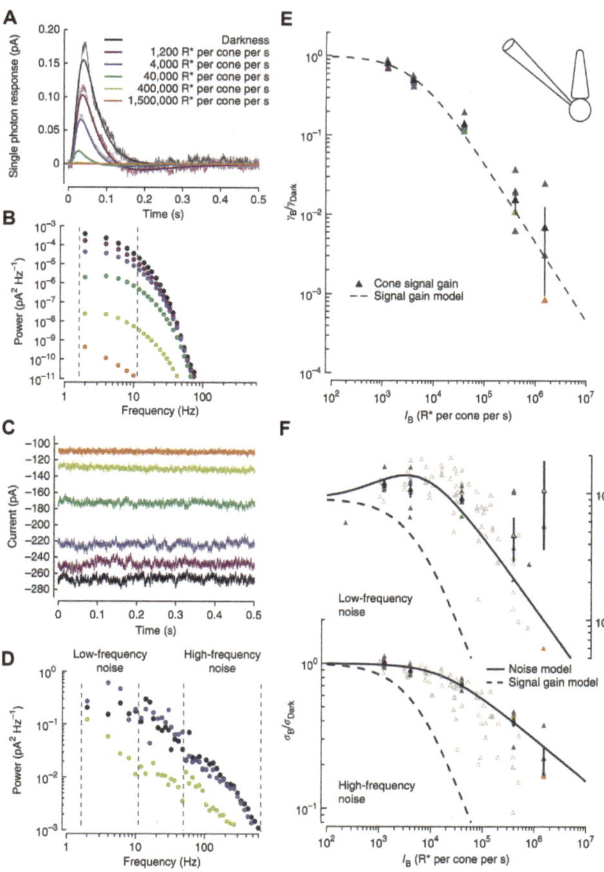

Fig. 2.6 Signal and noise adapt differently with luminance in cones. (**A**) Single-photon responses in primate cones measured at different background luminances as indicated. (**B**) Power spectra of the traces in (**A**). (**C**) Measurements of background noise at each luminance, colored as in (**A** and **B**). (**D**) Power spectra of three of the noise traces from (**C**). (**E**) Cone gain versus background luminance, expressed as a fraction of its value in darkness *(triangles)* along with a model fit to the data *(dashed line)*. (**F**) Gain of cone noise in the low- and high-frequency regimes as indicated in (**D**). *Solid line* is a noise gain model. *Dashed line* is the signal gain model from (**E**). *(Modified from Angueyra and Rieke (2013).)*

Adaptation on longer timescales

So far, we have focused on forms of luminance adaptation that are very rapid, on the timescale of single saccades. As described earlier, rapid adaptation is needed to account for the large dynamic range of natural visual scenes. But there are also predictable changes in luminance on longer timescales that engage different mechanisms. Environmental changes, like clearing cloud cover or walking into a cave, change luminance on the scale of tens of seconds to minutes. Of course, the rotation of the earth also changes luminance with a period of 24 h.

Layered on top of the five sites of fast adaptation described earlier are perhaps dozens of additional modulations that take place over seconds, minutes, and hours. We will point out some of these mechanisms in the following sections to highlight their remarkable diversity, from protein localization to functional circuit reconfiguration. Still, there are surely more that we will omit and others that remain to be discovered. Astounding capacity for change through multiple mechanisms is a general property of the central nervous system; some of the lessons we have learned from the retina have led the way in our understanding of adaptation and modulation more broadly in neuroscience.

Neuromodulators

Neuromodulators work throughout the nervous system to modify cellular and circuit function over timescales longer than those involved in fast synaptic transmission. Unsurprisingly, they play many and varied roles in retinal function but are perhaps best understood in the context of luminance adaptation. Dopamine is the best-characterized neuromodulator in the retina, and many of the other neuromodulators have been characterized in part by their relationship to dopamine.

Light stimulates dopamine synthesis in dopaminergic amacrine cells (DACs) of the inner plexiform layer (Iuvone et al., 1978), and DACs increase their firing as light levels increase (Gustincich et al., 1997). A principal site of action of dopamine is gap junctions, both in the outer and inner retina. Photoreceptor coupling is reduced by dopamine acting on the D4 receptor (Ribelayga et al., 2008). This dynamic regulation of photoreceptor coupling seems to increase the signal-to-noise ratio in dim lighting due to signal averaging across multiple rods while only modestly reducing the single-photon response (Jin and Ribelayga, 2016). Dopamine also decreases coupling between horizontal cells (Zhang et al., 2011) to modulate the outer-retinal component of surround suppression (see Chapter 8). In the inner retina, dopamine decouples AII ACs in photopic conditions, working to dephosphorylate the connexin 36 protein, which forms the gap junctions between AIIs (Kothmann et al., 2009). One functional consequence of AII decoupling is the reduction of the spatial extent of ON bipolar cell signals (through their gap junctions to the AII network) and OFF bipolar cells (through crossover inhibition), thus promoting high acuity vision (Hartveit and Veruki, 2012). Finally, a recent study showed that a decrease of surround inhibition in OFF bipolar cells occurs during light adaptation and is mediated by signaling through the D1 dopamine receptor (Mazade and Eggers, 2019). The reduced inhibition allows OFF bipolar cells to more strongly drive ganglion cell excitation in photopic conditions.

Adenosine, an ATP breakdown product, is another neuromodulator found in the retina, and its effects generally run counter to those of dopamine. The putative sources of adenosine, starburst ACs (Neal and Cunningham, 1994) and Müller glia (Newman, 2003), both secrete adenosine into the extracellular space. In contrast to dopamine, levels of adenosine peak during the night and are further modulated upward in conditions of low ambient light (Ribelayga and Mangel, 2005). While dopamine acts to reduce photoreceptor gap-junction connectivity, adenosine promotes gap-junction connectivity in high scotopic and mesopic conditions by inducing phosphorylation of connexin 36 (Li et al., 2014). Adenosine also attenuates spiking responses in intrinsically photosensitive (ip)RGCs in response to light via activation of the A1a receptor (Sodhi and Hartwick, 2014). This may act to reduce the transmittance of photic information to the suprachiasmatic nucleus during scotopic conditions, promoting proper circadian entrainment (see Chapter 3, Absolute luminance detection).

Nitric oxide (NO), synthesized in photoreceptors in the outer retina (Yoshida et al., 1995) and by a specific AC, called nNOS-2, in the inner retina (Jacoby et al., 2018), also plays a role in luminance adaptation. Increases in NO are generally associated with light adaptation through a variety of mechanisms, including decoupling of AII ACs from cone bipolar cells (BCs) (Mills and Massey, 1995), increasing the gain in cone synapses (Snellman and Nawy, 2004) while decreasing the gain in rod synapses (Kourennyi et al., 2004), and increasing the gain in cone BC synapses (Tooker et al., 2013; Vielma et al., 2014). Mice lacking neuronal nitric oxide synthase (nNOS) show deficiencies in their photopic electroretinogram (Wang et al., 2007). Unlike for dopamine, flickering light is more effective than constant light in eliciting NO synthesis in the inner retina, suggesting that NO plays a role in adaptation over shorter timescales than does dopamine (Eldred and Blute, 2005; Yoshida et al., 1995).

Other neuromodulators in the retina that have demonstrated effects, either directly or indirectly on luminance adaptation, include somatostatin (Akopian et al., 2000; Cervia et al., 2008), endocannabinoids (Middleton et al., 2019; Miraucourt et al., 2016; Yazulla, 2008), neuropeptide Y (Bruun and Ehinger, 1993; Santos-Carvalho et al., 2015; Sinclair et al., 2004), and acetylcholine through metabotropic receptors (Bonaventure et al., 1989; Jardon et al., 1992; Strang et al., 2010).

Feedback from the brain

One familiar source of luminance adaptation involves controlling the amount of light let into the eye in the first place (also see Chapter 3). The pupillary light reflex (PLR) modulates pupil diameter in response to changes in luminance, increasing diameter when the luminance level drops to allow more photons into the eye and decreasing diameter upon increases in luminance. This modulation relies on ipRGCs, and in particular, the M1 subtype, which expresses the photopigment melanopsin (Fig. 2.7) (Chen et al., 2011). M1 RGC axons project directly to the olivary pretectal nucleus, where luminance information is processed. This processed information is, in turn, routed to the parasympathetic Edinger-Westphal nuclei,

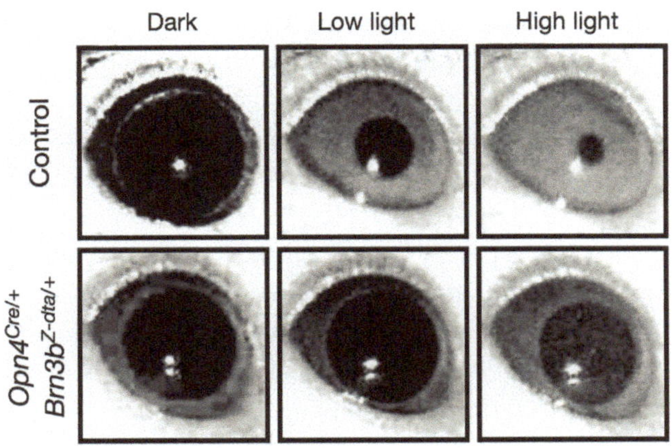

Fig. 2.7 The pupillary light reflex depends on a subset of ipRGCs. Images of a mouse eye showing the dramatic changes in pupil diameter in different luminance conditions. *Bottom row* shows the diminished pupillary light reflex in a mouse lacking melanopsin. *(Modified from Chen et al. (2011).)*

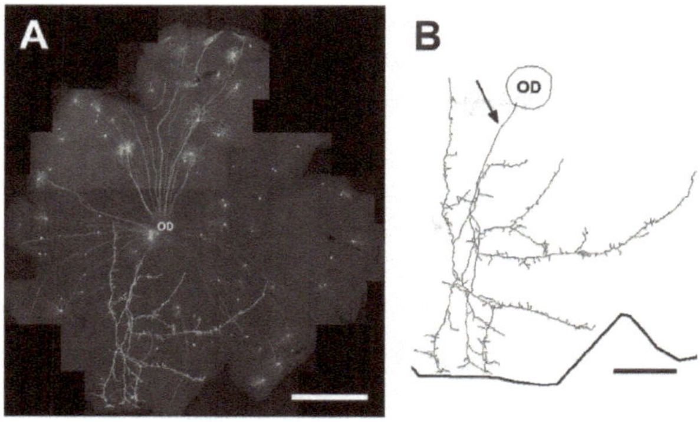

Fig. 2.8 An axon from the brain in the mouse retina. (**A**) Image of a retina with sparse expression of YFP in a subset of retinal neurons showing a single retinofugal axon. (**B**) Tracing of the axon. Scale bar = 0.5 mm. *(From Gastinger et al. (2006).)*

which innervate the iris sphincter muscles, thereby modulating pupil diameter (He et al., 2013). Another pathway for the PLR does not require the brain at all. An intrinsic PLR has been identified in a wide range of mammals, driven by melanopsin expression in the cells of the iris sphincter muscle itself (Xue et al., 2011).

The brain's influence on luminance adaptation is not limited to the PLR. In various species, including humans, axons project in the opposite direction along the optic nerve: from the brain to the retina (Fig. 2.8). Called "retinofugal" projections, these axons originate in nuclei expressing serotonin or histamine neurons (Gastinger et al., 2006; Lawrence and Studholme, 2014). One proposed action of serotonin in the retina is to promote light detection during the animal's active phase. Neurons in the dorsal raphe nucleus, where these axons originate, fire at higher rates during the active phase in both nocturnal and diurnal animals (Nakahama et al., 1981; Trulson and Jacobs, 1979). Agonists of 5-HT1 serotonin receptors reduce horizontal cell-mediated inhibition measured at the level of RGCs (Mangel and Brunken, 1992), thus enhancing the photic sensitivity of these cells to scotopic conditions. Stimulation of the H1 histamine receptor in guinea pigs strongly reduces dopamine release from DACs (Gastinger et al., 2004), which would help elicit dark adaptation at night.

Pigment regeneration

On the timescale of minutes, a key limitation in regaining visual sensitivity when transitioning from light to dark (often called "dark adaptation") is the availability of chromophores, the visual pigment molecules. Once it has been activated, 11-*cis*-retinal must be reisomerized to the all-*trans* form before it can once again integrate with an opsin and become light sensitive. The processes by which pigment recycling occurs are collectively called the retinoid cycles. Classic work in retinas detached from the retinal pigment epithelium (RPE) demonstrated that rhodopsin failed to regenerate, but cone opsins were able to regenerate via a mechanism within the retina (Bruch Goldstein and Wolf, 1973). The retinoid cycle in rods involves the release of all-*trans* retinal from the opsin and its conversion to all-*trans* retinol, its related alcohol. Molecules of all-*trans* retinol are in turn picked up by cells of the RPE, converted back to 11-*cis*-retinal, and reabsorbed by the rods (Wright et al., 2015).

The retinoid cycle in cones is substantially faster than in rods (Wang and Kefalov, 2009). Not only does this allow for continued signaling in bright conditions, but it also provides a mechanism for dark adaptation following exposure to bright light. The retinoid cycle in cones relies on Müller glial cells, which, like RPE cells, absorb all-*trans*-retinol released by rods and cones, isomerizing it from all-*trans*-retinol to 11-*cis*-retinol. This pathway's cone specificity stems from the inability of rods to oxidize 11-*cis*-retinol to 11-*cis*-retinal, which keeps rods from using the isomerized retinoids under daylight conditions, keeping them available for cones. The molecular pathway within Müller glia that is responsible for cone pigment regeneration was recently identified (Morshedian et al., 2019).

Receptor composition in RGCs

At synapses between ON BCs and both ON and ON-OFF RGCs, a switch in receptor composition has been reported to play a role in luminance adaptation (Jones et al., 2012). AMPA receptors mediate fast, excitatory synaptic transmission at these synapses and exist in two varieties, calcium-permeable and calcium-impermeable. At scotopic levels, EPSCs in ON RGCs are evoked primarily through calcium-impermeable AMPARs. As light intensity increases, though, EPSCs are evoked through calcium-permeable AMPARs. Also, at higher light levels, perisynaptic NMDA receptors are activated by glutamate spillover and facilitate a transition to calcium-impermeable AMPA receptor expression, which takes place in minutes. The effect of this receptor switch is to reduce gain at these synapses.

Subcellular localization of transducin and arrestin

Arrestin is a cytoplasmic protein found in rods and cones that quenches phototransduction by blocking photoactivated rhodopsin's interaction with its G-protein, transducin (see Chapter 1). Given this function, one might expect to find it in rod outer segments associated with rhodopsin. However, in the dark, arrestin is localized to rod inner segments, with very little expressed in the outer segments. When light levels reach about 1000 isomerizations/s, arrestin migrates to the outer segments (Calvert et al., 2006; Peterson et al., 2003). Conversely, transducin is found in the outer segments in the dark, migrating to the inner segments when light levels increase above 10,000 isomerizations/s (Fig. 2.9) (Sokolov et al., 2002). This translocation of transducin out of the outer segments in high light levels increases the operating range of rods, which are saturated at about this light level, by reducing the transducin-mediated amplification of light.

Other circuit reconfigurations

As he was first describing the retinal circuits that give rise to RGC receptive fields, Kuffler (1953) speculated that, even with fixed wiring, functional reconfigurations in retinal circuits could allow specialized computations in different visual conditions, like different luminance levels. Almost 70 years later, we are still only beginning to understand the functional reconfigurations that retinal circuits undergo with luminance. It has long been known that, under

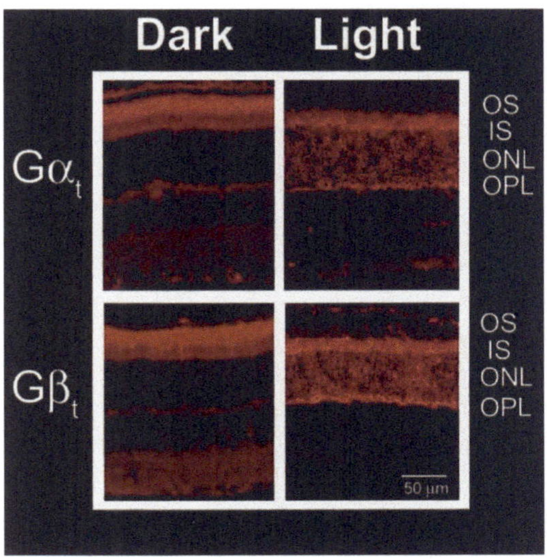

Fig. 2.9 Light-dependent translocation of transducin within photoreceptors. Images of antibody staining against the alpha *(top)* and beta *(bottom)* subunits of transducin in dark and light conditions. Layers in the retinal section are indicated: rod outer segments (OS), inner segments (IS), outer nuclear layer (ONL), outer plexiform layer (OPL). *(From Sokolov et al. (2002).)*

low light levels, the suppressive surround of some ganglion cells weakens or even nearly disappears (Barlow et al., 1957). While changes in the outer retina, as described earlier, certainly play a role in this phenomenon, more recent work has suggested that some ACs switch on only within a certain range of luminances (Farrow et al., 2013). Texture sensitivity, a retinal computation we will explore in Chapter 7, also depends on mean illumination through a mechanism that involves control of the resting potential of ON cone BCs by the AII ACs to which they are electrically coupled (Grimes et al., 2014).

Landmark paper: Dunn, F.A., and Rieke, F. (2008). Single-photon absorptions evoke synaptic depression in the retina to extend the operational range of rod vision. Neuron *57*(6), 894–904.

Discussion questions:

1. Fig. 2A–D uses paired recordings between RBCs and AII ACs to show a reduction in gain with background light. Why do the authors need an additional experiment to prove that the gain change is localized to this synapse? How does the voltage-clamp experiment support the conclusion that the gain change is localized to this synapse?
2. If the results in Fig. 2 prove that gain control is localized to the RBC to AII AC synapse, why do the authors additionally check for the role of inhibition in Fig. 3? While they did not find a role for inhibition, explain how the alternative result would have been possible. How can a mechanism be confined to a single synapse but still involve inhibition? Hint: Think about the anatomical specializations of bipolar cell synapses.
3. The first paragraph in the section "Gain Modulated by Single Photons: Paired-Flash Depression" describes the reasoning that predicted that single photons reduce gain at the RBC to AII AC synapse. Write an equation that captures this argument and explain it in your own words.
4. What experiment suggested that the gain control mechanism is located on the presynaptic side of the RBC to AII AC synapse? Which figure panel shows these data?
5. How do the authors explain the benefit of a gain control mechanism based on single-photon responses that are inherently noisy? Why would longer temporal integration be inappropriate? What about larger spatial integration?

Landmark paper: Angueyra, J.M., and Rieke, F. (2013). Origin and effect of phototransduction noise in primate cone photoreceptors. Nature Neuroscience *16*(11), 1692–1700.

Discussion questions:

1. Why does exposing a cone to very bright light (as in Fig. 1a) isolate the instrumental and inner segment noise? What does the measurement tell you about the ability of these authors to measure the phototransduction noise in cones?
2. Explain the rationale for using the power spectrum from the average dim-flash response to measure the low-frequency end of the noise power spectrum. About how much of the noise in cones could be attributed to this component? How is this different from what you learned about rods in Chapter 1?
3. Why did the authors use voltage clamp instead of current clamp to measure cone noise?
4. How does the two-electrode recording method work? Explain how the controls in Fig. 2 validate the method.
5. Which frequencies were most affected in the experiments shown in Figs. 3 and 4, respectively? What main conclusions could the authors draw from these differences?
6. Explain how Figs. 5–7 make the argument that signal and noise adapt similarly in rods but differently in cones and how that affects predictions of threshold-vs-intensity curves in perception.

Programming exercise

To access the programming exercise, please visit https://www.elsevier.com/books-and-journals/book-companion/9780128198964.

This chapter began with a discussion of how the dynamic range in natural scenes in daylight poses a challenge for luminance adaptation. Rapid adaptation within cones plays an important role in preventing saturation under these conditions. This exercise examines the optimization of the timecourse of cone adaptation based on real eye movement data from human subjects.

Load the file "DOVES_traces.mat." These data come from the DOVES database of eye movements (Van Der Linde et al., 2009). It includes eye movement traces of 29 subjects looking at calibrated natural images from a different database (Van Hateren and Van der Schaaf, 1998). I have extracted the luminance traces from the nearest pixel to the center of gaze for each of the 29 subjects for a single image. The variable "imTrace" contains these light intensity traces sampled at 200 Hz. The variable "imTrace_highPass" contains the same traces high-pass filtered at 5 Hz to remove the mean luminance changes from full saccades but retain the local sampling from fixational eye movements.

The goal of this exercise is to explore how a cone with a limited dynamic range can use adaptation to encode these large-dynamic-range intensity sequences. The included function "coneModel" implements a simple "optimal" adaptation strategy based on the highly influential work on optimal encoding by Simon Laughlin (1981), in which he showed that a response function that approximated the cumulative distribution of the stimulus variable was optimal for information transmission. The model has two parameters. First, "dynamicRange" is the number of discrete response values the model can take. The images themselves are 8-bit, so they span 256 values. Start by setting dynamicRange to 25, so that the cone capacity is about 10 times less than that of the image. The other parameter, "timeWindow," is the amount of time (in ms) the cone samples take to estimate the luminance distribution. Start with a value of 50 ms.

1. Plot some of the traces in "imTrace." Can you tell where the saccades are? Try out the model like this: "coneTrace = coneAdapModel (imTrace{1}, 25, 50). Plot "coneTrace" along with the original image trace to see how well they match. Explain which aspects of the trace are captured well and which are captured less well.
2. Now try using the included "compareTraces" function to explore some more details about how the model is encoding the stimulus. It is called like this: "[cc, errPerConeBin] = compareTraces(coneTrace, imTrace {1}, imTrace_highPass{1}, 50, 1)." The correlation coefficient (cc) and the error per cone bin (errPerConeBin) are two error metrics in matching the cone response to the high-pass filtered luminance trace. The last column of the resulting figure plots the actual intensity vs. the model cone response. How is this plot different for the original vs. the high-pass filtered version? What does this tell you about what aspects of the luminance trace the cone is encoding and what aspects it is failing to encode?
3. Vary dynamic range from all the way down to 2 (binary responses) up to 50 to see how this parameter affects the model. Plot the dependence of both error metrics on dynamic range. Is this relationship what you expected? Why or why not?
4. Now vary the time window from 0 (no adaptation) to 500 ms. Again, plot the dependence of the error metrics on time. What is the shape of this relationship? How do you explain the peak?
5. Explore both model parameters systematically by making a heatmap of each error metric over a matrix where you vary both "dynamicRange"

and "timeWindow." Average the error across the traces from all 29 subjects. What do these heatmaps reveal about the relationship between these two variables?

References

Akopian, A., Johnson, J., Gabriel, R., Brecha, N., and Witkovsky, P. (2000). Somatostatin modulates voltage-gated K+ and Ca2+ currents in rod and cone photoreceptors of the salamander retina. Journal of Neuroscience *20*(3), 929–936.

Anastassov, I.A., Wang, W., and Dunn, F.A. (2019). Synaptogenesis and synaptic protein localization in the postnatal development of rod bipolar cell dendrites in mouse retina. Journal of Comparative Neurology *527*(1), 52–66.

Angueyra, J.M., and Rieke, F. (2013). Origin and effect of phototransduction noise in primate cone photoreceptors. Nature Neuroscience *16*(11), 1692–1700.

Barlow, H.B. (1957). Increment thresholds at low intensities considered as signal/noise discriminations. The Journal of Physiology *136*(3), 469–488.

Barlow, H.B., Fitzhugh, R., and Kuffler, S.W. (1957). Change of organization in the receptive fields of the cat's retina during dark adaptation. The Journal of Physiology *137*(3), 338–354.

Bonaventure, N., Jardon, B., Sahel, J., and Wioland, N. (1989). Neurotransmission in the frog retina: possible physiological and histological correlations. Documenta Ophthalmologica *72*(1), 71–82.

Bruch Goldstein, E., and Wolf, B.M. (1973). Regeneration of the green-rod pigment in the isolated frog retina. Vision Research *13*(3), 527–534.

Bruun, A., and Ehinger, B. (1993). NPY-induced neurotransmitter release from the rabbit and chicken retina. Acta Ophthalmologica *71*(5), 590–596.

Calvert, P.D., Strissel, K.J., Schiesser, W.E., Pugh, E.N., and Arshavsky, V.Y. (2006). Light-driven translocation of signaling proteins in vertebrate photoreceptors. Trends in Cell Biology *16*(11), 560–568.

Cervia, D., Casini, G., and Bagnoli, P. (2008). Physiology and pathology of somatostatin in the mammalian retina: a current view. Molecular and Cellular Endocrinology *286*(1–2), 112–122.

Chen, S.K., Badea, T.C., and Hattar, S. (2011). Photoentrainment and pupillary light reflex are mediated by distinct populations of ipRGCs. Nature *476*(7358), 92–96.

Donner, K., Copenhagen, D.R., and Reuter, T. (1990). Weber and noise adaptation in the retina of the toad *Bufo marinus*. Journal of General Physiology *95*(4), 733–753.

Dunn, F.A., and Rieke, F. (2008). Single-photon absorptions evoke synaptic depression in the retina to extend the operational range of rod vision. Neuron *57*(6), 894–904.

Dunn, F.A., Doan, T., Sampath, A.P., and Rieke, F. (2006). Controlling the gain of rod-mediated signals in the mammalian retina. Journal of Neuroscience *26*(15), 3959–3970.

Dunn, F.A., Lankheet, M.J., and Rieke, F. (2007). Light adaptation in cone vision involves switching between receptor and post-receptor sites. Nature *449*(7162), 603–606.

Eldred, W.D., and Blute, T.A. (2005). Imaging of nitric oxide in the retina. Vision Research *45*(28), 3469–3486.

Farrow, K., Teixeira, M., Szikra, T., Viney, T.J., Balint, K., Yonehara, K., and Roska, B. (2013). Ambient illumination toggles a neuronal circuit switch in the retina and visual perception at cone threshold. Neuron *78*(2), 325–338.

Field, G.D., and Rieke, F. (2002). Nonlinear signal transfer from mouse rods to bipolar cells and implications for visual sensitivity. Neuron *34*(5), 773–785.

Frazor, R.A., and Geisler, W.S. (2006). Local luminance and contrast in natural images. Vision Research *46*(10), 1585–1598.

Gastinger, M.J., Yusupov, R.G., Glickman, R.D., and Marshak, D.W. (2004). The effects of histamine on rat and monkey retinal ganglion cells. Visual Neuroscience *21*(6), 935–943.

Gastinger, M.J., Tian, N., Horvath, T., and Marshak, D.W. (2006). Retinopetal axons in mammals: emphasis on histamine and serotonin. Current Eye Research *31*(7–8), 655–667.

Grimes, W.N., Schwartz, G.W., and Rieke, F. (2014). The synaptic and circuit mechanisms underlying a change in spatial encoding in the retina. Neuron *82*(2), 460–473.

Gustincich, S., Feigenspan, A., Wu, D.K., Koopman, L.J., and Raviola, E. (1997). Control of dopamine release in the retina: a transgenic approach to neural networks. Neuron *18*(5), 723–736.

Hartveit, E., and Veruki, M.L. (2012). Electrical synapses between AII amacrine cells in the retina: function and modulation. Brain Research *1487*, 160–172.

He, L., Wendt, M., and Glasser, A. (2013). Long-term reproducibility of Edinger-Westphal stimulated accommodation in rhesus monkeys. Experimental Eye Research *113*, 80–86.

Ingram, N.T., Sampath, A.P., and Fain, G.L. (2016). Why are rods more sensitive than cones? Journal of Physiology *594*(19), 5415–5426.

Iuvone, P.M., Galli, C.L., Garrison-Gund, C.K., and Neff, N.H. (1978). Light stimulates tyrosine hydroxylase activity and dopamine synthesis in retinal amacrine neurons. Science *202*(4370), 901–902.

Jacoby, J., Nath, A., Jessen, Z.F., and Schwartz, G.W. (2018). A self-regulating gap junction network of amacrine cells controls nitric oxide release in the retina. Neuron *100*(5), 1149–1162.e5.

Jardon, B., Bonaventure, N., and Scherrer, E. (1992). Possible involvement of cholinergic and glycinergic amacrine cells in the inhibition exerted by the ON retinal channel on the OFF retinal channel. European Journal of Pharmacology *210*(2), 201–207.

Jin, N.G., and Ribelayga, C.P. (2016). Direct evidence for daily plasticity of electrical coupling between rod photoreceptors in the mammalian retina. The Journal of Neuroscience *36*(1), 178–184.

Jones, R.S., Carroll, R.C., and Nawy, S. (2012). Light-induced plasticity of synaptic ampa receptor composition in retinal ganglion cells. Neuron *75*(3), 467–478.

Kothmann, W.W., Massey, S.C., and O'Brien, J. (2009). Dopamine-stimulated dephosphorylation of connexin 36 mediates AII amacrine cell uncoupling. Journal of Neuroscience *29*(47), 14903–14911.

Kourennyi, D.E., Liu, X.D., Hart, J., Mahmud, F., Baldridge, W.H., and Barnes, S. (2004). Reciprocal modulation of calcium dynamics at rod and cone photoreceptor synapses by nitric oxide. Journal of Neurophysiology *92*(1), 477–483.

Kuffler, S.W. (1953). Discharge patterns and functional organization of mammalian retina. Journal of Neurophysiology *16*(1), 37–68.

Laughlin, S. (1981). A simple coding procedure enhances a neuron's information capacity. Zeitschrift fuer Naturforschung, C: Journal of Biosciences *36*(9–10), 910–912.

Lawrence, P.M., and Studholme, K.M. (2014). Retinofugal projections in the mouse. Journal of Comparative Neurology *522*(16), 3733–3753.

Li, H., Chuang, A.Z., and O'brien, J. (2014). Regulation of photoreceptor gap junction phosphorylation by adenosine in zebrafish retina. Visual Neuroscience *31*(3), 237–243.

Mangel, S.C., and Brunken, W.J. (1992). The effects of serotonin drugs on horizontal and ganglion cells in the rabbit retina. Visual Neuroscience *8*(3), 213–218.

Mazade, R.E., and Eggers, E.D. (2019). Inhibitory components of retinal bipolar cell receptive fields are differentially modulated by dopamine D1 receptors. Visual Neuroscience *37*, E01.

Middleton, T.P., Huang, J.Y., and Protti, D.A. (2019). Cannabinoids modulate light signaling in ON-sustained retinal Ganglion cells of the mouse. Frontiers in Neural Circuits *13*, 37.

Mills, S.L., and Massey, S.C. (1995). Differential properties of two gap junctional pathways made by AII amacrine cells. Nature *377*(6551), 734–737.

Miraucourt, L.S., Tsui, J., Gobert, D., Desjardins, J.-F., Schohl, A., Sild, M., Spratt, P., Castonguay, A., De Koninck, Y., Marsh-Armstrong, N., et al. (2016). Endocannabinoid signaling enhances visual responses through modulation of intracellular chloride levels in retinal ganglion cells. eLife *5*, e15932.

Morshedian, A., Kaylor, J.J., Ng, S.Y., Tsan, A., Frederiksen, R., Xu, T., Yuan, L., Sampath, A.P., Radu, R.A., Fain, G.L., et al. (2019). Light-driven regeneration of cone visual pigments through a mechanism involving RGR opsin in Müller glial cells. Neuron *102*(6), 1172–1183.e5.

Nakahama, H., Shima, K., Yamamoto, M., and Aya, K. (1981). Regularity of the spontaneous discharge of neurons in the nucleus raphe dorsalis of the cat. Neuroscience Letters *23*(2), 161–165.

Neal, M., and Cunningham, J. (1994). Modulation by endogenous ATP of the light-evoked release of ACh from retinal cholinergic neurones. British Journal of Pharmacology *113*(4), 1085–1087.

Newman, E.A. (2003). Glial cell inhibition of neurons by release of ATP. Journal of Neuroscience *23*(5), 1659–1666.

Peterson, J.J., Tam, B.M., Moritz, O.L., Shelamer, C.L., Dugger, D.R., McDowell, J.H., Hargrave, P.A., Papermaster, D.S., and Smith, W.C. (2003). Arrestin migrates in photoreceptors in response to light: a study of arrestin localization using an arrestin-GFP fusion protein in transgenic frogs. Experimental Eye Research *76*(5), 553–563.

Ribelayga, C., and Mangel, S.C. (2005). A circadian clock and light/dark adaptation differentially regulate adenosine in the mammalian retina. Journal of Neuroscience *25*(1), 215–222.

Ribelayga, C., Cao, Y., and Mangel, S.C. (2008). The circadian clock in the retina controls rod-cone coupling. Neuron *59*(5), 790–801.

Rieke, F., and Rudd, M.E. (2009). The challenges natural images pose for visual adaptation. Neuron *64*(5), 605–616.

Santos-Carvalho, A., Ambrósio, A.F., and Cavadas, C. (2015). Neuropeptide Y system in the retina: from localization to function. Progress in Retinal and Eye Research *47*, 19–37.

Schneeweis, D.M., and Schnapf, J.L. (1999). The photovoltage of macaque cone photoreceptors: adaptation, noise, and kinetics. The Journal of Neuroscience *19*(4), 1203–1216.

Schwartz, G.W., and Rieke, F. (2013). Controlling gain one photon at a time. eLife *2*, e00467.

Sinclair, J.R., Jacobs, A.L., and Nirenberg, S. (2004). Selective ablation of a class of amacrine cells alters spatial processing in the retina. Journal of Neuroscience *24*(6), 1459–1467.

Snellman, J., and Nawy, S. (2004). cGMP-dependent kinase regulates response sensitivity of the mouse on bipolar cell. Journal of Neuroscience *24*(29), 6621–6628.

Sodhi, P., and Hartwick, A.T.E. (2014). Adenosine modulates light responses of rat retinal ganglion cell photoreceptors througha cAMP-mediated pathway. Journal of Physiology *592*(19), 4201–4220.

Sokolov, M., Lyubarsky, A.L., Strissel, K.J., Savchenko, A.B., Govardovskii, V.I., Pugh, E.N., and Arshavsky, V.Y. (2002). Massive light-driven translocation of transducin between the two major compartments of rod cells: a novel mechanism of light adaptation. Neuron *34*(1), 95–106.

Strang, C.E., Renna, J.M., Amthor, F.R., and Keyser, K.T. (2010). Muscarinic acetylcholine receptor localization and activation effects on ganglion response properties. Investigative Ophthalmology and Visual Science *51*(5), 2778–2789.

Tooker, R.E., Lipin, M.Y., Leuranguer, V., Rozsa, E., Bramley, J.R., Harding, J.L., Reynolds, M.M., and Vigh, J. (2013). Nitric oxide mediates activity-dependent plasticity of retinal bipolar cell output via S-nitrosylation. Journal of Neuroscience *33*(49), 19176–19193.

Trulson, M.E., and Jacobs, B.L. (1979). Raphe unit activity in freely moving cats: correlation with level of behavioral arousal. Brain Research *163*(1), 135–150.

Tsukamoto, Y., Morigiwa, K., Ueda, M., and Sterling, P. (2001). Microcircuits for night vision in mouse retina. Journal of Neuroscience *21*(21), 8616–8623.

Tuten, W.S., Harmening, W.M., Sabesan, R., Roorda, A., and Sincich, L.C. (2017). Spatiochromatic interactions between individual cone photoreceptors in the human Retina. Journal of Neuroscience *37*(39), 9498–9509.

Van Der Linde, I., Rajashekar, U., Bovik, A.C., and Cormack, L.K. (2009). DOVES: a database of visual eye movements. Spatial Vision *22*(2), 161–177.

Van Hateren, J.H., and Van der Schaaf, A. (1998). Independent component filters of natural images compared with simple cells in primary visual cortex. Proceedings of the Royal Society B: Biological Sciences *265*(1394), 359–366.

Vielma, A.H., Agurto, A., Valdés, J., Palacios, A.G., and Schmachtenberg, O. (2014). Nitric oxide modulates the temporal properties of the glutamate response in type 4 OFF bipolar cells. PLoS ONE *9*(12), e114330.

Wang, J.S., and Kefalov, V.J. (2009). An alternative pathway mediates the mouse and human cone visual cycle. Current Biology *19*(19), 1665–1669.

Wang, G.Y., van der List, D.A., Nemargut, J.P., Coombs, J.L., and Chalupa, L.M. (2007). The sensitivity of light-evoked responses of retinal ganglion cells is decreased in nitric oxide synthase gene knockout mice. Journal of Vision 7(14). https://doi.org/10.1167/7.14.7.

Wright, C.B., Redmond, T.M., and Nickerson, J.M. (2015). A history of the classical visual cycle. Progress in Molecular Biology and Translational Science *134*, 433–448.

Xue, T., Do, M.T.H., Riccio, A., Jiang, Z., Hsieh, J., Wang, H.C., Merbs, S.L., Welsbie, D. S., Yoshioka, T., Weissgerber, P., et al. (2011). Melanopsin signalling in mammalian iris and retina. Nature *479*(7371), 67–72.

Yazulla, S. (2008). Endocannabinoids in the retina: from marijuana to neuroprotection. Progress in Retinal and Eye Research *27*(5), 501–526.

Yoshida, A., Pozdnyakov, N., Dang, L., Orselli, S.M., Reddy, V.N., and Sitaramayya, A. (1995). Nitric oxide synthesis in retinal photoreceptor cells. Visual Neuroscience *12*(3), 493–500.

Zhang, A.J., Jacoby, R., and Wu, S.M. (2011). Light- and dopamine-regulated receptive field plasticity in primate horizontal cells. Journal of Comparative Neurology *519*(11), 2125–2134.

Light detection is not only for vision. The different light environments within this room have profound effects on mood and sleep. An evolutionarily ancient set of retinal circuits encodes absolute luminance to keep our physiology synchronized with slow luminance changes.

CHAPTER 3

Absolute luminance detection

Gregory William Schwartz[a,b,c]
[a]Departments of Ophthalmology and Physiology, Feinberg School of Medicine, Northwestern University, Chicago, IL, United States
[b]Department of Neurobiology, Weinberg College of Arts and Sciences, Northwestern University, Chicago, IL, United States
[c]Northwestern Interdepartmental Neuroscience Program (NUIN), Northwestern University, Chicago, IL, United States

The previous chapter described five retinal mechanisms for fast luminance adaptation and potentially dozens more that act on timescales from seconds to hours. Clearly, the retina is highly specialized for adapting to mean luminance and transmitting contrast: positive and negative change from the mean over both space (see Chapter 8) and time. But what if there is a need to retain information about absolute luminance? Is that information completely lost at the level of the retinal output? Reflexes, like the pupillary light reflex (PLR) and more complex behaviors, like circadian photoentrainment, depend on a measurement of absolute luminance averaged over temporal and spatial scales at which luminance adaptation acts to cancel the signal. Thus, counter to our intuition from human-made devices, like cameras and photomultipliers, contrast detection is the default in the retina. A highly specialized pathway is needed to circumvent some of the luminance adaptation mechanisms and relay absolute luminance information.

The source of the absolute luminance signal in the retina was a mystery until about 20 years ago when it was discovered that a subset of retinal ganglion cells (RGCs) represents a third class of photoreceptors in the vertebrate retina (Berson et al., 2002; Hattar et al., 2002; Provencio et al., 2000). These cells, called intrinsically photosensitive (ip)RGCs, contain the photopigment melanopsin, making them *directly* responsive to light. While contributions from other RGC types have not been conclusively excluded, our current understanding is that absolute luminance signals are relayed to the brain via ipRGCs. While all ipRGCs receive inputs from rods and cones that are subject to most (but perhaps not all) of the circuit mechanisms for fast luminance adaptation, their own phototransduction cascade works on longer spatial and temporal scales to support the encoding of absolute luminance. A more thorough treatment of ipRGCs, including details of their phototransduction, upstream circuits, downstream targets, and behavioral consequences, can be

Retinal Computation
https://doi.org/10.1016/B978-0-12-819896-4.00004-4

found in an excellent contemporary review by Michael Do (2019). Here, I will focus specifically on what is known and unknown about the mechanisms by which ipRGCs encode absolute luminance.

IpRGC types

Six different ipRGC types have been identified in mouse and named M1–M6 (Fig. 3.1A and B) (Aranda and Schmidt, 2020; Lee and Schmidt, 2018; Quattrochi et al., 2019; Sand et al., 2012; Schmidt et al., 2008, 2011; Sondereker et al., 2020). Homologs to some of these types have been identified in other mammalian species (Fig. 3.1C), and, in particular, primates, including humans, have at least the M1 and M2 types (Dacey et al., 2005; Mure et al., 2019; Nasir-Ahmad et al., 2019; Neumann et al., 2011). M1 ipRGCs contain much more melanopsin than the other ipRGC types, and they are both necessary and sufficient (among RGCs) for most aspects of the PLR and circadian photoentrainment (Altimus et al., 2008; Chen et al., 2011; Guler et al., 2008; Hatori et al., 2008). M4 ipRGCs are synonymous with ON alpha RGCs (Ecker et al., 2010; Estevez et al., 2012; Schmidt et al., 2014), a well-studied RGC type in the mouse (Pang et al., 2003). While melanopsin drives a very small current

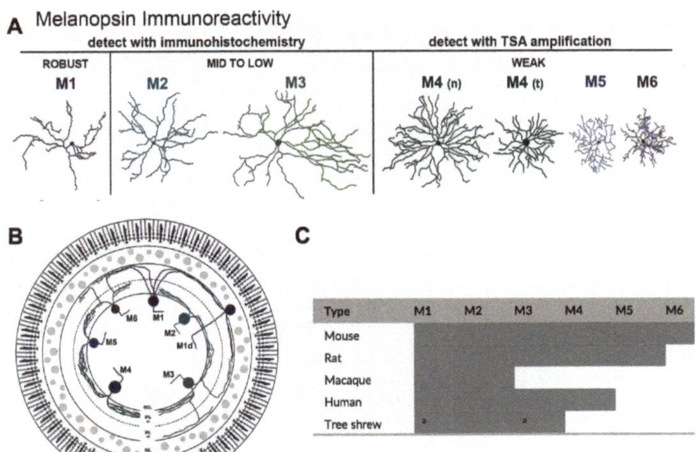

Fig. 3.1 IpRGC types. (**A**) En face morphologies of ipRGC types in mouse separated into classes by their melanopsin immunoreactivity. Both nasal (n) and temporal (t) M4 examples are shown. (**B**) Schematic of the stratification of each ipRGC type. (**C**) Table with shaded regions indicating which ipRGC types have been identified in which species. *(Modified from Sondereker et al. (2020).)*

in these RGCs, it has sustained effects on excitability (Sonoda et al., 2018). Careful studies have revealed many components of the phototransduction pathway in M1 and M4 ipRGCs, and remarkably, they are very different, but less is known about the role of melanopsin in the other ipRGC types and whether they support absolute luminance encoding. Therefore, I will focus on M1 and M4 ipRGCs.

Behavioral roles for absolute luminance detection

Before delving into the retinal mechanisms of absolute luminance detection, the varied behavioral roles of this computation are worth a brief discussion (reviewed by Warthen and Provencio, 2012; Schmidt et al., 2011; Sondereker et al., 2020). Light influences many aspects of physiology apart from visual perception, including sleep and arousal (Chen et al., 2011; Guler et al., 2008), body temperature (Rupp et al., 2019), melatonin production (Legates et al., 2012), development of the early visual system (Chew et al., 2017; Kirkby and Feller, 2013; Renna et al., 2011), and retinal vasculature (Rao et al., 2013), mood (Fernandez et al., 2018), and reflexes like the PLR (Chen et al., 2011; Guler et al., 2008). IpRGCs have traditionally been associated with these nonperceptual influences of light, collectively called "non-image-forming vision." However, recent work has demonstrated a role for melanopsin phototransduction in M4 ipRGCs, and perhaps in others as well, in "image-forming" vision (Schmidt et al., 2011, 2014; Sonoda et al., 2018). Retrograde effects of ipRGCs within retinal circuits further entangle absolute luminance detection with other retinal computations. Some M1 ipRGCs actually send axons into the inner plexiform layer (IPL) where they release glutamate onto dopaminergic amacrine cells to stimulate dopamine release (Prigge et al., 2016). Dopamine has a variety of effects, including on luminance adaptation (Chapter 2) and surround suppression (Chapter 8), though the specific links between ipRGCs and the various effects of dopamine in the retina remain largely unresolved. Axon collaterals of M1 ipRGCs have even been found in the outer retina where they could interact with rod and cone outputs (Joo et al., 2013).

The separation of the influences of melanopsin and ipRGCs into "image-forming" and "non-image-forming" functions may represent a false dichotomy (Sondereker et al., 2020). The computation of absolute luminance detection, by definition, is non-image-forming since it averages over space and time at scales much greater than those of objects in the visual environment. However, melanopsin may not be used exclusively for absolute luminance detection, and ipRGCs most certainly transmit signals about other aspects of visual stimuli. Additionally, while originally considered "daylight" detectors, only sensitive to very bright stimuli, it has become increasingly clear that ipRGCs can encode luminance information and influence behavior over a huge intensity range including in the scotopic range (Lee et al., 2019). There is still a great deal to learn about how melanopsin and absolute luminance detection interact with other retinal computations to influence both vision and other light-sensitive aspects of physiology and behavior.

Mechanisms for absolute luminance detection in the retina

M1 ipRGCs

Spatial integration

Large spatial integration is a natural consequence of melanopsin's distribution throughout the entire plasma membrane of ipRGCs. Whereas the collecting area in the outer segments of rods and cones in mice is ~1 μm^2 and 0.67 μm^2, respectively (reviewed by Luo et al., 2008), M1 ipRGCs are among the largest RGC types in mice, with a dendritic diameter ~350 μm, ~12 degrees of visual angle (Berson et al., 2010). M1 ipRGCs in primates are even larger, often over 1 mm in dendritic diameter (Liao et al., 2016). But because the primate eye is so much larger than that of the mouse, they cover a smaller region of visual space, ~5 degrees. Within the tiny lateral profile of rods and cones, thousands of membrane disks stack to support the extremely high number of opsin molecules, ~10^8 per cell (reviewed by Luo et al., 2008). By contrast, each M1 ipRGC has ~10^4 molecules of melanopsin total, spread out along the huge lateral extent of its soma, axon, and dendrites (Do et al., 2009). Thus, M1 ipRGCs appear to sacrifice spatial precision in photon capture to average over a large area in visual space.

Temporal integration

Temporal integration in M1 ipRGCs begins with the single-photon response of melanopsin. While the integration time of the single-photon response is ~80 ms in cones (Cao et al., 2014) and ~200 ms in rods (Mendez et al., 2000), it is dramatically longer in M1 ipRGCs: ~8 s (Do et al., 2009). Unlike in rods and cones, the details of the shutoff mechanism of melanopsin phototransduction remain unclear. Experiments have shown that repeating flashes of the same intensity causes a progressive increase in firing rate in M1 ipRGCs (Emanuel and Do, 2015). In steady illumination, the shutoff mechanism that truncates the single-photon response likely becomes saturated, leading to persistent stability of melanopsin in its active state (Do, 2019). Melanopsin is also very resistant to bleaching. Since both its active state and its two silent states are thermodynamically stable, following activation it does not need to unbind and rebind a chromophore to produce another response (Emanuel and Do, 2015). While chromophores are not indefinitely stable within melanopsin, there is a robust pigment regeneration mechanism that transports 11-*cis* retinal from the retinal pigmented epithelium through Müller glia to ipRGCs (Zhao et al., 2016).

In addition to their own phototransduction pathway, ipRGCs, like other RGCs, receive input from rods and cones via bipolar cells (BCs). Despite the location of their dendrites at the outer border of the IPL, excitatory synaptic input to M1 ipRGCs comes exclusively from ON BCs that make ectopic synapses in what is otherwise the OFF layer (Fig. 3.2A and B) (Dumitrescu et al., 2009; Hoshi et al., 2009). Ultrastructural experiments

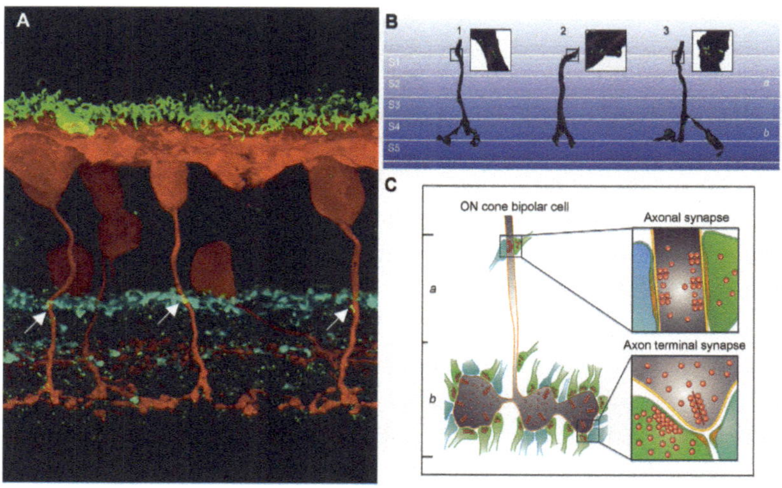

Fig. 3.2 Ectopic synapses from ON BCs make monad synapses near the outer border of the IPL. (**A**) Retinal section showing ON bipolar cells *(red)* with ribbons *(green)* on the axon stalk as it passes through the outermost portion of the IPL *(arrows)*. (**B**) EM reconstructions of ON bipolar cells with insets showing ribbons in sublamina 1 of the IPL. (**C**) Schematic of monad (axonal) ectopic synapses and dyad (axon terminal) synapses. *(From (A) Hoshi et al. (2009) and (B, C) Kim et al. (2012).)*

have shown that the ectopic synapses onto ipRGCs have a different morphology than most BC synapses in the rest of the IPL. Instead of forming a dyad with two postsynaptic cells, where one is typically an amacrine cell that provides local (reciprocal) inhibitory feedback, the ectopic BC synapses onto ipRGCs are monads. The monad synapses not only lack reciprocal inhibition, but they tend to contain more synaptic ribbons than other BC synapses (Fig. 3.2C) (Kim et al., 2012; Sabbah et al., 2018). The extra ribbons may endow the monads with larger and more quickly replenished vesicle pools to avoid some of the presynaptic depression mechanisms present at BC dyads. While this speculation from anatomy has not been demonstrated physiologically in the form of very long-lasting excitatory current in M1 ipRGCs (Zhao et al., 2017), the biophysical properties of the RGCs may also contribute to temporal integration. The input resistance in M1 ipRGCs is substantially higher than in the other RGC types in which it has been measured, and they rest very close to spike threshold (Do et al., 2009; Emanuel et al., 2017). Therefore, even the small, sustained tail of an otherwise transient excitatory current can cause a maintained increase in firing rate (Zhao et al., 2017). Rod-driven responses lasting tens of seconds were recently shown in M1 ipRGCs in melanopsin knockout animals (Lee et al., 2019).

Distributed coding

Like other RGCs, and neurons in general, ipRGCs have a limited dynamic range of spike output. M1 ipRGCs, in particular, appear to have a very limited dynamic range of spike output. Thus, it had been difficult to reconcile the large dynamic range of absolute luminance-sensitive behavior (at least 6 orders of magnitude) with the firing range of ipRGCs (~2 orders of magnitude or less for M1). Milner and Do (2017) introduced a fascinating solution to this apparent paradox. These authors measured melanopsin-driven absolute luminance tuning curves of M1 ipRGCs in mice across 8 orders of magnitude of light intensity. The majority of cells did not increase firing monotonically with luminance but instead had unimodal luminance-response functions, each peaking at a different intermediate light intensity (Fig. 3.3A–C). They went on to show that the falling phase of the M1 ipRGC tuning curves was caused by depolarization block. Beyond a threshold, which varied substantially within the M1 ipRGC population, further increases in luminance caused a steady depolarization that inactivated voltage-gated Na^+ channels, resulting in a sharp decrease in spike rate (Fig. 3.3D–F). Diversity in the biophysical properties of M1 ipRGCs

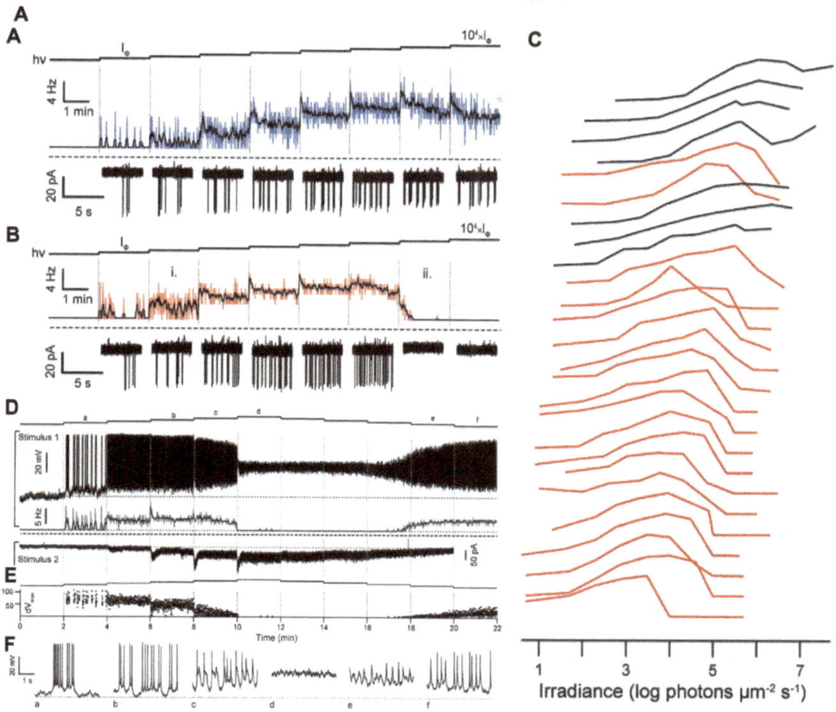

Fig. 3.3 M1 ipRGCs have heterogeneous luminance tuning curves and use depolarization block to stop firing in bright light. (**A**) Firing rates of an M1 ipRGC in response to steps of increasing luminance from threshold to 10,000 times brighter. (**B**) Same as (**A**) but for a different M1 ipRGC with a nonmonotonic relationship between luminance and firing rate. (**C**) Luminance-response curves for M1 ipRGCs ordered by their peak responses. Red traces are unimodal cells, black traces are monotonic cells. (**D**) *Top*, perforated patch-clamp recording showing depolarization block in an M1 ipRGC at high luminance. *Middle*, firing rate from the cell shown above. *Bottom*, voltage-clamp recording from the same RGC to a second luminance ramp following dark adaptation. (**E**) Maximum slope for each action potential in (**D**). (**F**) Magnified views of the trace from (**D**) at the labeled times. *(Modified from Milner and Do (2017).)*

supported the diversity in their luminance tuning curves (Emanuel et al., 2017). This distributed code for luminance among the M1 ipRGC population represents another specialization of these cells. Response homogeneity is presumably an important optimization principle for other RGC types, so that they can encode the same stimulus features across visual space. No such spatial constraint exists for M1 ipRGCs; instead of spreading their tuning curves over space, they spread them over the luminance range (Milner and Do, 2017).

Not only do they vary in their biophysical properties, but M1 ipRGCs also vary in the extent of their synaptic input (Lee et al., 2019). Heterogeneity in the M1 ipRGC population also extends into the brain, where different subpopulations innervate distinct brain regions, presumably to support behaviors that require absolute luminance detection in different intensity ranges (Fernandez et al., 2018; Guler et al., 2008; Rupp et al., 2019). There is even a population of M1 ipRGCs that release GABA at their postsynaptic targets in the brain (Sonoda et al., 2020). Selective elimination of GABAergic ipRGCs actually *increases* the sensitivity of the PLR and circadian photoentrainment (Fig. 3.4), suggesting that one of the roles

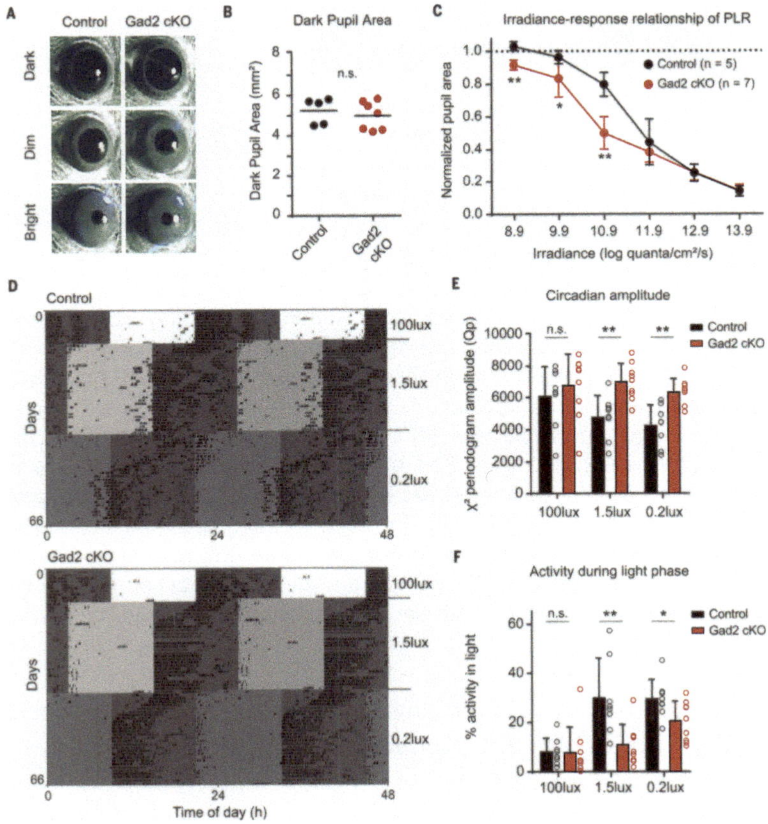

Fig. 3.4 GABAergic ipRGCs influence behavioral sensitivity to light. (**A**) Pupil constriction in WT mice and Gad2 cKO mice in which RGCs do not release GABA. (**B**) Dark pupil area in WT and Gad2 cKO mice was not statistically different. (**C**) Pupil area as a function of luminance for WT and Gad2 cKO mice. (**D**) Actograms of control and Gad2 cKO mice. Ticks indicate running time. (**E** and **F**) Quantification of circadian amplitude and activity levels for WT and Gad2 cKO mice. *(From Sonoda et al. (2020).)*

of this subpopulation is to dampen the light sensitivity of these reflexes so that they are not engaged by small fluctuations in luminance (Sonoda et al., 2020).

M4 ipRGCs

Unlike M1 ipRGCs, ON alpha RGCs were identified long before it was known that they contained melanopsin and thus gained the name M4. These RGCs receive robust input through traditional photoreceptor pathways, they are very sensitive to contrast, and they project to the dorsal lateral geniculate nucleus (dLGN) where they have been associated with the conscious perception of contrast (Román Rosón et al., 2019; Rompani et al., 2017; Schmidt et al., 2014). As mentioned earlier, melanopsin expression is substantially lower in M4 ipRGCs than in M1 ipRGCs. So what is the M4 ipRGC doing in a chapter on absolute luminance detection?

Tiffany Schmidt et al. (2014) showed that M4 ipRGCs had slow and sustained light responses in the absence of rod and cone inputs (Fig. 3.5A and B) and that melanopsin transduction increased their contrast sensitivity when rod and cone signals were intact (Sonoda et al., 2018). Mice lacking melanopsin or mice with many M4 ipRGCs ablated showed deficits in a perceptual test of contrast sensitivity (Fig. 3.5C) (Schmidt et al., 2014). Thus, even with a small fraction of the melanopsin present in M1 ipRGCs, M4 ipRGCs can integrate light signals over a long time period to influence behavior.

The magnitude of the melanopsin-dependent effects on M4 ipRGCs initially seemed inconsistent with the small (~50 pA) intrinsic currents,

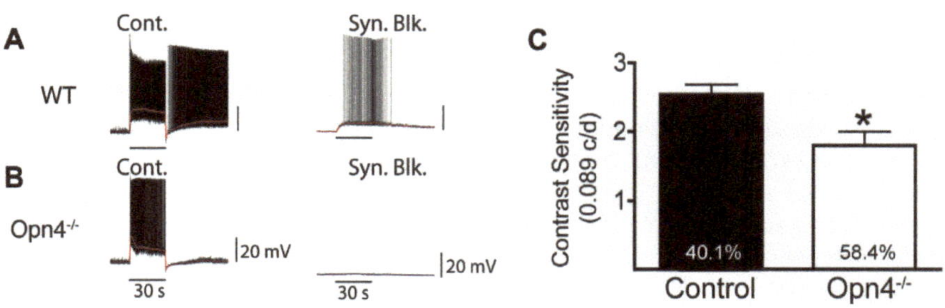

Fig. 3.5 M4 ipRGCs have intrinsic light responses and their melanopsin expression influences behavioral contrast sensitivity. (**A**) Response of an M4 ipRGC in control conditions *(left)* and in synaptic blockers *(right)*. (**B**) Same as (**A**) for a melanopsin knockout animal. (**C**) Behavioral measurement of contrast sensitivity for control and melanopsin knockout animals. *(Modified from Schmidt et al. (2014).)*

compared to the ~2 nA excitatory synaptic currents from bipolar cells. Sonoda et al. (2018) showed that the dominant phototransduction mechanism in M4 ipRGCs was different than in M1 ipRGCs (Fig. 3.6). While melanopsin phototransduction in M1 ipRGCs leads to the opening of nonspecific cation channels, the phototransduction cascade in M4 ipRGCs instead closes leak K^+ channels. The closure of a sufficient number of these channels can cause the RGC to depolarize, but even a modest closure of channels substantially increases input resistance. Thus, upon the activation of melanopsin, even by steady, dim light, M4 ipRGCs become more excitable (Fig. 3.6A and B). With increased resistance, small excitatory currents from bipolar cells responding to low contrast are sufficient to drive spike responses in the RGC. The time-course of melanopsin-driven resistance changes in M4 ipRGCs has not been tested systematically, but it showed no apparent adaptation over at least 10 min.

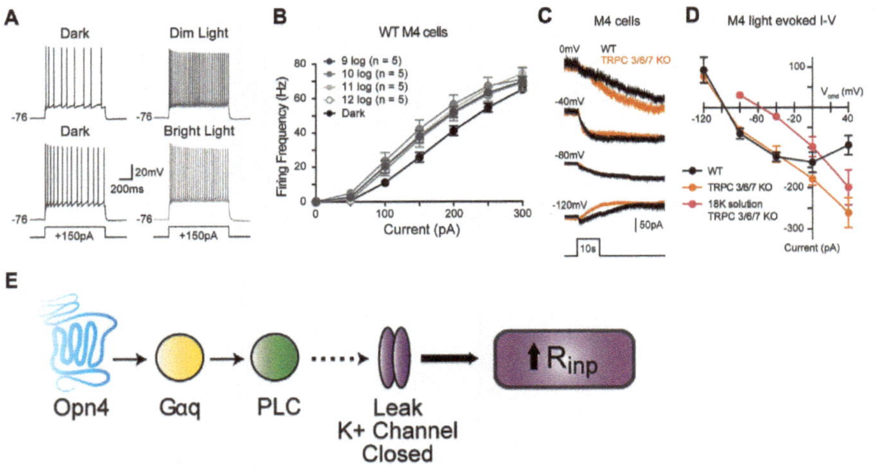

Fig. 3.6 Melanopsin phototransduction in M4 ipRGCs closes leak potassium channels to increase excitability. (**A**) Depolarization-evoked spike responses in M4 ipRGCs in darkness, dim and bright light. (**B**) Firing rate as a function of current injected in darkness and at four levels of background illumination. (**C**) Voltage-clamp recordings of light responses from an M4 ipRGC in a WT mouse *(black)* and in a triple knockout lacking the CNG channels required for phototransduction in M1 ipRGCs *(orange)*. (**D**) Current-voltage relationship for M4 ipRGC transduction currents in WT *(black)*, triple knockouts *(orange)*, and triple knockouts in a solution that altered the potassium reversal potential *(magenta)*. (**E**) Schematic of the phototransduction cascade in M4 ipRGCs. *(Modified from Sonoda et al. (2018).)*

Additional "non-visual" opsins

In addition to melanopsin, encoded by the Opn4 gene, there are at least two other noncanonical opsins, neuropsin (Opn5) and encephalopsin (Opn3), found in the mammalian retina, including in humans (Kojima et al., 2011; Nieto et al., 2011). Although neuropsin-dependent light responses have yet to be demonstrated in mammalian neurons within the intact retina, several functions for the pigment in the eye have been shown. Neuropsin plays a role in vascular development in the retina (Nguyen et al., 2019), it is necessary for entertaining the local circadian oscillator in the retina and cornea (Buhr et al., 2015; Van Gelder and Buhr, 2016), and it is upregulated and light sensitive in the cornea itself following a wound (Díaz et al., 2020). Even less is known about the role of encephalopsin in light sensation, but there is a report of expression in Müller glia in chicken driving a light response (Rios et al., 2019). Encephalopsin appears to be expressed broadly in the retina and even in some parts of the brain (Kato et al., 2016). The study of these additional noncanonical opsins is a rapidly emerging field, so there will surely be discoveries in the coming years about their role, if any, in absolute luminance detection in the retina and beyond and how they influence light-sensation behavior.

Conclusions and future directions

Though the field is only 20 years old, it is remarkable how much has been learned about ipRGCs, from the molecular pathways of phototransduction to their circuits within the retina, their projections in the brain, and especially, their roles in behavior. Genetic tools in mice have been particularly well utilized in this field to link specific RGC types and their targets in the brain to well-defined behaviors. Of course, there is much more to be discovered. The details of melanopsin phototransduction have only been worked out in two of the six ipRGC types, and since they are completely different, it will be important to study the pathways in the other four types. We are only beginning to appreciate the complexity of the interactions between melanopsin signaling and the rod and cone pathways that converge in all ipRGCs. The sixth ipRGC type was published in 2019 (Quattrochi et al., 2019), so there still may be others yet to be found. The central projections of ipRGCs are extensive and vary by type, but they have not been fully characterized (Do, 2019). Finally, we have only begun to explore the roles of neuropsin and encephalopsin in light sensation and whether cell types beyond M1–M6 play a role in conveying absolute luminance signals within the retina or to the brain.

Landmark paper: Milner, E.S., and Do, M.T.H. (2017). A population representation of absolute light intensity in the mammalian retina. Cell *171*(4), 865–876.

Discussion questions:

1. What are two advantages of recording at the axon rather than the soma for the experiments shown in Fig. 1? What is being demonstrated in panel G and why?
2. In Fig. 2, what is the difference between the black and the red tuning curves in A? Explain the difference between how the tuning curves are being plotted in panel A vs. panel B. Why do the authors show both?
3. Which figure(s) in the paper report intrinsic (melanopsin-driven) light responses alone and which also include the rod and cone inputs? How are the methods different? Are you convinced that the results agree?
4. M1 ipRGCs only receive ON input and melanopsin depolarizes them, so why are there firing patterns in Fig. 5 that look like OFF responses?
5. For each of the following, what is the trend with light increase in an M1 ipRGC undergoing depolarization block: (A) inward current, (B) subthreshold membrane potential, (C) maximum slope of actual potentials, (D) spike rate?
6. What is the purpose of Fig. 7? Explain in your own words the biophysical mechanism of this effect.

Landmark paper: Sonoda, T., Lee, S.K., Birnbaumer, L., and Schmidt, T.M. (2018). Melanopsin phototransduction is repurposed by ipRGC subtypes to shape the function of distinct visual circuits. Neuron *99*(4), 754–767.e4.

Discussion questions:

1. How do the authors measure contrast sensitivity in M4 ipRGCs? Based on the results of Fig. 1 alone, what is the caveat with the interpretation that melanopsin activation in M4 ipRGCs enhances their contrast sensitivity?
2. Explain the experiment shown in Fig. 2. Does this address the caveat from question 1? Why or why not?
3. According to the data in Fig. 4, what is the dimmest light intensity that causes a melanopsin-driven depolarization in M4 ipRGCs? Is this inconsistent with the data in Fig. 1? Why or why not? What data in Fig. 5 support your conclusion?

4. Fig. 6 shows the effect of melanopsin activation on the intrinsic excitability of M1 ipRGCs. Was this considered in Milner and Do (2017)? How could it influence the interpretation of their data? How would it affect the depolarization-block mechanism in these cells?
5. What is the most direct evidence in this paper that melanopsin phototransduction in M4 ipRGCs closes K^+ channels?
6. What is the nucleated patch recording technique and why was it used in this study?
7. What is the timecourse of the photocurrent in M4 ipRGCs? How does this compare to the single-photon response in M1 ipRGCs?

Programming exercise

To access the programming exercise, please visit https://www.elsevier.com/books-and-journals/book-companion/9780128198964.

As mentioned earlier, M1 ipRGCs appear to have much more melanopsin protein than do M4 ipRGCs. Does this mean that they are more light sensitive? This exercise explores this question based on the different transduction pathways and different functional roles of M1 and M4 ipRGCs.

The file "melanopsin_LIF_models.zip" contains the code you need for this simulation. It implements a simple neuronal model called a leaky-integrate-and-fire (LIF) model to simulate spiking activity in M1 and M4 ipRGCs. The M1 model uses the single-photon response current from Do et al. (2009) each time a photon is absorbed by melanopsin. The M4 model instead implements a time-dependent change in resistance with a longer time constant. The models also have different starting input resistances consistent with measurements: 200 MΩ for the M1 and 100 MΩ for the M4.

1. Run the "setup" and "base model" section of the script "M1_script.m." Choose a short simulation time (~1 s) so you can see the voltage trajectory and then a longer time (~600 s) so you can get a good estimate of the firing rate. How does the firing rate depend on the amount of noise in the input current and on the resistance of the cell?
2. Now generate the melanopsin-generated current with the following command: "Imel = melActivationCurrent(simulation_time, sampleRate, mel_activity):" where mel_activity is the rate of photon absorptions in units of events per second. Plot "Imel" for different activity rates. What is the lowest rate at which the melanopsin-driven events overlap substantially in time?

3. Run the full script. It plots the firing rate as a function of the melanopsin activity rate with a dotted line to indicate baseline. Note the logarithmic scale. What value of melanopsin activity would you consider the sensitivity threshold? How could you be more quantitative about this selection than just looking at the curve?
4. Now open "M4_script.m" and note which parameters are different in the setup. Make the baseline current and run the LIF model. What is the baseline firing rate for the M4 model? Why is it so different than for the M1?
5. Now run the section with "melResistanceChange." Plot "Rmel" for different values of "mel_activity_rate." What is the lowest rate at which the melanopsin-driven events overlap substantially in time? How does this compare to what you observed for the currents in the M1 model?
6. Run the next section of the script and plot (I) so you can see the structure of the experiment. Zoom in to see the current pulse series. This simulates a set of 10 currents of increasing intensity to see how the cell responds. This is like the contrast-response function, as increasing contrast will cause increasing synaptic input from bipolar cells. Each series has 10 pulses. How many of them can you make out from the noise? Does this change from the first to the last set?
7. Run the final section of the M4_script. Does the M4 model do a good job encoding absolute luminance? Why or why not? How does the model contrast-response function change from the first to the last minute of the experiment? Is this similar to data in Sonoda et al. (2018)?
8. Explore a range of different melanopsin activity rates. How does the activity rate change your answer to question 7? If we assume that M4 ipRGCs have 100 times less melanopsin than M1 ipRGCs, in these models, which cell type is more sensitive? How do you define sensitivity? Does it depend on the length of the simulation? Why?

References

Altimus, C.M., Güler, A.D., Villa, K.L., McNeill, D.S., LeGates, T.A., and Hattar, S. (2008). Rods-cones and melanopsin detect light and dark to modulate sleep independent of image formation. Proceedings of the National Academy of Sciences of the United States of America *105*(50), 19998–20003.

Aranda, M.L., and Schmidt, T.M. (2020). Diversity of intrinsically photosensitive retinal ganglion cells: circuits and functions. Cellular and Molecular Life Sciences. https://doi.org/10.1007/s00018-020-03641-5.

Berson, D.M., Dunn, F.A, and Takao, M. (2002). Phototransduction by retinal ganglion cells that set the circadian clock. Science *295*(1998), 1070–1073.

Berson, D.M., Castrucci, A.M., and Provencio, I. (2010). Morphology and mosaics of melanopsin-expressing retinal ganglion cell types in mice. Journal of Comparative Neurology *518*(13), 2405–2422.

Buhr, E.D., Yue, W.W.S., Ren, X., Jiang, Z., Liao, H.W.R., Mei, X., Vemaraju, S., Nguyen, M.T., Reed, R.R., Lang, R.A., et al. (2015). Neuropsin (OPN5)-mediated photoentrainment of local circadian oscillators in mammalian retina and cornea. Proceedings of the National Academy of Sciences of the United States of America *112*(42), 13093–13098.

Cao, L.H., Luo, D.G., and Yau, K.W. (2014). Light responses of primate and other mammalian cones. Proceedings of the National Academy of Sciences of the United States of America *111*(7), 2752–2757.

Chen, S.K., Badea, T.C., and Hattar, S. (2011). Photoentrainment and pupillary light reflex are mediated by distinct populations of ipRGCs. Nature *476*(7358), 92–96.

Chew, K.S., Renna, J.M., McNeill, D.S., Fernandez, D.C., Keenan, W.T., Thomsen, M.B., Ecker, J.L., Loevinsohn, G.S., Vandunk, C., Vicarel, D.C., et al. (2017). A subset of iprgcs regulates both maturation of the circadian clock and segregation of retinogeniculate projections in mice. eLife *6*, e22861.

Dacey, D.M., Liao, H.W., Peterson, B.B., Robinson, F.R., Smith, V.C., Pokomy, J., Yau, K.W., and Gamlin, P.D. (2005). Melanopsin-expressing ganglion cells in primate retina signal colour and irradiance and project to the LGN. Nature *433* (7027), 749–754.

Díaz, N.M., Lang, R.A., Van Gelder, R.N., and Buhr, E.D. (2020). Wounding induces facultative Opn5-dependent circadian photoreception in the murine cornea. Investigative Ophthalmology & Visual Science *61*(6), 37.

Do, M.T.H. (2019). Melanopsin and the intrinsically photosensitive retinal ganglion cells: biophysics to behavior. Neuron *104*(2), 205–226.

Do, M.T.H., Kang, S.H., Xue, T., Zhong, H., Liao, H.W., Bergles, D.E., and Yau, K.W. (2009). Photon capture and signalling by melanopsin retinal ganglion cells. Nature *457*(7227), 281–287.

Dumitrescu, O.N., Pucci, F.G., Wong, K.Y., and Berson, D.M. (2009). Ectopic retinal ON bipolar cell synapses in the OFF inner plexiform layer: contacts with dopaminergic amacrine cells and melanopsin ganglion cells. Journal of Comparative Neurology *517*(2), 226–244.

Ecker, J.L., Dumitrescu, O.N., Wong, K.Y., Alam, N.M., Chen, S.K., LeGates, T., Renna, J.M., Prusky, G.T., Berson, D.M., and Hattar, S. (2010). Melanopsin-expressing retinal ganglion-cell photoreceptors: cellular diversity and role in pattern vision. Neuron *67*(1), 49–60.

Emanuel, A.J., and Do, M.T.H. (2015). Melanopsin tristability for sustained and broadband phototransduction. Neuron *85*(5), 1043–1055.

Emanuel, A.J., Kapur, K., and Do, M.T.H. (2017). Biophysical variation within the M1 type of ganglion cell photoreceptor. Cell Reports *21*(4), 1048–1062.

Estevez, M.E., Michelle Fogerson, P., Ilardi, M.C., Borghuis, B.G., Chan, E., Weng, S., Auferkorte, O.N., Demb, J.B., and Berson, D.M. (2012). Form and function of the M4 cell, an intrinsically photosensitive retinal ganglion cell type contributing to geniculocortical vision. Journal of Neuroscience *32*(39), 13608–13620.

Fernandez, D.C., Fogerson, P.M., Lazzerini Ospri, L., Thomsen, M.B., Layne, R.M., Severin, D., Zhan, J., Singer, J.H., Kirkwood, A., Zhao, H., et al. (2018). Light affects mood and learning through distinct retina-brain pathways. Cell *175*(1), 71–84.e18.

Guler, A.D., Ecker, J.L., Lall, G.S., Haq, S., Altimus, C.M., Liao, H.-W., Barnard, A.R., Cahill, H., Badea, T.C., Zhao, H., et al. (2008). Melanopsin cells are the principal conduits for rod-cone input to non-image-forming vision. Nature *453*(7191), 102–105.

Hatori, M., Le, H., Vollmers, C., Keding, S.R., Tanaka, N., Schmedt, C., Jegla, T., and Panda, S. (2008). Inducible ablation of melanopsin-expressing retinal ganglion cells reveals their central role in non-image forming visual responses. PLoS One *3*(6), e2451.

Hattar, S., Liao, H.W., Takao, M., Berson, D.M., and Yau, K.W. (2002). Melanopsin-containing retinal ganglion cells: architecture, projections, and intrinsic photosensitivity. Science *295*(5557), 1065–1070.

Hoshi, H., Liu, W.L., Massey, S.C., and Mills, S.L. (2009). ON inputs to the OFF layer: bipolar cells that break the stratification rules of the retina. Journal of Neuroscience *29*(28), 8875–8883.

Joo, H.R., Peterson, B.B., Dacey, D.M., Hattar, S., and Chen, S.K. (2013). Recurrent axon collaterals of intrinsically photosensitive retinal ganglion cells. Visual Neuroscience *30*(4), 175–182.

Kato, M., Sugiyama, T., Sakai, K., Yamashita, T., Fujita, H., Sato, K., Tomonari, S., Shichida, Y., and Ohuchi, H. (2016). Two opsin 3-related proteins in the chicken retina and brain: a TMT-type opsin 3 is a blue-light sensor in retinal horizontal cells, hypothalamus, and cerebellum. PLoS One *11*(11), e0163925.

Kim, H.L., Jeon, J.H., Koo, T.H., Lee, U.Y., Jeong, E., Chun, M.H., Moon, J.I., Massey, S.C., and Kim, I.B. (2012). Axonal synapses utilize multiple synaptic ribbons in the mammalian retina. PLoS One 7(12), e52295.

Kirkby, L.A., and Feller, M.B. (2013). Intrinsically photosensitive ganglion cells contribute to plasticity in retinal wave circuits. Proceedings of the National Academy of Sciences of the United States of America *110*(29), 12090–12095.

Kojima, D., Mori, S., Torii, M., Wada, A., Morishita, R., and Fukada, Y. (2011). UV-sensitive photoreceptor protein OPN5 in humans and mice. PLoS One *6*(10), e26388.

Lee, S.K., and Schmidt, T.M. (2018). Morphological identification of melanopsin-expressing retinal ganglion cell subtypes in mice. Methods in Molecular Biology *1753*, 275–287.

Lee, S.K., Sonoda, T., and Schmidt, T.M. (2019). M1 intrinsically photosensitive retinal ganglion cells integrate rod and melanopsin inputs to signal in low light. Cell Reports *29*(11), 3349–3355.e2.

Legates, T.A., Altimus, C.M., Wang, H., Lee, H.K., Yang, S., Zhao, H., Kirkwood, A., Weber, E.T., and Hattar, S. (2012). Aberrant light directly impairs mood and learning through melanopsin-expressing neurons. Nature *491*(7425), 594–598.

Liao, H.W., Ren, X., Peterson, B.B., Marshak, D.W., Yau, K.W., Gamlin, P.D., and Dacey, D.M. (2016). Melanopsin-expressing ganglion cells on macaque and human retinas form two morphologically distinct populations. Journal of Comparative Neurology *524*(14), 2845–2872.

Luo, D.G., Kefalov, V., and Yau, K.W. (2008). Phototransduction in rods and cones. In The Senses: A Comprehensive Reference (Elsevier), Vol. 1, pp. 269–301.

Mendez, A., Burns, M.E., Roca, A., Lem, J., Wu, L.W., Simon, M.I., Baylor, D.A., and Chen, J. (2000). Rapid and reproducible deactivation of rhodopsin requires multiple phosphorylation sites. Neuron *28*(1), 153–164.

Milner, E.S., and Do, M.T.H. (2017). A population representation of absolute light intensity in the mammalian retina. Cell *171*(4), 865–876.

Mure, L.S., Vinberg, F., Hanneken, A., and Panda, S. (2019). Functional diversity of human intrinsically photosensitive retinal ganglion cells. Science *366*(6470), 1251–1255.

Nasir-Ahmad, S., Lee, S.C.S., Martin, P.R., and Grünert, U. (2019). Melanopsin-expressing ganglion cells in human retina: morphology, distribution, and synaptic connections. Journal of Comparative Neurology *527*(1), 312–327.

Neumann, S., Haverkamp, S., and Auferkorte, O.N. (2011). Intrinsically photosensitive ganglion cells of the primate retina express distinct combinations of inhibitory neurotransmitter receptors. Neuroscience *199*,24–31.

Nguyen, M.T.T., Vemaraju, S., Nayak, G., Odaka, Y., Buhr, E.D., Alonzo, N., Tran, U., Batie, M., Upton, B.A., Darvas, M., et al. (2019). An opsin 5-dopamine pathway mediates light-dependent vascular development in the eye. Nature Cell Biology *21*(4), 420–429.

Nieto, P.S., Valdez, D.J., Acosta-Rodríguez, V.A., and Guido, M.E. (2011). Expression of novel opsins and intrinsic light responses in the mammalian retinal ganglion cell line RGC-5. Presence of Opn5 in the rat retina. PLoS One *6*(10), e26417.

Pang, J.-J.,Gao, F., and Wu, S.M. (2003). Light-evoked excitatory and inhibitory synaptic inputs to ON and OFF alpha ganglion cells in the mouse retina. Journal of Neuroscience *23*(14), 6063–6073.

Prigge, C.L., Yeh, P.T., Liou, N.F., Lee, C.C., You, S.F., Liu, L.L., McNeill, D.S., Chew, K.S., Hattar, S., Chen, S.K., et al. (2016). M1 ipRGCs influence visual function through retrograde signaling in the retina. Journal of Neuroscience *36*(27), 7184–7197.

Provencio, I, Rodriguez, I.R., Jiang, G, Hayes, W.P., Moreira, E.F., and Rollag, M.D. (2000). A novel human opsin in the inner retina. The Journal of Neuroscience: The Official Journal of the Society for Neuroscience *20*(2), 600–605.

Quattrochi, L.E., Stabio, M.E., Kim, I., Ilardi, M.C., Michelle Fogerson, P., Leyrer, M.L., and Berson, D.M. (2019). The M6 cell: a small-field bistratified photosensitive retinal ganglion cell. Journal of Comparative Neurology *527*(1), 297–311.

Rao, S., Chun, C., Fan, J., Kofron, J.M., Yang, M.B., Hegde, R.S., Ferrara, N., Copenhagen, D.R., and Lang, R.A. (2013). A direct and melanopsin-dependent fetal light response regulates mouse eye development. Nature *494*(7436), 243–246.

Renna, J.M., Weng, S., and Berson, D.M. (2011). Light acts through melanopsin to alter retinal waves and segregation of retinogeniculate afferents. Nature Neuroscience *14*(7), 827–829.

Rios, M.N., Marchese, N.A., and Guido, M.E. (2019). Expression of non-visual opsins Opn3 and Opn5 in the developing inner retinal cells of birds. Light-responses in Müller glial cells. Frontiers in Cellular Neuroscience *13*, 376.

Román Rosón, M., Bauer, Y., Kotkat, A.H., Berens, P., Euler, T., and Busse, L. (2019). Mouse dLGN receives functional input from a diverse population of retinal ganglion cells with limited convergence. Neuron *102*(2), 462–476.e8.

Rompani, S.B., Müllner, F.E., Wanner, A., Zhang, C., Roth, C.N., Yonehara, K., and Roska, B. (2017). Different modes of visual integration in the lateral geniculate nucleus revealed by single-cell-initiated transsynaptic tracing. Neuron *93*(4), 767–776.e6.

Rupp, A.C., Ren, M., Altimus, C.M., Fernandez, D.C., Richardson, M., Turek, F., Hattar, S., and Schmidt, T.M. (2019). Distinct ipRGC subpopulations mediate light's acute and circadian effects on body temperature and sleep. eLife *8*, e44358.

Sabbah, S., Papendorp, C., Koplas, E., Beltoja, M., Etebari, C., Gunesch, A.N., Carrete, L., Kim, M.T., Manoff, G., Bhatia-Lin, A., et al. (2018). Synaptic circuits for irradiance coding by intrinsically photosensitive retinal ganglion cells. BioRxiv, 442954.

Sand, A., Schmidt, T.M., and Kofuji, P. (2012). Diverse types of ganglion cell photoreceptors in the mammalian retina. Progress in Retinal and Eye Research *31*(4), 287–302.

Schmidt, T.M., Taniguchi, K., and Kofuji, P. (2008). Intrinsic and extrinsic light responses in melanopsin-expressing ganglion cells during mouse development. Journal of Neurophysiology *100*(1), 371–384.

Schmidt, T.M., Chen, S.K., and Hattar, S. (2011). Intrinsically photosensitive retinal ganglion cells: many subtypes, diverse functions. Trends in Neurosciences *34*(11), 572–580.

Schmidt, T.M., Alam, N.M., Chen, S., Kofuji, P., Li, W., Prusky, G.T., and Hattar, S. (2014). A role for melanopsin in alpha retinal ganglion cells and contrast detection. Neuron *82*(4), 781–788.

Sondereker, K.B., Stabio, M.E., and Renna, J.M. (2020). Crosstalk: the diversity of melanopsin ganglion cell types has begun to challenge the canonical divide between image-forming and non-image-forming vision. Journal of Comparative Neurology *528*(12), 2044–2067.

Sonoda, T., Lee, S.K., Birnbaumer, L., and Schmidt, T.M. (2018). Melanopsin phototransduction is repurposed by ipRGC subtypes to shape the function of distinct visual circuits. Neuron *99*(4), 754–767.e4.

Sonoda, T., Li, J.Y., Hayes, N.W., Chan, J.C., Okabe, Y., Belin, S., Nawabi, H., and Schmidt, T.M. (2020). A noncanonical inhibitory circuit dampens behavioral sensitivity to light. Science *368*(6490), 527–531.

Van Gelder, R.N., and Buhr, E.D. (2016). Ocular photoreception for circadian rhythm entrainment in mammals. Annual Review of Vision Science *2*(1), 153–169.

Warthen, D.M., and Provencio, I. (2012). The role of intrinsically photosensitive retinal ganglion cells in nonimage-forming responses to light. Eye and Brain *4*, 43–48.

Zhao, X., Pack, W., Khan, N.W., and Wong, K.Y. (2016). Prolonged inner retinal photoreception depends on the visual retinoid cycle. Journal of Neuroscience *36*(15), 4209–4217.

Zhao, X., Reifler, A.N., Schroeder, M.M., Jaeckel, E.R., Chervenak, A.P., and Wong, K.Y. (2017). Mechanisms creating transient and sustained photoresponses in mammalian retinal ganglion cells. The Journal of General Physiology *149*(3), 335–353.

PART 2

Contrast

Contrast sensitivity sets a critical limit on our perception in daylight. This sea turtle is drawn with a gradient of contrast; more spatial details emerge as the contrast increases.

CHAPTER 4

Contrast sensitivity

Gregory William Schwartz[a,b,c]
[a]Departments of Ophthalmology and Physiology, Feinberg School of Medicine, Northwestern University, Chicago, IL, United States
[b]Department of Neurobiology, Weinberg College of Arts and Sciences, Northwestern University, Chicago, IL, United States
[c]Northwestern Interdepartmental Neuroscience Program (NUIN), Northwestern University, Chicago, IL, United States

Contrast is the central property of a visual scene that allows us to distinguish different surfaces. The fur of an animal in a tree reflects less light than the surrounding branches, and it appears darker, or it reflects more light, and it appears lighter. Similarly, these letters are easily discriminable from the background of the page, whether you are reading this book by a dim lamp, under fluorescent indoor lighting, outside in the bright sun, or on your computer at any given luminance setting on the screen. Adaptation, to both luminance and contrast, enables vision to work in different lighting conditions, and these adaptation computations are covered in other chapters. Here, our focus is on the underlying computation that sets the contrast sensitivity limit in the steady-state. Like photon detection (Chapter 1), there is a clear survival advantage to maximizing contrast sensitivity. An animal that can reliably detect a difference in luminance 100-fold lower than the background (1% contrast) can see predators, prey, and other food sources that would be invisible to competitors.

Humans have an exquisite ability to detect contrast as low as 0.5% (De Valois et al., 1974). Other animals have somewhat lower contrast sensitivity, as measured by just-detectable contrast: 1% in macaque (De Valois et al., 1974), 0.8% in squirrel monkey (Merigan, 1976), 1% in cat (Bisti and Maffei, 1974), 1% in rabbit (Pak, 1984), 1.7% in mouse (Histed et al., 2012), 1.8% in goldfish (Northmore and Dvorak, 1979). As with luminance detection (Section 1), signal-to-noise ratio (SNR) and dynamic range are important considerations. First, we will look at a well-conserved retinal ganglion cell (RGC) type in which synaptic conductances are tuned to maximize SNR for contrast sensitivity. Then, we will consider the problem of dynamic range and the speculation that a distributed code helps solve this problem.

Retinal Computation
https://doi.org/10.1016/B978-0-12-819896-4.00005-6

The OFF delta (OFF sustained alpha) RGC

The OFF delta RGC type in guinea pig and cat is homologous to the OFF sustained alpha in mouse (Manookin et al., 2008; Pang et al., 2003; Van Wyk et al., 2009). This cell is of special interest because of its remarkable contrast sensitivity, the highest reported among RGCs in the mammalian retina (Manookin et al., 2008). Thus we can speculate that OFF delta RGCs set the contrast sensitivity limit of vision, though this remains to be demonstrated experimentally. Nonetheless, the circuit mechanisms that contribute to high contrast sensitivity in this RGC are reasonably well understood, and many of the principles of this computation can be applied throughout the central nervous system.

A cell that sits at the inflection point of its tuning curve

The principle of dynamic range is ubiquitous in retinal computation, and we have already seen an example (in Chapter 3) of how neurons transmit the most information about a stimulus variable in the steepest part of their tuning curves. Contrast tuning curves for retinal neurons can be one of four general shapes: ON, OFF, ON-OFF, or suppressed-by-contrast (Fig. 4.1). By definition, these curves must pass through the origin because zero contrast elicits zero change in response.

As an OFF RGC, the OFF delta cell has a monotonically decreasing contrast tuning curve, and like most monotonic neuronal tuning curves, it can

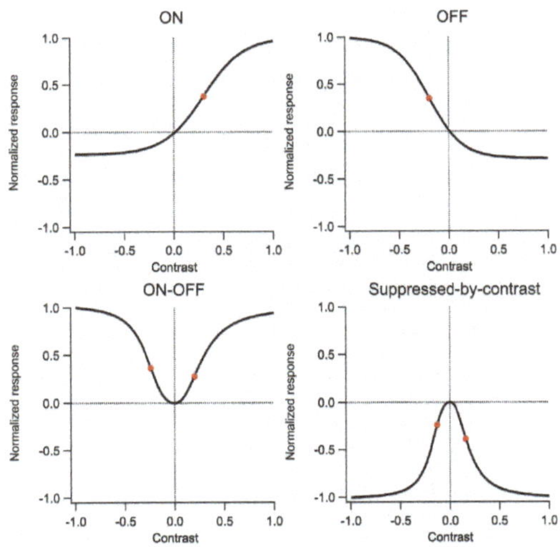

Fig. 4.1 The four classes of contrast tuning curves. Examples of the four classes of contrast tuning curves are plotted. Red markers indicate inflection points where contrast sensitivity is maximized.

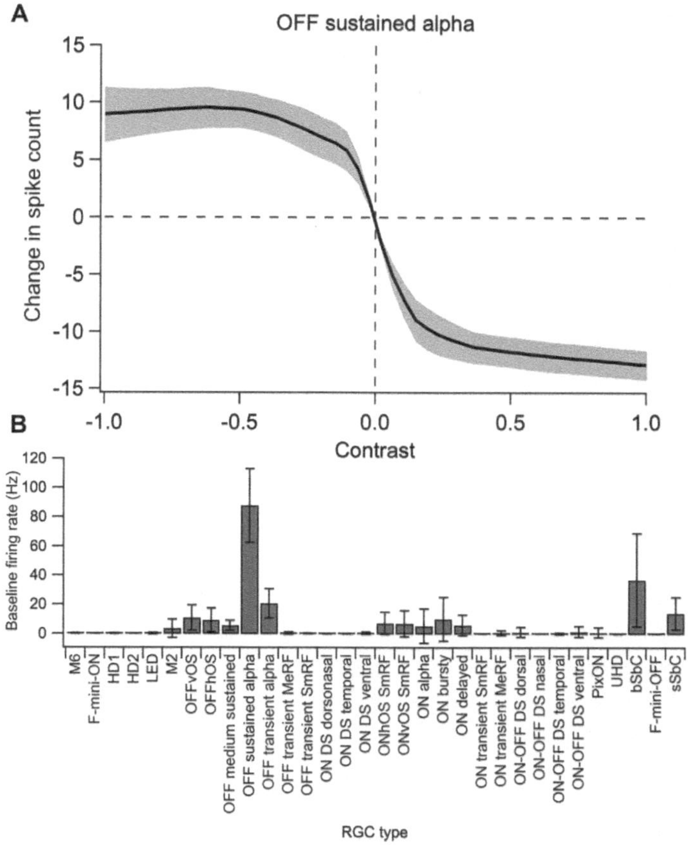

Fig. 4.2 OFF sustained alpha RGCs have high contrast sensitivity and high baseline firing rate. (**A**) Contrast response function of an OFF sustained alpha RGC from mouse. Shaded region represents SEM. Data from Sophia Wienbar. (**B**) Baseline firing rates of different RGC types. Error bars are SD.

be described by a sigmoidal curve. The steepest point on such a curve is the inflection point, where the second derivative is zero (red points in Fig. 4.1). Therefore, to achieve maximal sensitivity to small positive and negative contrasts, an ideal ON or OFF RGC would place the inflection point of its tuning curve at zero contrast. OFF delta RGCs come remarkably close to this idealization (Fig. 4.2A).

Such a tuning curve creates a situation that is unique among RGCs and rare among sensory neurons generally: the background firing rate sits in the middle of the cell's dynamic range. For the OFF delta RGC, this means that at zero contrast, it must maintain a steady firing rate of nearly 100 Hz, midway between zero and its maximum rate of ~200 Hz (Murphy and Rieke, 2011). This is the highest maintained firing rate among RGCs (Fig. 4.2B), and it certainly comes with an energetic cost. That the OFF delta maintains a high spike rate suggests that its optimization for contrast sensitivity was an important enough selective pressure in evolution to offset the metabolic cost of so many spikes.

A conserved microcircuit

Several elements of the circuit upstream of OFF delta RGCs in the retina have been identified. The cells providing synaptic input have been best studied in the OFF sustained alpha RGC of the mouse. Excitation comes mostly from type 2 bipolar cells and a recently discovered glutamatergic monopolar interneuron (GluMI) (Della Santina et al., 2016). As with most RGCs, inhibition to the OFF delta/OFF sustained alpha likely comes from several amacrine cells, most of which have not been identified. However, one prominent inhibitory input is known. The AII amacrine cell provides glycinergic inhibition onto the dendrites of OFF delta RGCs in guinea pig and OFF sustained alpha RGCs in mouse (Di Marco et al., 2013; Manookin et al., 2008). Originally identified as a critical component of the rod bipolar circuit described in Chapter 1, AII amacrine cells are ON cells because they depolarize for positive contrast, and they are very sensitive to small contrasts because they collect (indirectly) from many photoreceptors by virtue of their extensive network of gap junctions. So positive contrast leads to increased direct inhibition onto OFF delta RGCs via AII amacrine cells as well as decreased excitation (Fig. 4.3). This relationship where excitation and inhibition work together

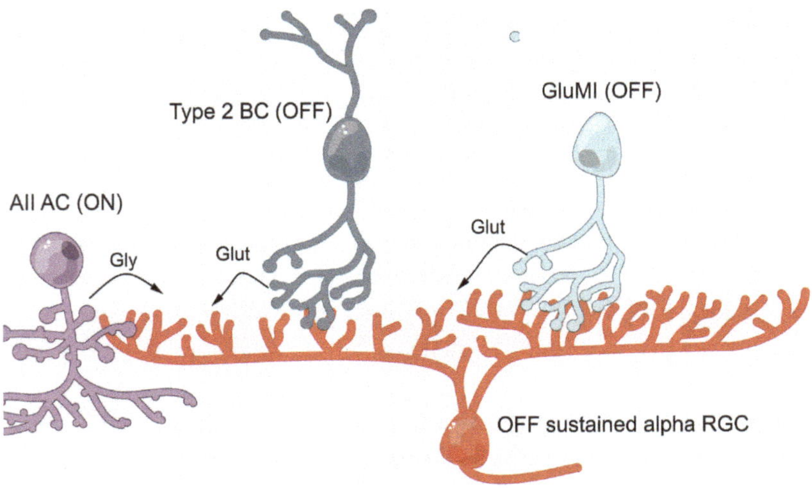

Fig. 4.3 Synaptic inputs to the OFF sustained alpha RGC. Known elements of the push-pull circuit upstream of OFF sustained alpha RGCs are shown. Presynaptic cells are labeled by their ON or OFF polarity. Type 2 BCs and GluMIs provide OFF excitation via glutamate while AII ACs provide ON inhibition via glycine.

to control the firing rate of a neuron is called "push-pull," and it is a common motif in sensory neurons, notably in the lateral geniculate nucleus one synapse downstream of the RGCs (Hirsch et al., 2015).

Push-pull mechanism to maximize SNR

The push-pull mechanism has an intuitive advantage for improving sensitivity because excitation increases at the same time inhibition decreases (or vice versa), thus exerting a large effect on membrane voltage. Excitation and inhibition working together in this way to change the firing rate of a neuron is more effective than either one alone. But what is the most advantageous way to combine excitation and inhibition to maximize sensitivity? The answer is to add each conductance with a weight proportional to its SNR. This strategy is called maximal-ratio combining in telecommunications (Fig. 4.4) (Brennan, 1959). Do OFF delta RGCs combine excitation and inhibition according to the maximal-ratio combining principle?

This was the question investigated by Homann and Freed (2017) in the landmark paper for this chapter. The authors recorded synaptic currents in OFF delta RGCs of the guinea pig, flashed dark spots on a gray background, and measured signal and noise variances to compute SNR (Fig. 4.5). By voltage clamping the RGC at different holding potentials between the excitatory (E) and inhibitory (I) reversal potential, they could manipulate the E:I balance. For a high contrast spot matched to the size of the receptive field center, SNR was maximized for a membrane potential of −53 mV, which was, within measurement error, indistinguishable from the resting potential of the RGC. Even more remarkably, the authors repeated the experiment over a range of contrasts and spot sizes, and while both manipulations had profound and varying

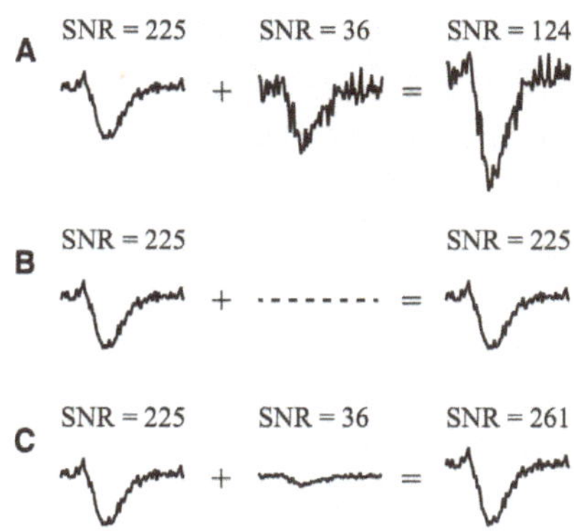

Fig. 4.4 Maximal-ratio combining to maximize SNR. Examples of combining signals with different SNR. (**A**) Signals with different SNR are combined with equal weights. (**B**) The signal with lower SNR is excluded. (**C**) The signals are added with weights proportional to their SNR. *(From Homann and Freed (2017).)*

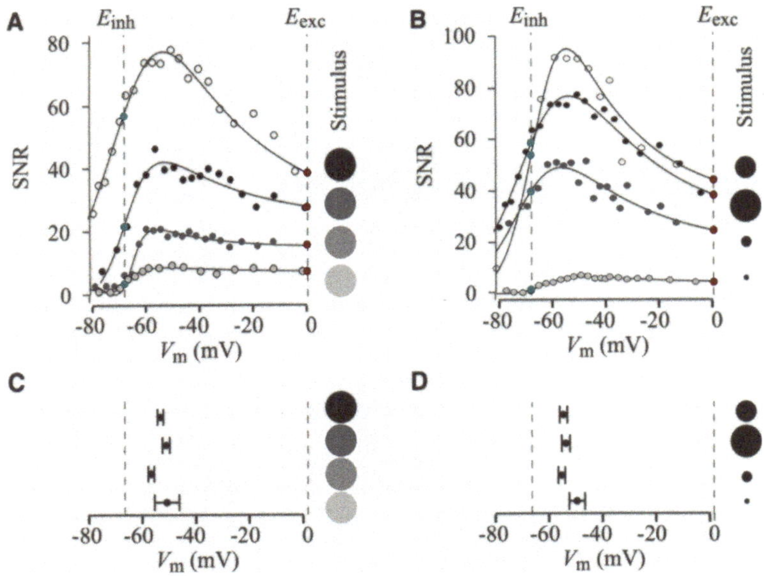

Fig. 4.5 OFF delta RGCs weight their excitation and inhibition according to maximal-ratio combining. (**A**) SNR of currents recorded in OFF delta RGCs at different holding potentials. Different lines are for different contrast stimuli as indicated. Dotted lines mark the excitatory and inhibitory reversal potentials. (**B**) Same as (**A**) but for spots of different sizes at a fixed contrast. (**C** and **D**) Membrane potential at which SNR is maximized. *(From Homann and Freed (2017).)*

effects on the signal and noise of excitation and inhibition, the optimal membrane potential to maximize SNR remained close to the resting potential. Thus, the weights of excitation and inhibition to OFF delta RGCs are dynamically adjusted to maintain high SNR across different stimuli, allowing the cell to represent small contrasts reliably and robustly.

A distributed code for contrast

We have seen that the OFF delta RGC has a contrast tuning curve with its steepest portion aligned to small contrasts, and we have seen how its synaptic inputs maximize SNR, but this RGC alone cannot account for contrast sensitivity in perception. Re-examining the contrast tuning curve (Fig. 4.2A) reveals a problem. While the curve is steep around zero contrast, it is flat for large positive and negative contrasts. The firing rate of the OFF delta RGC cannot distinguish 50% contrast from 100% contrast (either positive or negative), even though this is a trivial distinction perceptually. As we will see in the next chapter, contrast adaptation is part of the way RGCs adjust their dynamic range to the contrast distribution in the visual scene, but there is another solution that does not depend on adaptation and thus allows the retina to encode multiple contrast ranges simultaneously.

Contrast sensitivity distributed among RGCs

Analogous to how absolute luminance is encoded in the M1 RGC population (Chapter 3), the solution is to distribute tuning curves among RGCs so that each range of contrasts falls into the steep part of the tuning curve for at least one cell type. Indeed different RGC types vary dramatically in the shapes of their contrast tuning curves (Fig. 4.6). The weak hypothesis that different RGC types contribute to contrast sensitivity in different ranges is almost certainly true, but the stronger hypothesis that the contrast turning curves of some or all RGCs are optimized to represent the contrast distribution of natural scenes remains untested.

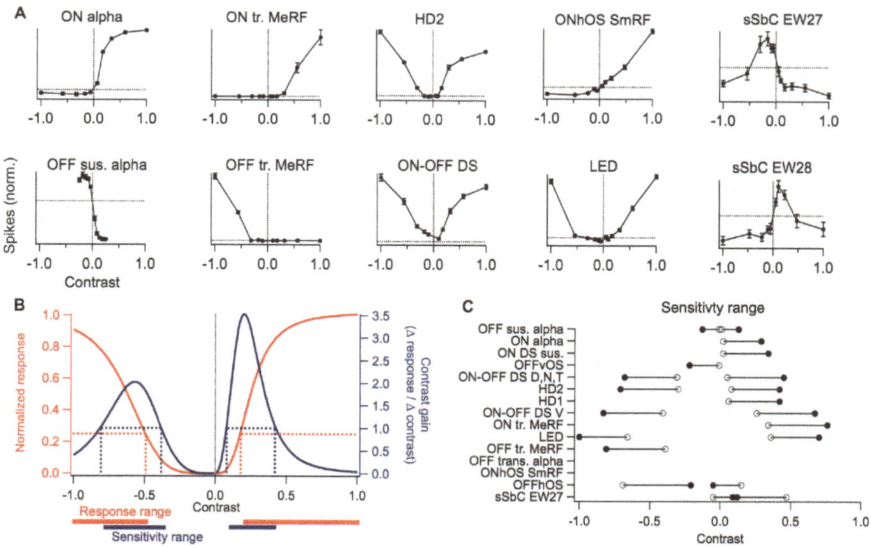

Fig. 4.6 Diversity of contrast response functions among mouse RGCs. (**A**) Contrast tuning curves from ten different RGC types. Error bars are SEM across trials ($n = 5$–15 trials). (**B**) Functional fit to the normalized contrast tuning curve for the ON-OFF DS cell from (**A**) *(red)* and the first derivative, the contrast gain *(blue)*. *Dotted lines* indicate threshold crossings of 25% of the maximum response *(red)* and a contrast gain of 1 *(blue)*. *Solid lines* beneath the plot indicate the response range and the sensitivity range. (**C**) Sensitivity range for an example RGC of each type. *Open circles* indicate the bottom of the sensitivity range beyond which contrast gain is between 0 and 1. *Closed circles* indicate the top of the sensitivity range beyond which contrast gain is between −1 and 0.

To test this strong hypothesis, one first needs to measure the contrast tuning curve of each RGC type. Many contrast tuning curves have been measured, but two issues have so far prevented their systematic study. First, for reasons both physiological, like contrast and luminance adaptation and dependence on stimulus size, and non-physiological, like excessive bleaching or rod dysfunction, contrast tuning functions measured from the same RGC type can be highly variable (Sethuramanujam et al., 2016). Second, the atlas of RGC types has been evolving rapidly over the last decade, and only now are we reaching consensus on the identities of the ~40 types in the mouse retina (Baden et al., 2016; Bae et al., 2018; Tran et al., 2019). This is a fertile area for future research because it is not known which RGCs contribute to contrast perception. While many RGC types are identified according to their trigger features, contrast sensitivity is ubiquitous. Therefore, the contrast code is likely multiplexed in the spike trains of RGCs along with other stimulus features.

Contrast sensitivity distributed among bipolar cells

Before signals reach RGCs they must pass through the interneurons of the retina, the bipolar and amacrine cells. At the first stage of vision, photoreceptors mostly lack a substrate for distributed contrast encoding. Rods and cones occupy different ranges of luminance sensitivity (see Chapter 2) and cone types differ in wavelength sensitivity (see Chapter 16), but there is not enough diversity in the photoreceptor population for different cell types to have the steepest part of their contrast response functions aligned to different contrast regimes. Parallel processing begins at the next synapse at the photoreceptor output. Due in part to their different complements of glutamate receptors, bipolar cells receiving input from the same cones are classified as either ON or OFF type, and they are further subdivided by response kinetics into more transient and more sustained types.

Burkhardt and Fahey (1998) measured contrast response functions in bipolar cells of the tiger salamander retina. Their data showed wide variability in the curves of different cells, suggesting a distributed code for contrast at the level of bipolar cells. This group went on to show, in a larger sample of bipolar cell recordings, that the distribution of dynamic range in the bipolar cell population was well aligned to the contrast range of natural images in both land and aquatic environments (Fig. 4.7) (Burkhardt, 2011; Burkhardt et al., 2006). The 15 bipolar cell types of the mouse retina have been identified and registered in morphological, transcriptomic, and functional

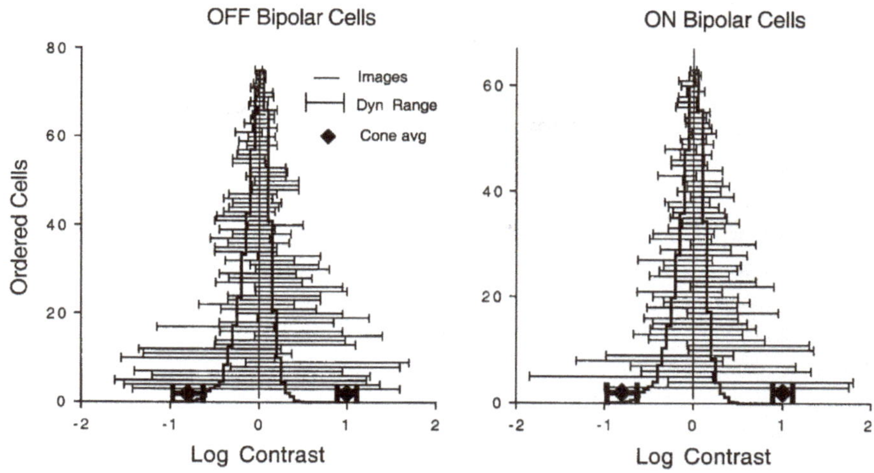

Fig. 4.7 The dynamic range of salamander bipolar cells is matched to natural scenes. The dynamic range (defined as 10%–90% response range) for each bipolar cell is superimposed on the contrast distribution of natural scenes. Left panel is OFF bipolar cells and right panel is ON-bipolar cells. The contrast scale is logarithmic base 10, so −1 is a step to 0.1 times the background and +1 is a step to 10 times the background. *Dark symbols* show the dynamic range of cones recorded in the same conditions. *(From Burkhardt (2011).)*

studies (Franke et al., 2017; Kim et al., 2014; Shekhar et al., 2016). While functional data at the level of glutamate imaging suggest a diversity of contrast response functions among types (Franke et al., 2017), a systematic study relating contrast sensitivity in mammalian bipolar cells to the contrast distributions of natural scenes is lacking.

Open questions

Amacrine cells were conspicuously absent from the discussion of distributed coding of contrast. Due to their extreme diversity, we are far from understanding the range of contrast tuning curves present in amacrine cells. Perhaps additional amacrine cell circuits contribute to achieving high contrast sensitivity and distributing sensitivity for different contrast ranges among different RGCs.

Another key area for future research is the connection between contrast sensitivity at the level of RGCs and the level of visual behavior. We began by citing the contrast sensitivity limits of vision in different species, but these experiments were limited to particular behavior tasks, like detecting a grating pattern. Conscious detection tasks like these are only a small fraction of the repertoire of visual behavior, and many RGC types (the majority in mice (Ellis et al., 2016)) are not part of the direct thalamocortical pathway most associated with conscious perception. For the most part, the contrast sensitivity limits of different visual behaviors are not known. Perhaps different populations of RGCs encode contrast in the appropriate range for the behaviors they serve. The distributed code for contrast among RGCs may not be a single code but rather a distinct distributed code for each retinorecipient brain region, tuned to the behaviors supported by that brain region.

Landmark paper: Homann, J., and Freed, M.A. (2017). A mammalian retinal ganglion cell implements a neuronal computation that maximizes the SNR of its postsynaptic currents. Journal of Neuroscience *37*(6), 1468–1478.

Discussion questions:

1. What point is demonstrated in Fig. 6? Why was the perforated patch technique necessary for the experiments in panel F?
2. The authors conclude that the contribution of (dis)inhibition to the OFF delta is large relative to the contribution of excitation for low contrast and small radius stimuli. Explain how the data shown in Fig. 4A and 4B support this conclusion. Can you relate this finding to properties of the cell types that provide synaptic input to OFF delta and OFF sustained alpha RGCs?
3. Why did the authors use two different internal solutions in the experiments shown in Fig. 5B? How did using two solutions help control for the effect of electrochemical driving force on excitatory and inhibitory conductances?

Programming exercise

To access the programming exercise, please visit https://www.elsevier.com/books-and-journals/book-companion/9780128198964.

The first part of this chapter made the simplifying assumption that the most sensitive range in a contrast tuning curve was around its inflection point. This is not necessarily true in cases where the noise varies as a function of the response amplitude. Multiplicative noise is such an example, and it is common in neuroscience. The variance of spike counts usually scales with their mean. This can be modeled with a noise term whose amplitude depends on the signal amplitude.

This exercise explores how multiplicative and additive noise contribute to the discriminability of contrast and how the optimal weighting of excitation and inhibition in a push-pull system depend on the properties of the noise.

Load the file "tuning_curves.mat." It contains two variables: "tuning_curve_exc" and "tuning_curve_inh." Both are 201 × 2 matrices where the first column is contrast (−1,1) and the second column is the normalized response. These are simulated tuning curves for excitatory and inhibitory input, respectively, to an OFF delta RGC.

The script "contrast_sensitivity_SNR.m" contains the code to get you started with the simulation.

1. Start by varying only testContrast1, and use a high value of multiplicative noise, like 1.5. Plot the SNR of excitation and inhibition across the full range of contrasts. Find the inflection point on each tuning curve and see how it compares to the peak SNR of each channel.
2. Now try comparisons by varying both testContrast1 and testContrast2, again across the full range of contrast. Make a heatmap of discrimination performance (measured by d′) across different combinations of both contrast values. If you set a d′ threshold of 1 as "just detectable," what is the contrast discriminability of this model cell? How does discriminability vary across the contrast range?
3. Repeat part 2 with lower and higher values of multiplicative noise. How does this change the result? Can you explain why?
4. Now try keeping the multiplicative noise at 1.5 and increasing the additive noise. How does this affect discriminability and change where the maximal sensitivity range is located?
5. Add an additional noise term that comes after excitation and inhibition are combined. This would be something like noise in spike generation within the RGC. Is maximal-ratio combining still the optimal strategy? Why or why not? Use a figure to back up your answer.

References

Baden, T., Berens, P., Franke, K., Román Rosón, M., Bethge, M., and Euler, T. (2016). The functional diversity of retinal ganglion cells in the mouse. Nature *529*(7586), 345–350.

Bae, J.A., Mu, S., Kim, J.S., Turner, N.L., Tartavull, I., Kemnitz, N., Jordan, C.S., Norton, A.D., Silversmith, W.M., Prentki, R., et al. (2018). Digital museum of retinal ganglion cells with dense anatomy and physiology. Cell *173*(5), 1293–1306.e19.

Bisti, S., and Maffei, L. (1974). Behavioural contrast sensitivity of the cat in various visual meridians. The Journal of Physiology *241*(1), 201–210.

Brennan, D.G. (1959). Linear diversity combining techniques. Proceedings of the IRE *47*(6), 1075–1102.

Burkhardt, D.A. (2011). Contrast processing by on and off bipolar cells. Visual Neuroscience *15*, 219–229.
Burkhardt, D.A., and Fahey, P.K. (1998). Contrast enhancement and distributed encoding by bipolar cells in the retina. Journal of Neurophysiology *80*(3), 1070–1081.
Burkhardt, D.A., Fahey, P.K., and Sikora, M.A. (2006). Natural images and contrast encoding in bipolar cells in the retina of the land- and aquatic-phase tiger salamander. Visual Neuroscience *23*(1), 35–47.
De Valois, R.L., Morgan, H., and Snodderly, D.M. (1974). Psychophysical studies of monkey Vision-III. Spatial luminance contrast sensitivity tests of macaque and human observers. Vision Research *14*(1), 75–81.
Della Santina, L., Kuo, S.P., Yoshimatsu, T., Okawa, H., Suzuki, S.C., Hoon, M., Tsuboyama, K., Rieke, F., and Wong, R.O.L. (2016). Glutamatergic monopolar interneurons provide a novel pathway of excitation in the mouse retina. Current Biology *26*(15), 2070–2077.
Di Marco, S., Protti, D.A., and Solomon, S.G. (2013). Excitatory and inhibitory contributions to receptive fields of alpha-like retinal ganglion cells in mouse. Journal of Neurophysiology *110*(6), 1426–1440.
Ellis, E.M., Gauvain, G., Sivyer, B., and Murphy, G.J. (2016). Shared and distinct retinal input to the mouse superior colliculus and dorsal lateral geniculate nucleus. Journal of Neurophysiology *116*(2), 602–610.
Franke, K., Berens, P., Schubert, T., Bethge, M., Euler, T., and Baden, T. (2017). Inhibition decorrelates visual feature representations in the inner retina. Nature *542*(7642), 439–444.
Hirsch, J.A., Wang, X., Sommer, F.T., and Martinez, L.M. (2015). How inhibitory circuits in the thalamus serve vision. Annual Review of Neuroscience *38*, 309–329.
Histed, M.H., Carvalho, L.A., and Maunsell, J.H.R. (2012). Psychophysical measurement of contrast sensitivity in the behaving mouse. Journal of Neurophysiology *107*(3), 758–765.
Homann, J., and Freed, M.A. (2017). A mammalian retinal ganglion cell implements a neuronal computation that maximizes the SNR of its postsynaptic currents. Journal of Neuroscience *37*(6), 1468–1478.
Kim, J.S., Greene, M.J., Zlateski, A., Lee, K., Richardson, M., Turaga, S.C., Purcaro, M., Balkam, M., Robinson, A., Behabadi, B.F., et al. (2014). Space-time wiring specificity supports direction selectivity in the retina. Nature *509*(7500), 331–336.
Manookin, M.B., Beaudoin, D.L., Ernst, Z.R., Flagel, L.J., and Demb, J.B. (2008). Disinhibition combines with excitation to extend the operating range of the OFF visual pathway in daylight. Journal of Neuroscience *28*(16), 4136–4150.
Merigan, W.H. (1976). The contrast sensitivity of the squirrel monkey (*Saimiri sciureus*). Vision Research *16*(4), 375–379.
Murphy, G.J., and Rieke, F. (2011). Electrical synaptic input to ganglion cells underlies differences in the output and absolute sensitivity of parallel retinal circuits. The Journal of Neuroscience *31*(34), 12218–12228.
Northmore, D.P.M., and Dvorak, C.A. (1979). Contrast sensitivity and acuity of the goldfish. Vision Research *19*(3), 255–261.
Pak, M.A. (1984). Ocular refraction and visual contrast sensivity of the rabbit, determined by the VECP. Vision Research *24*(4), 341–345.
Pang, J.-J.,Gao, F., and Wu, S.M. (2003). Light-evoked excitatory and inhibitory synaptic inputs to ON and OFF alpha ganglion cells in the mouse retina. Journal of Neuroscience *23*(14), 6063–6073.
Sethuramanujam, S., McLaughlin, A.J., DeRosenroll, G., Hoggarth, A., Schwab, D.J., and Awatramani, G.B. (2016). A central role for mixed acetylcholine/GABA transmission in direction coding in the retina. Neuron *90*(6), 1243–1256.
Shekhar, K., Lapan, S.W., Whitney, I.E., Tran, N.M., Macosko, E.Z., Kowalczyk, M., Adiconis, X., Levin, J.Z., Nemesh, J., Goldman, M., et al. (2016). Comprehensive

classification of retinal bipolar neurons by single-cell transcriptomics. Cell *166*(5), 1308–1323.e30.

Tran, N.M., Shekhar, K., Whitney, I.E., Jacobi, A., Benhar, I., Hong, G., Yan, W., Adiconis, X., Arnold, M.E., Lee, J.M., et al. (2019). Single-cell profiles of retinal ganglion cells differing in resilience to injury reveal neuroprotective genes. Neuron *104*(6), 1039–1055.e12.

Van Wyk, M., Wässle, H., and Taylor, W.R. (2009). Receptive field properties of ON- and OFF-ganglion cells in the mouse retina. Visual Neuroscience *26*(3), 297.

Contrast is an important depth cue in art. The three sections of this coral reef look like they are progressively further away because they are drawn at decreasing levels of contrast. Retinal circuits adapt rapidly to contrast with each eye movement to bring out spatial details. Because of contrast adaptation, diving on this reef, you would be able to identify corals at various distances.

CHAPTER 5

Contrast adaptation and sensitization

Gregory William Schwartz[a,b,c]
[a]Departments of Ophthalmology and Physiology, Feinberg School of Medicine, Northwestern University, Chicago, IL, United States
[b]Department of Neurobiology, Weinberg College of Arts and Sciences, Northwestern University, Chicago, IL, United States
[c]Northwestern Interdepartmental Neuroscience Program (NUIN), Northwestern University, Chicago, IL, United States

In Chapter 2 (Luminance adaptation), we reviewed the many mechanisms the retina uses to adjust gain based on the mean luminance in the visual scene. Of course, there are statistics of the light distribution beyond the mean that could also influence gain in the retina. The second moment of a distribution is its variance. When referring to the deviations in light intensity from its mean over space or time, vision scientists use the term "contrast." Contrast provides a great deal of information about objects in the environment because the relative intensity of objects remains nearly constant despite large variations in average lighting conditions.

As with luminance, the principle of efficient coding guides our understanding of how the input-output function of retinal circuits adapts to contrast. To maintain sensitivity to contrast, which is a key constraint in vision across species (see Chapter 4), neurons must have high contrast gain, in other words, a steep slope in the function relating contrast to response amplitude. On the other hand, since neurons have limited dynamic range, high contrast gain comes with the risk of saturation.

The tradeoff between high gain and saturation is more nuanced, however, for contrast than for luminance adaptation. While increases in luminance (beyond the absolute sensitivity limit) only decrease gain in the retina (see Chapter 2), increases in contrast can engage either gain decreases (adaptation) or increases (sensitization), and the balance of these opposing mechanisms depends on cell types and stimulus conditions. Contrast adaptation in the retina has been a topic of study for several decades, beginning with the work of Shapley and Victor in the 1970s (Shapley and Victor, 1978) (reviewed by Meister and Berry, 1999; Shapley, 1997). An excellent review

Retinal Computation
https://doi.org/10.1016/B978-0-12-819896-4.00006-8

by Jonathan Demb (2008) captured the state of the field in 2008. Contrast sensitization was discovered more recently (Kastner and Baccus, 2011), but progress has been rapid in showing that this computation exists across species and in identifying the necessary circuit components.

Contrast has a large dynamic range in natural scenes

As with luminance, contrast changes in predictable ways in the visual world at many different timescales, and the shortest relevant timescale is that of a single saccade. Within a daylight scene, contrast can vary within the field of view of a retinal ganglion cell (RGC) by more than a factor of 1000 on subsequent fixations (Frazor and Geisler, 2006). Photographers face this issue when trying to capture a large image, like that of a landscape, with the camera sensor's limited dynamic range. A 24-bit digital camera only has a range of 256 (2^8) values in each of its three color channels. If this range were optimized for the high contrasts of highlights and shadows on a cliff, it would not capture the subtle light variations that give clouds their texture. Conversely, if contrast gain were increased to capture details in the sky, dark shadows and bright highlights on the cliff would saturate the sensor. Some photographers take multiple exposures of the same scene with different gain settings and then assemble a composite image digitally to approximate the contrast gain control that occurs naturally and seamlessly in vision (Fig. 5.1).

Mechanisms of contrast adaptation in the retina

Contrast adaptation occurs through a collection of mechanisms at different synapses and cells throughout retinal circuits. Each of these different mechanisms and locations confers certain computational advantages and limitations that I will explore in the following sections. Unlike for luminance adaptation, where key mechanisms occur in phototransduction in both rods and cones (see Chapter 2), contrast adaptation begins at the synapse between photoreceptors and bipolar cells.

Photoreceptor to bipolar cell synapses

By the year 2000, it was clear that contrast adaptation occurs within the retina at multiple timescales (Fig. 5.2) (Chander and Chichilnisky, 2001; Kim and Rieke, 2001; Smirnakis et al., 1997), but the sites of the adaptive

Fig. 5.1 Multiple exposure photography blends different contrast ranges in different regions of an image. Exposures capturing the low and high contrast ranges *(top)* are combined into a composite *(bottom)* in which different regions have different contrast encoding. The retina performs a similar computation seamlessly within each saccade. *(Retrieved from https://www.digitalphotomentor.com/exposure-blending-in-photoshop-a-beginners-tutorial/.)*

mechanisms were unclear. Fred Rieke (2001) sought to determine the first site of contrast adaptation by recording from photoreceptors, horizontal cells, and bipolar cells (BCs) in the salamander retina. The light stimulus was a noise sequence that alternated between low and high contrast, similar to the temporal contrast adaptation stimuli used in previous studies. Rieke found that BCs adapted to the stimulus by decreasing gain following the low-high transition with a time constant of 2.1 s and increasing gain following the high-low transition over a longer timescale (6.2 s). Adaptation could be explained by a change in the gain of the photoreceptor to BC synapse. Voltage responses of photoreceptors and horizontal cells showed no such adaptation, pointing to the dendrites of BCs as the location of the first contrast adaptation mechanism in the retina (Fig. 5.3). Adaptation was observed both at rod and cone light levels, and OFF BCs showed more contrast adaptation than ON BCs, in part due to a calcium-dependent mechanism unique to the OFF pathway. This result was consistent with earlier results showing

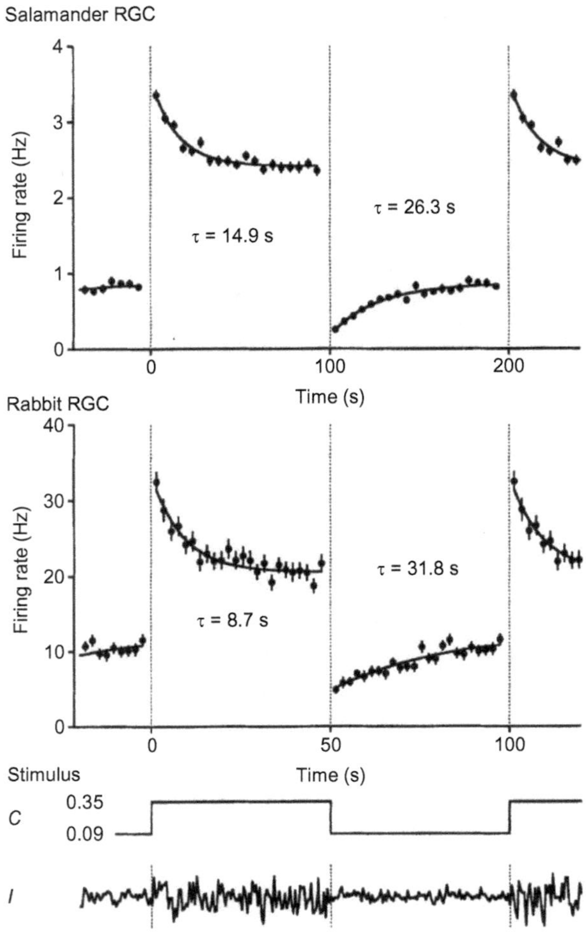

Fig. 5.2 Contrast adaptation in the firing rate of RGCs. Firing rate of a salamander RGC *(top)* and a rabbit RGC *(middle)* in response to an alternating contrast stimulus *(bottom)*. Time constants of exponential fits following low-high contrast and high-low contrast transitions are indicated. *(Modified from Smirnakis et al. (1997).)*

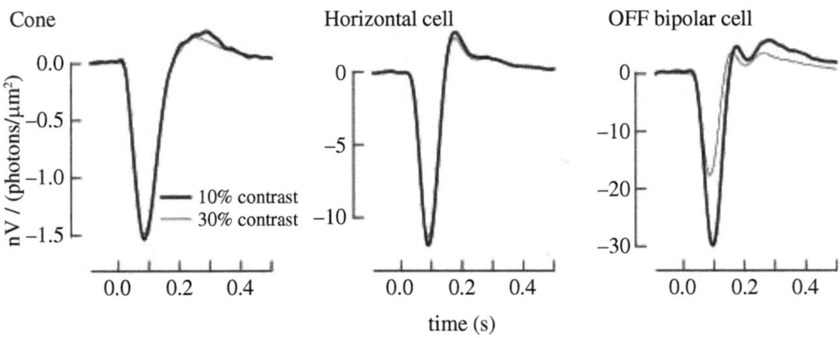

Fig. 5.3 The first site of contrast adaptation is in bipolar cells. Responses of cones, horizontal cells, and OFF bipolar cells of the salamander retina to 10% and 30% contrast flashes. OFF bipolar cells are the first site where contrast adaptation is observed. *(From Rieke (2001).)*

stronger contrast adaptation in OFF than ON RGCs (Chander and Chichilnisky, 2001).

The next year, Baccus and Meister (2002) confirmed that BCs were the first site of contrast adaptation. They used linear-nonlinear (LN) models to capture both changes in gain and response kinetics following contrast transitions. In addition to decreased gain and faster kinetics following a low-high contrast switch, they introduced an additional physiological mechanism, prolonged hyperpolarization, which subsequent studies would show occurs in downstream circuit elements as well (Manookin and Demb, 2006).

Synaptic depression in bipolar cells

Contrast adaptation was also measured in voltage-clamp recordings of synaptic inputs to RGCs (Beaudoin et al., 2008; Kim and Rieke, 2001; Manookin and Demb, 2006). These measurements yielded two key insights. First, contrast adaptation was insensitive to pharmacological block of inhibition with GABA and glycine receptor antagonists (Brown and Masland, 2001; Manookin and Demb, 2006), suggesting that amacrine cells (ACs) are not required. Second, the input currents displayed two distinct time scales of adaptation. While the first timescale of ~1–2 s was consistent with the outer retinal mechanism described earlier, prolonged hyperpolarization in the RGC membrane potential could last for up to 10 s, suggesting an additional mechanism in the inner retina (Fig. 5.4A–C) (Manookin and Demb, 2006).

To isolate the inner retinal mechanism, Jarsky et al. (2011) circumvented the photoreceptors and used paired recordings between synaptically coupled rod bipolar cells (RBCs) and AII ACs. The authors injected white noise currents into RBCs and alternated between high and low standard deviation current injections. Currents in postsynaptic AII ACs showed a gain change consistent with contrast adaptation (Fig. 5.4D and E). A mechanistic model of vesicle depletion at the RBC ribbon synapses explained the gain control phenomenon.

Beaudoin et al. (2008) investigated the functional properties and contributions of different gain control mechanisms across light levels. Both excitation and inhibition to OFF transient alpha RGCs in guinea pig decreased gain following low to high contrast transitions, but only the excitatory pathway displayed faster kinetics at high contrast. At scotopic luminance, the authors measured little contrast adaptation in synaptic inputs to the RGC, so the third mechanism, detailed next, dominated the computation.

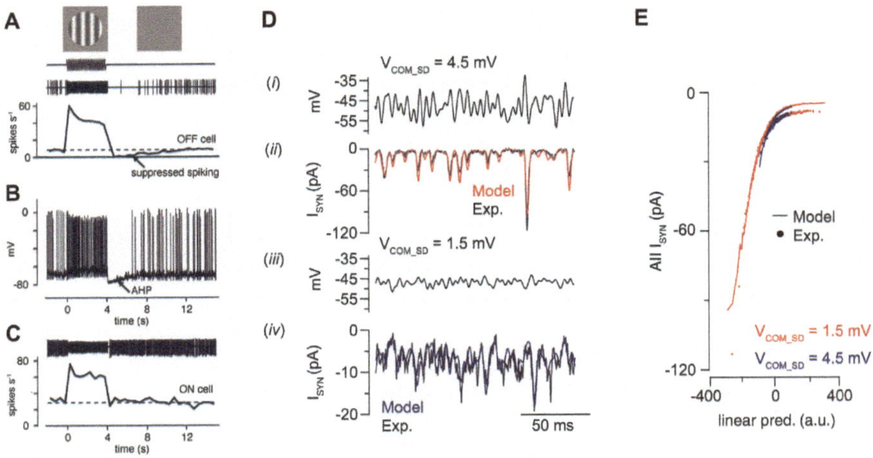

Fig. 5.4 Contrast adaptation at bipolar cell output synapses. (**A**) Spike recording of an OFF RGC in guinea pig retina shows a prolonged period of suppression following its response to a contrast-reversing grating. (**B**) Intracellular (current-clamp) recording from the cell in (**A**) shows a membrane hyperpolarization after the high contrast stimulus. (**C**) An ON RGC from the same retina does not show suppression following the high contrast stimulus. (**D**) Linear-nonlinear (LN) modeling of the rod bipolar cell to AII AC synapse. Top traces (i and iii) show current injected into the rod bipolar cell: either high variance (i) or low variance (iii). Bottom traces (ii and iv) show resulting currents measured in the postsynaptic AII AC *(black)* and the LN model fit *(red, blue)*. (**E**) Static nonlinearity of the LN model for high *(red)* and low *(blue)* variance current injections into the rod bipolar cell. *(Modified from (A–C) Manookin and Demb (2006) and (D, E) Jarsky et al. (2011).)*

Intrinsic mechanism in RGC spike generation

Unlike the first two contrast adaptation mechanisms in upstream BCs, several studies identified a third mechanism that was intrinsic to the RGC itself (Beaudoin et al., 2008; Kim and Rieke, 2001). In a strategy analogous to the one used by Jarsky et al. to isolate the RBC to AII AC synapse, Kim and Rieke (2003) used direct current injections in RGCs to determine whether the spike generator could adapt to the variance of the cell's membrane voltage. Indeed, the gain of the relationship between membrane voltage and spike probability was adjusted dynamically based on the variance of the past current trace. These authors went on to show that the cellular mechanism for this phenomenon was slow inactivation of the voltage-gated Na^+ channels that generate action potentials (Fig. 5.5). Notably, "slow" inactivation of Na^+ channels has a time constant of ~100–200 ms, so the intrinsic mechanism of contrast adaptation in RGCs is considerably faster than either of the

Fig. 5.5 High variance current injections into an RGC inactivate sodium channel. (**A**) Current injection paradigm for salamander RGCs. Two seconds of a white noise stimulus at different amplitude (variance) values was followed by a brief test pulse. (**B**) Responses to the test pulse following noise at each amplitude. *(From Kim and Rieke (2003).)*

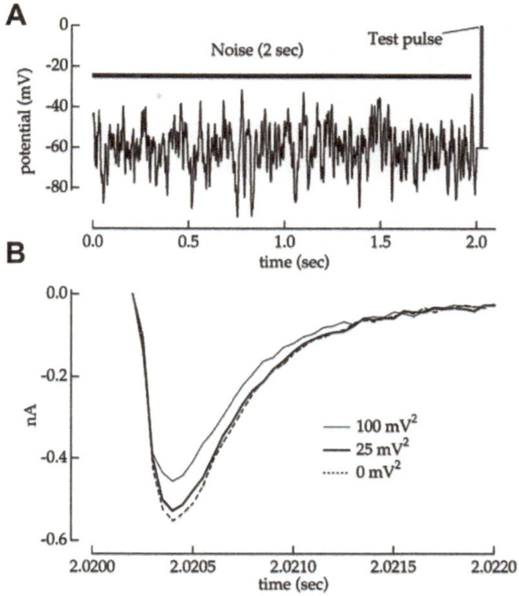

upstream circuit mechanisms. Kim and Rieke also pointed out the generality of this mechanism throughout the central nervous system. Many neurons need to adjust their gain based on the variance of their input currents, and most spiking neurons contain Na^+ channels that undergo inactivation.

Spatial and temporal scales of contrast adaptation

Since the two slower mechanisms of contrast adaptation occur in the inputs and outputs of BCs, respectively, one would expect the slow component of the computation to have a spatial scale more closely aligned to that of the BC receptive field (RF) than to that of the RGC RF. Brown and Masland (2001) first investigated this question by presenting a stimulus to the rabbit retina that alternated between high and low contrast in small subregions of the RGC RF. Indeed, adaptation was independent at different locations within the RF center, and the RF center and surround also adapted to contrast independently. A more recent study re-examined this question with a larger sample of mouse RGCs recorded on a multi-electrode array and found a diverse set of spatial adaptation patterns (Fig. 5.6) (Khani and Gollisch, 2017). While some RGCs showed local contrast adaptation,

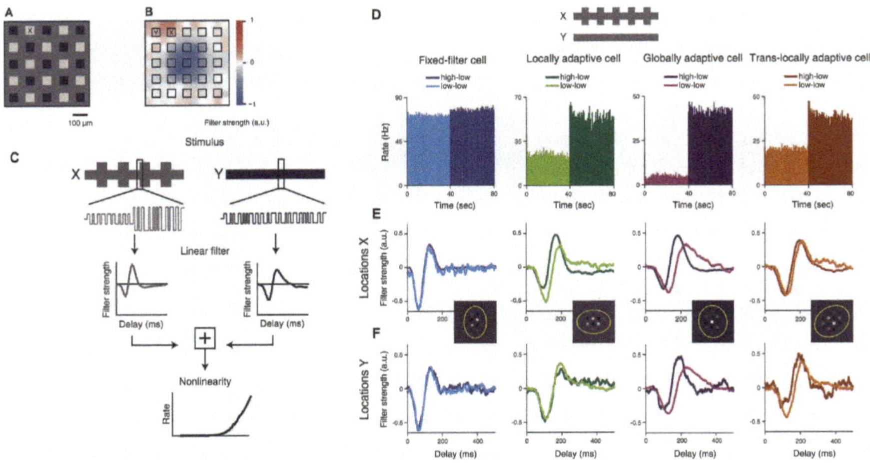

Fig. 5.6 Mouse RGCs show diverse spatial patterns of contrast adaptation. (**A**) Stimulus paradigm: alternate squares are marked "X" and "Y." X squares undergo periodic contrast transitions while Y squares do not. (**B**) Size of a typical RGC RF relative to the squares. (**C**) Schematic of LN model construction from the low-high contrast transition period. (**D**) Four different classes of responses to the stimulus: mean firing rate across the low-high contrast transition period. (**E**) Linear filters at X positions for examples of the 4 RGC classes measured during periods when both X and Y squares were low contrast *(lighter colors)* and following the transition to high contrast in X squares *(dark colors)*. (**F**) Same as (**E**) but linear filters measured from the positions of Y squares. *(Modified from Khani and Gollisch (2017).)*

consistent with a location in BCs, other RGCs showed global contrast adaptation that transferred from one RF location to another. A third category of RGCs had almost no contrast adaptation. Most surprisingly, a fourth category, called "trans-adapting" RGCs, actually showed greater contrast adaptation at RF locations in which the stimulus contrast remained fixed than in adjacent locations where it alternated between high and low contrast. Some of this functional diversity may be explained by the interplay of adaptation and sensitization, as described in the following section.

The three adaptation mechanisms described previously have time constants around 100–200 ms (RGC spike generation), 2–6 s (BC dendrites), and ~10 s (BC output synapses), so the overall time course of contrast adaptation might be expected to follow a weighted sum of three exponentials in this range. In fact, beyond the high-speed gain control in the first 100–200 ms, the time course of contrast adaptation is well fit by a single exponential, and remarkably, the time constant itself depends on the statistics of the preceding stimulus. Wark et al. (2009) used the familiar stimulus

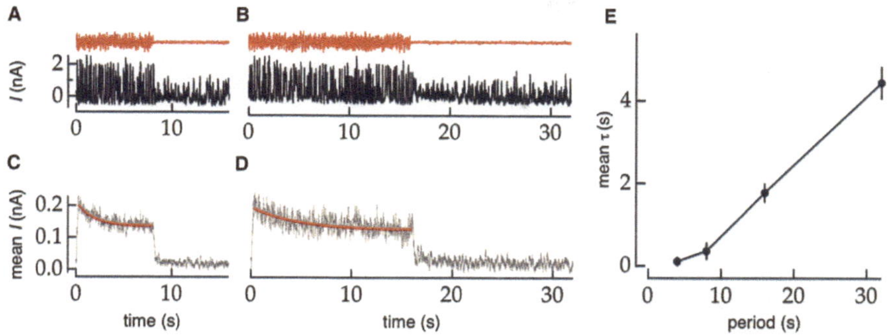

Fig. 5.7 The time constant of contrast adaptation itself adapts to the stimulus switching period. (**A**) Light stimulus switching from high to low contrast *(red)* and corresponding inhibitory conductance from an OFF transient RGC in mouse *(black)*. (**B**) Same as (**A**) but for a longer switching period. (**C** and **D**) Mean currents over trials for the conditions above with an exponential fit at the low-high contrast transition *(red)*. (**E**) Time constant of the exponential decay of current (averaged over all recorded cell types) as a function of switching period. *(From Wark et al. (2009).)*

paradigm of alternating low and high temporal contrast, but they varied the switching period from 4 s to 32 s. Within the range from 8 s to 32 s, the time constant of contrast adaptation changed linearly with the switching period (Fig. 5.7). They modeled this phenomenon from the perspective of Bayesian statistics. As evidence accumulates over a long period at one contrast, the prior distribution becomes narrow, so it takes more samples from the new distribution to achieve certainty about the contrast change, and thus adaptation proceeds slowly. The prior distribution remains broader for shorter switching periods, and faster adaptation is the optimal solution to the inference problem.

The Bayesian framework made several additional predictions that the authors confirmed experimentally. More discriminable changes in contrast (those with higher signal-to-noise ratio) drive faster adaptation, and adaptation from low to high contrast is faster than adaptation from high to low contrast. Finally, the authors tested whether higher-order moments of the light distribution (skewness and kurtosis) drive adaptation. While they confirmed the results of a previous study that the retina does not adapt to statistical moments beyond the variance (Bonin et al., 2006), the higher-order moments did impact the time course of contrast adaptation, consistent with the Bayesian inference framework. For equivalent switches in contrast, bimodal distributions (high kurtosis) are more discriminable than Gaussian distributions (low kurtosis), and indeed they drove faster adaptation (Wark et al., 2009).

Contrast sensitization

Contrast adaptation is effective in preventing saturation because it decreases gain at transitions from low to high contrast, but consider the opposite transition from high to low contrast. Gain begins in a low state and slowly recovers. Because of the statistical properties of contrast inference, this recovery is slower than the gain decrease at low to high transitions (Wark et al., 2009). Does this mean that RGCs remain insensitive and unable to encode fluctuations in light intensity for many seconds after a transition from high to low contrast? Using the example at the beginning of this chapter, adapting RGCs help you see the high contrast in the cliffs after you have been looking at the sky, but how do you see the low contrasts in the clouds after you have been looking at the cliffs?

The answer was unknown until Kastner and Baccus (2011) discovered a different form of change in response to contrast switches. Populations of RGCs in salamander, mouse, and rabbit retina responded to high-low contrast switches with a prolonged *elevation* in gain, opposite to the adapting RGCs (Fig. 5.8A). They called this phenomenon "sensitization" and went on to show that it occurs for both luminance and contrast switches in a subset of RGCs. Population data revealed a bimodal distribution; RGCs were cleanly separated into adapting and sensitizing groups (Fig. 5.8B). The two populations work together to encode temporal fluctuations in light across contrast transitions by dividing the contrast range. Adapting cells encode strong signals and use adaptation to prevent saturation; sensitizing cells encode weak signals and use sensitization to compensate for the slow gain recovery in adapting cells after a transition to low contrast.

Mechanism and spatial properties of contrast sensitization

The general mechanism of contrast sensitization turns out to be simple and elegant: a sign flip implemented by ACs. Sensitizing RGCs, and/or the BCs that provide excitation to them, are tonically inhibited by ACs. These ACs are, in turn, driven by BCs that undergo adaptation. Thus, as the adapting RGCs become hyperpolarized by a tonic decrease in excitation, the sensitizing RGCs become depolarized by a tonic decrease in inhibition either onto upstream BCs or onto their own dendrites (Fig. 5.9). Kastner and Baccus (2011) proposed such a model in their paper that first described sensitization, and they later demonstrated the mechanism experimentally by manipulating sensitization circuits via current injections in ACs (Kastner et al., 2019). This sign-flip mechanism of sensitization was also

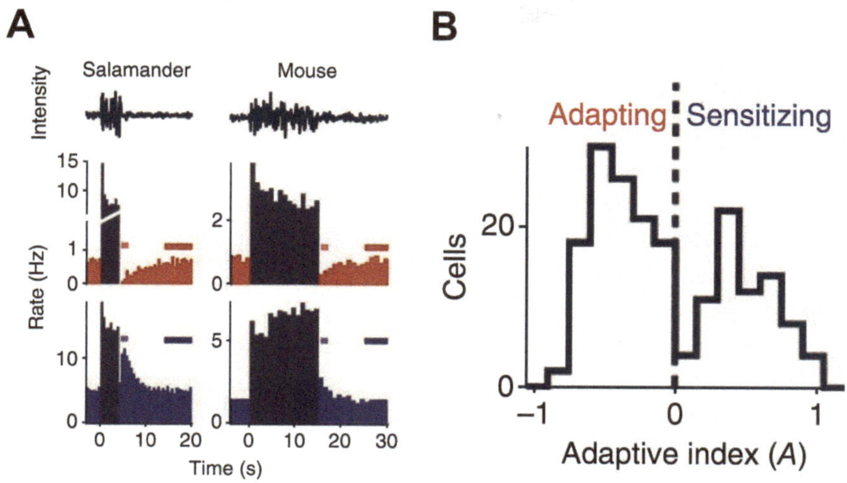

Fig. 5.8 Separate populations of RGCs show contrast adaptation and contrast sensitization. (**A**) *Top*, stimuli including contrast transitions. *Middle*, firing rate of example RGCs from salamander *(left)* and mouse *(right)* that show contrast adaptation indicated by a lower firing rate immediately after the high-low contrast transition *(light red bar)* than at steady-state low contrast *(dark red bar)*. *Bottom*, the firing rate of example sensitizing RGCs that have a higher firing rate immediately after the high-low contrast transition *(light blue bar)* than at steady-state low contrast *(dark blue bar)*. (**B**) Histogram of adaptive index (comparing time periods in (**A**), see text) for all RGCs measured in the study. *(Modified from Kastner and Baccus (2011).)*

confirmed in zebrafish retina in vivo using glutamate and calcium imaging (Nikolaev et al., 2013). Adapting and sensitizing glutamate signals from BCs were measured in different sublayers of the inner plexiform layer, and adapting and sensitizing calcium signals were measured from different populations of postsynaptic ACs and RGCs.

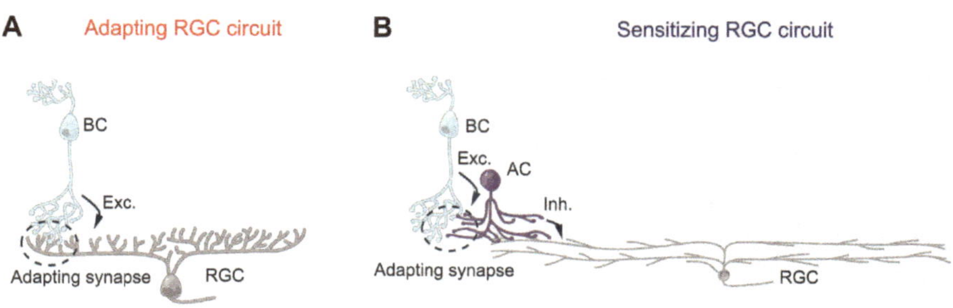

Fig. 5.9 Simple circuits for contrast adaptation and sensitization. (**A**) Schematic of a BC with an adapting synapse onto an adapting RGC. (**B**) An intervening AC makes the RGC sensitizing because tonic inhibition adapts.

So far, I have described sensitization as a mirror-symmetric computation to adaptation, but the fact that ACs implement the sign flip introduces a potential asymmetry in the spatial domain. AC RFs vary enormously from nearly the size of BCs on the small end to substantially larger than RGCs on the large end. Thus, while the spatial scale of adaptation is aligned to the small RFs of BCs, sensitization could take place on a range of larger scales. Kastner and Baccus (2013) investigated the spatial scale of contrast adaptation and sensitization using a low-contrast checkerboard stimulus in which a single square was temporarily switched to high contrast (Fig. 5.10). The position of the high-contrast square was varied throughout the experiment to build a spatial map of the sign and amplitude of the RGC's response to contrast switches. While adaptation was generally confined to the RF center, sensitization could extend far into the RF surround. A subset of RGCs, called "adapting OFF" cells, even had a center-surround organization to their contrast responses. The RF center showed contrast adaptation while the RF surround showed contrast sensitization. A model with small (BC) adapting subunits and larger (AC) sensitizing subunits could account for all the spatial profiles they observed simply by varying the strength of inhibition.

Separate populations of adapting and sensitizing cells enable the retina to encode both low and high contrasts with minimal delay at transitions (Kastner and Baccus, 2011), but what is the computational role of an RGC with an adapting center and a sensitizing surround? Kastner and Baccus (2013) suggested that this type of RF embodies a prediction about future object position. The spike response to a moving object within an RGC's RF decay rapidly after the object stops moving, but contrast adaptation and sensitization can last up to tens of seconds. Thus, even long after motion stops, the sensitizing RF surround stores information about where it may move next with its increased gain. The surround forms a halo of heightened sensitivity around the previous object location so that the RGC is poised to respond quickly to subsequent motion.

Since their discovery in 2011, sensitizing RGCs were found in zebrafish, salamander, mouse, and rabbit, implying that this computation is a general property of the vertebrate retina. Appleby and Manookin (2019) extended this result to the primate retina. These authors showed that midget RGCs, the numerically dominant RGC types of the primate retina, exhibit strong contrast sensitization to full-field stimuli. They went on to show that, like the "adapting OFF" cells in the salamander, midget RGCs had weak contrast adaptation in the RF center along with strong

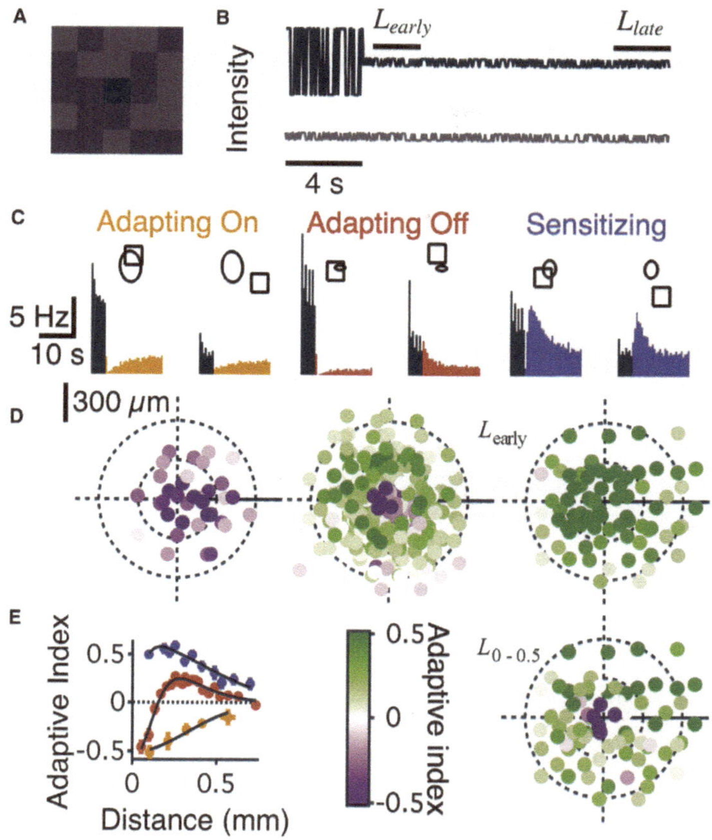

Fig. 5.10 Adaptation and sensitization can co-occur in the same RGCs in different parts of the receptive field. (**A**) Stimulus in which one square periodically switches to high contrast. (**B**) Intensity trace for the switching square *(top)* and the rest of the squares *(bottom)*. (**C**) Examples of firing rates for three different types of RGCs following the high-low contrast transition. Contrast-switching square position relative to RF center (ellipse) is indicated above each trace. (**D**) Adaptive index (colormap) for every cell, grouped into the three types above. Position of each point indicates the contrast-switching square position relative to the RF center (at the origin). Adaptive index is shown for two different time points for sensitizing RGCs *(rightmost column)*. (**E**) Mean adaptive index as a function of distance from RF center for the three classes of RGCs. *(Form Kastner and Baccus (2013).)*

sensitization throughout the RF surround (Fig. 5.11A–C). Voltage-clamp recordings revealed sensitization in the excitatory input to midget RGCs, localizing the site of the computation to AC inhibition onto the terminals of midget BCs. Midget RGCs make up the vast majority of the signal transmitted from the primate fovea, so they are likely optimized for

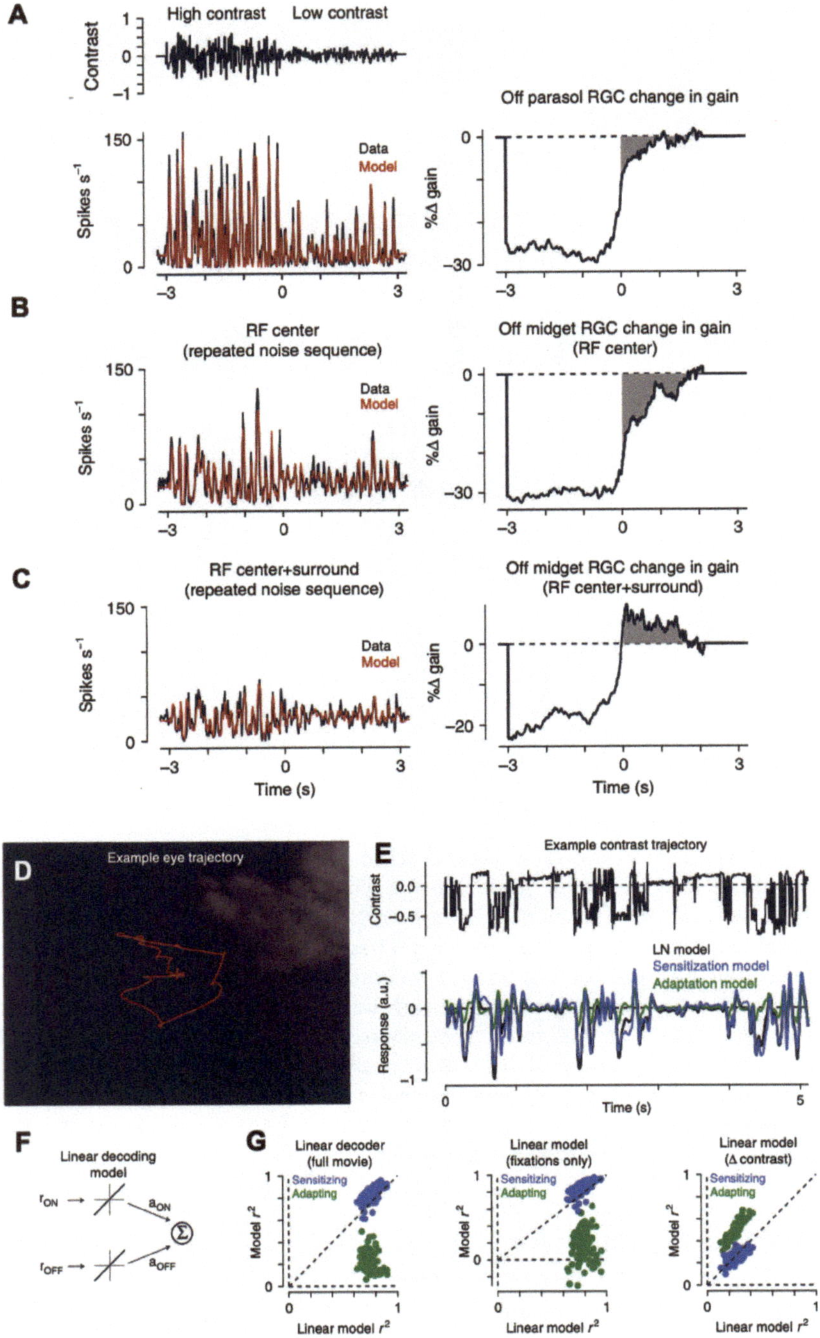

Fig. 5.11 Midget RGCs in primate have center adaptation and surround sensitization, and these properties improve the encoding of natural scenes. (**A**) *Left*, spike response *(black)* from an OFF parasol RGC and gain-modulated LN model *(red)* for a

(continued)

Fig. 5.11—cont'd contrast-switching stimulus *(top)*. *Right*, gain of the model over time relative to the high-low contrast transition. *Shaded region* shows adaptation. (**B**) Same as (**A**) for an OFF midget RGC for stimuli restricted to the RF center. (**C**) Same as (**B**) but for full-field stimulation of an OFF midget RGC. In this condition, the RGC shows contrast sensitization (positive shaded region on gain plot). (D) Example eye trajectory from the DOVES database used as a stimulus (see text). (**E**) Corresponding contrast trace for the trajectory in (**D**). (**F**) Schematic of two-channel ON-OFF linear decoder of contrast. (**G**) Performance of adapting *(green)* and sensitizing *(blue)* models compared to the linear model (x-axis) in encoding the naturalistic contrast trace as in (**E**) *(left)*, a trace containing only fixation periods *(middle)*, or the first derivative of the contrast trace *(right)*. *(Modified from Appleby and Manookin (2019).)*

encoding fluctuations in light intensity across saccades in natural scenes. Appleby and Manookin used image sequences from a database of human saccades across natural scenes to model how midget RGCs could encode the light fluctuations. They found that models incorporating sensitization outperformed adapting models, suggesting an important role for sensitization in foveal vision (Fig. 5.11D–G).

Conclusions and future directions

Contrast adaptation and sensitization are remarkable success stories in research on retinal computation. The concept of fast and slow forms of contrast adaptation was not even developed until the late 1990s, and contrast sensitization was discovered in 2011. Now the field has detailed measurements and predictive models of three different circuit mechanisms for contrast adaptation, a theoretical framework for how the computation solves the statistical inference problem (though the details remain controversial (Młynarski and Hermundstad, 2018)), and evidence that sensitization plays a key role in luminance encoding in the midget RGCs that drive vision in the human fovea.

Of course, there is always more to learn. In particular, I hope that, in the next decade, our field leverages this success in building the framework for understanding mechanisms of contrast adaptation and sensitization to reveal interactions among specific circuit components. Despite our ability to localize contrast adaptation and sensitization mechanisms to particular synapses, we still know very little about how these computations play out in different cell types. Which receptor types and downstream signaling molecules cause photoreceptor synapses onto BCs to adapt while HCs do not? How do both ON and OFF BCs adapt when their receptor types and signaling pathways are so different? Which of the ~40 different functionally identified mouse RGC types are contrast adapting and which are sensitizing? Which types fall into the four different spatial adaptation categories identified by Khani and Gollisch (2017): nonadapting, globally adapting, locally adapting, and trans-adapting? Which amacrine cell types are responsible for the adaptation sign-flip operation that causes sensitization? Are they morphologically and molecularly homologous from zebrafish all the way to primates? Along with greater detail about cell types, future work on contrast adaptation and sensitization may develop more theories about the functional consequences of these computations in behavior, and genetic tools in mice are bringing us closer to being able to manipulate mechanisms specifically in the retina and measure the behavioral consequences.

Landmark paper: Wark, B., Fairhall, A., and Rieke, F. (2009). Timescales of inference in visual adaptation. Neuron *61*(5), 750–761.

Discussion questions:

1. How do the authors explain the short adaptation time scales in Fig. 1E that do not match the linear relationship? Is there a different site of adaptation at work? How could one test this experimentally?
2. How do the three panels in Fig. 6 each show a different prediction of the inference model? Explain how kurtosis (panel C) effects the adaptation rate in the inference model even though the RGCs do not adapt to a change in kurtosis. Where in the paper is this prediction tested?

Landmark paper: Kastner, D.B., and Baccus, S.A. (2011). Coordinated dynamic encoding in the retina using opposing forms of plasticity. Nature Neuroscience *14*(10), 1317–1322.

Discussion questions:

1. How do adapting and sensitizing cells respond to changes in luminance? Where is this result in the paper, and how do the authors explain it?
2. What is the significance of Fig. 2E,F? For context, cell type classification has been historically difficult in salamander. The four classes in panel D are groups, not individual cell types.
3. How does Fig. 3 support the argument that adapting and sensitizing cells work together to encode a large range of contrasts? Even within the adapting and sensitizing groups, is there substantial diversity in contrast encoding? Why might this be beneficial?

4. Fig. 5 makes the argument about cooperativity between adapting and sensitizing RGCs using information theory. Explain this argument. Do you find it more or less convincing than the argument based on the data in Fig. 3? Why?

Programming exercise

To access the programming exercise, please visit https://www.elsevier.com/books-and-journals/book-companion/9780128198964.

This exercise explores the inference problem in contrast adaptation as investigated by Wark et al. (2009). How many samples of a different contrast distribution does it take to determine the distribution has changed?

1. Make a Gaussian random stimulus to represent the light intensity delivered to the neuron over time. Your stimulus should have 5 s with a coefficient of variation of 0.20 followed by 5 s with a coefficient of variation of 0.36 (units are arbitrary—call them Lumens if you like). The sampling rate is 100 Hz. Plot the stimulus and note the time of the contrast change (e.g., with a dotted line).
2. Look up the documentation on the MATLAB function "vartest2." Why is this an appropriate statistical test for this situation?
3. Assume that your neuron collects samples from a total of 1 s before the present time point. At each time point, it compares the first half of its collection window (−1 s to −0.5 s) to the second half (−0.5 s to present). Run "vartest2" for each (valid) time point in the stimulus. Collect all the p-values and plot the p-value over time. Hint: Use a logarithmic y-axis so you can see the small values better.
4. According to this test, at what time point does the neuron detect a contrast change for a p-value threshold of 0.05? Is this correct? What is a reasonable p-value threshold to reject the null hypothesis? (Hint: How many comparisons are you making here?). With your "corrected" p-value threshold, does the neuron now detect the contrast change correctly?
5. Run the simulation 20 times with different random seeds and get the mean and standard error of the latency (in milliseconds) to the detection of the contrast change.
6. Repeat the simulation, but this time change the order: high contrast (0.36) first followed by low contrast (0.20). How did your p-values change? If you make the sampling interval of the neuron longer than 1 s (still comparing the first and seconds halves of the interval), can you get comparable p-values in this situation to the ones you got in part 4?

7. Explain your intuition behind why the p-values were different when all you did was change the order. What does this mean for the speed at which an "optimal" visual neuron adjusts to contrast?

References

Appleby, T.R., and Manookin, M.B. (2019). Neural sensitization improves encoding fidelity in the primate retina. Nature Communications *10*(1), 4017.

Baccus, S.A., and Meister, M. (2002). Fast and slow contrast adaptation in retinal circuitry. Neuron *36*(5), 909–919.

Beaudoin, D.L., Manookin, M.B., and Demb, J.B. (2008). Distinct expressions of contrast gain control in parallel synaptic pathways converging on a retinal ganglion cell. Journal of Physiology *586*(22), 5487–5502.

Bonin, V., Mante, V., and Carandini, M. (2006). The statistical computation underlying contrast gain control. The Journal of Neuroscience *26*(23), 6346–6353.

Brown, S.P., and Masland, R.H. (2001). Spatial scale and cellular substrate of contrast adaptation by retinal ganglion cells. Nature Neuroscience *4*(1), 44–51.

Chander, D., and Chichilnisky, E.J. (2001). Adaptation to temporal contrast in primate and salamander retina. Journal of Neuroscience *21*(24), 9904–9916.

Demb, J.B. (2008). Functional circuitry of visual adaptation in the retina. Journal of Physiology *586*(18), 4377–4384.

Frazor, R.A., and Geisler, W.S. (2006). Local luminance and contrast in natural images. Vision Research *46*(10), 1585–1598.

Jarsky, T., Cembrowski, M., Logan, S.M., Kath, W.L., Riecke, H., Demb, J.B., and Singer, J.H. (2011). A synaptic mechanism for retinal adaptation to luminance and contrast. Journal of Neuroscience *31*(30), 11003–11015.

Kastner, D.B., and Baccus, S.A. (2011). Coordinated dynamic encoding in the retina using opposing forms of plasticity. Nature Neuroscience *14*(10), 1317–1322.

Kastner, D.B., and Baccus, S.A. (2013). Spatial segregation of adaptation and predictive sensitization in retinal ganglion cells. Neuron *79*(3), 541–554.

Kastner, D.B., Ozuysal, Y., Panagiotakos, G., and Baccus, S.A. (2019). Adaptation of inhibition mediates retinal sensitization. Current Biology *29*(16), 2640–2651.e4.

Khani, M.H., and Gollisch, T. (2017). Diversity in spatial scope of contrast adaptation among mouse retinal ganglion cells. Journal of Neurophysiology *118*(6), 3024–3043.

Kim, K.J., and Rieke, F. (2001). Temporal contrast adaptation in the input and output signals of salamander retinal ganglion cells. Journal of Neuroscience *21*(1), 287–299.

Kim, K.J., and Rieke, F. (2003). Slow Na+ inactivation and variance adaptation in salamander retinal ganglion cells. Journal of Neuroscience *23*(4), 1506–1516.

Manookin, M.B., and Demb, J.B. (2006). Presynaptic mechanism for slow contrast adaptation in mammalian retinal ganglion cells. Neuron *50*(3), 453–464.

Meister, M., and Berry, M.J. (1999). The neural code of the retina. Neuron *22*(3), 435–450.

Młynarski, W.F., and Hermundstad, A.M. (2018). Adaptive coding for dynamic sensory inference. eLife 7, e32055.

Nikolaev, A., Leung, K.M., Odermatt, B., and Lagnado, L. (2013). Synaptic mechanisms of adaptation and sensitization in the retina. Nature Neuroscience *16*(7), 934–941. 10.1038/nn.3408.

Rieke, F. (2001). Temporal contrast adaptation in salamander bipolar cells. Journal of Neuroscience *21*(23), 9445–9454.

Shapley, R. (1997). Retinal physiology: adapting to the changing scene. Current Biology 7(7), R421–R423.
Shapley, R.M., and Victor, J.D. (1978). The effect of contrast on the transfer properties of cat retinal ganglion cells. The Journal of Physiology *285*(1), 275–298.
Smirnakis, S.M., Berry, M.J., Warland, D.K., Bialek, W., and Meister, M. (1997). Adaptation of retinal processing to image contrast and spatial scale. Nature *386*(6620), 69–73.
Wark, B., Fairhall, A., and Rieke, F. (2009). Timescales of inference in visual adaptation. Neuron *61*(5), 750–761.

Contrast suppression is an unintuitive retinal computation in which a subset of cells, called suppressed-by-contrast (SbC), is most active in the *absence* of contrast. In the foreground is a seadragon with only high contrast that would strongly activate conventional cells. In the background is the ghostly SbC-dragon retaining only low contrasts.

CHAPTER 6

Contrast suppression

Gregory William Schwartz[a,b,c]
[a]Departments of Ophthalmology and Physiology, Feinberg School of Medicine, Northwestern University, Chicago, IL, United States
[b]Department of Neurobiology, Weinberg College of Arts and Sciences, Northwestern University, Chicago, IL, United States
[c]Northwestern Interdepartmental Neuroscience Program (NUIN), Northwestern University, Chicago, IL, United States

Contrast suppression is unique among retinal computations in that the field has largely discovered *how* before *why*. Many of the cellular and circuit mechanisms responsible for contrast suppression in the retina have been identified within the last decade, but why this computation evolved to be performed so early in the visual pathway and how it is used in the brain remain topics of speculation with little or no direct evidence.

Contrast suppression is the computation that produces retinal ganglion cells (RGCs) that reduce their tonic firing rate for both positive and negative contrasts. As discussed in Chapter 4, in addition to the three familiar ON, OFF, and ON-OFF polarity classes, there is a fourth class called suppressed-by-contrast (SbC) RGCs, which is comprised of cells with this response pattern. While they have typically received much less attention than the other three RGC classes, it is clear that SbC RGCs are conserved, at least across mammalian species, and that they represent a major information channel to the canonical pathway for visual perception through the dorsal lateral geniculate nucleus (dLGN) to primary visual cortex (V1).

Originally thought to be a single RGC type, I will argue that, like the other three polarity classes, the SbC class includes multiple functionally and morphologically distinct RGC types. At least three different types of SbC RGCs exist in the mouse; the differences in mechanism between the different types likely lead to functional specialization in ways we are only beginning to understand. Jason Jacoby and I wrote a recent review that covers some of the functional and morphological differences and circuit mechanisms of two of the SbC RGCs in mice (Jacoby and Schwartz, 2018). Here, I will additionally cover published work on rabbit SbC RGCs and currently unpublished work from my lab on a third (or possibly fourth) type of SbC RGC in mouse.

Retinal Computation
https://doi.org/10.1016/B978-0-12-819896-4.00007-X

Discovery of SbC RGCs in different species

SbC RGCs were first discovered in 1967 when Bob Rodieck (1967) found them in cat and Bill Levick (1967) found a similar RGC type in rabbit, which he called the "uniformity detector" (Fig. 6.1). In both cases, recordings were in vivo in anesthetized animals. Their existence in both species was confirmed by other groups using the same preparation (Amthor et al., 1989; Cleland and Levick, 1974; Rowe and Stone, 1976; Troy et al., 1989), and they were also found in the macaque (De Monasterio, 1978). These studies generally reported very few SbC RGCs among their quasi-random samples of RGCs, suggesting that they were a relatively rare cell type. The low numbers of reported cells also led to some controversy over whether they belonged to a single type or multiple types. David Mastronarde (1985) argued for the separation of cat SbC RGCs into "transient" and "sustained" subtypes based on the duration of their spike rate suppression. Despite this early work on their functional identification, SbC RGCs remained relatively unstudied for the next three decades. The first paper on the retinal mechanism of the SbC computation was not published until Sivyer et al. (2010) were able to record them ex vivo, and SbC RGCs in mouse were not identified until 2015 (Jacoby et al., 2015; Tien et al., 2015).

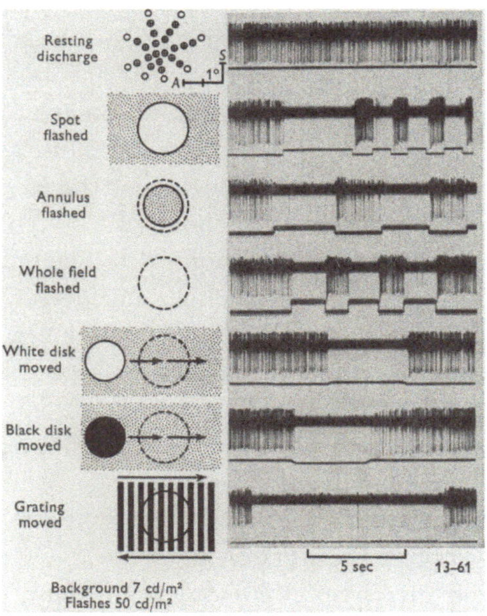

Fig. 6.1 Discovery of suppressed-by-contrast (uniformity detector) RGCs in rabbit. Example of a uniformity detector RGC. *Top,* receptive field map where *shaded points* indicate positions at which a flashed spot suppressed the steady firing seen to the right. Subsequent rows show the variety of stimuli that could suppress firing in this cell. No stimulus was able to increase firing above the tonic rate. *(From Levick (1967).)*

Table 6.1 Characteristics of SbC RGCs.

RGC name	Species	Suppression temporal scale	Suppression time vs. spot size	Suppression time vs. contrast	ON suppression mechanism	OFF suppression mechanism	References
Uniformity Detector (UD)	Rabbit	0.5–1 s	Negative correlation	?	Increase in inhibition (glycine)	Increase in inhibition	Sivyer et al. (2010) and Sivyer and Vaney (2010)
ON delayed/ transient SbC (tSbC)	Mouse	0.2–0.5 s	Negative correlation	Positive correlation	Increase in inhibition (GABA + glycine)	Increase in inhibition (GABA + glycine from VGluT3 AC) + decrease in excitation	Tien et al. (2015) and Mani and Schwartz (2017)
Sustained SbC (sSbC)	Mouse	up to 20 s	EW28: negative correlation; EW27: none	Positive correlation	Increase in inhibition (GABA from CRH-2 AC + glycine from AII AC)	Small increase in inhibition + decrease in excitation	Jacoby et al. (2015)
Bursty SbC (bSbC)	Mouse	0.5–1 s	None	Positive correlation	Increase in inhibition + decrease in excitation	Depolarization block	Wienbar and Schwartz (manuscript in prep.)

In this chapter, I will describe the functional and morphological characteristics and circuit mechanisms of the four (or five) SbC RGC types that have been well studied: one in rabbit, and three (or four) in mouse. Table 6.1 summarizes the characteristics for each type.

"Uniformity detector" RGC (rabbit)

Functional and morphological characteristics

Sivyer et al. (2010) managed to target SbC RGCs in rabbit based on their soma morphology and spontaneous firing patterns. They retained the "uniformity detector" name, but I will use SbC for consistency. These RGCs appeared to fall into a single functional and morphological class. They had a spontaneous firing rate of ~20 Hz that was suppressed to zero at both light onset and offset for 0.5–1 s before a slower recovery back to baseline (Fig. 6.2A–C). Suppression was maximal for a light stimulus near the size of the dendritic field, and larger stimuli produced a weaker suppressive response, indicating a center-surround receptive field structure.

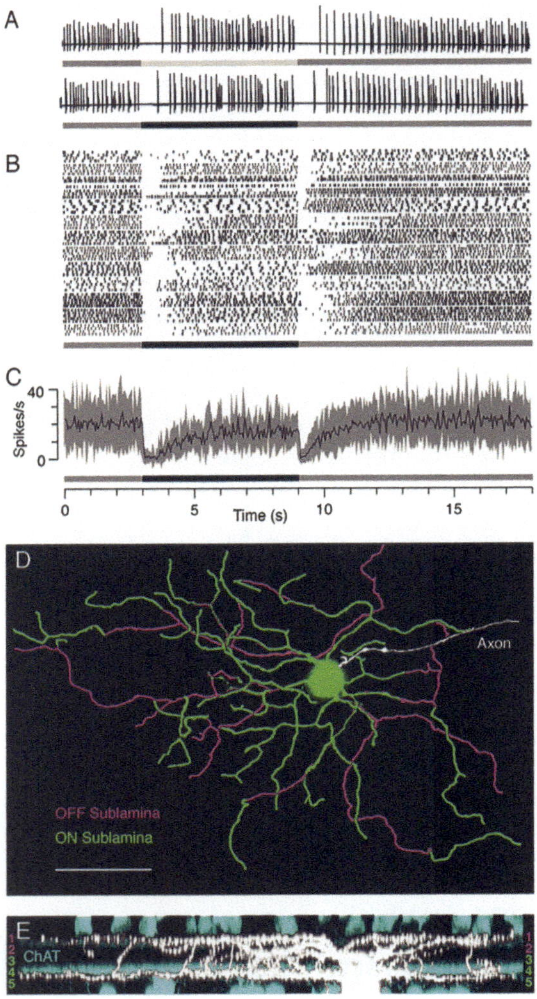

Fig. 6.2 Light responses and morphology of SbC RGCs in rabbit. (**A**) Cell-attached recordings of a rabbit SbC RGC responding to a positive *(top)* and negative *(bottom)* contrast spot. (**B**) Spike rasters of many SbC RGCs from the study. (**C**) Mean firing rate across cells. (**D**) Image of an SbC RGC in en face view (scale bar = 50 μm) and (**E**) in side view. *(From Sivyer et al. (2010).)*

Rabbit SbC RGCs had large, bistratified dendritic arbors (Sivyer et al., 2010; Sivyer and Vaney, 2010). Their most distinct morphological feature was the presence of "recursive" dendrites that snaked between the ON and OFF sublamina of the inner plexiform layer (IPL), which led some anatomists to call them "diving" cells (Fig. 6.2D and E) (Hoshi et al., 2009; Sivyer and Vaney, 2010). Neurobiotin injections revealed homotypic gap-junction coupling to other morphologically similar RGCs, as well as coupling to a single morphological class of wide-field GABAergic amacrine cells (ACs) (Sivyer and Vaney, 2010).

Mechanism of contrast suppression

Sivyer et al. (2010) used voltage clamp to record the synaptic conductances onto SbC RGCs and found that suppression at both ON and OFF is driven by a transient inhibitory conductance, with little or no contribution from excitation (Fig. 6.3). Pharmacology experiments revealed that the inhibition was carried mostly by glycine receptors. This was the first RGC type discovered that appeared to be driven nearly exclusively by inhibition. There was no significant light-evoked change in excitation at light ON or OFF. Nonetheless, there was a tonic excitatory conductance that could be reduced by glutamate receptor blockade. Recent work has identified a glutamatergic monopolar interneuron (GluMI) in mouse that is a candidate mechanism for tonic excitation that is not increased at light ON or OFF (Della Santina et al., 2016).

A distinct physiological feature of SbC RGCs in the Sivyer et al. (2010) study was their "complex" spikes. Sodium spikes of various sizes would ride on a lower frequency plateau potential. The first spikes that occurred following a light or dark-induced hyperpolarization were larger and had a higher maximum slope. The authors demonstrated that the small and variable maintained spikes were a result of sodium channel inactivation, which was relieved by light-evoked hyperpolarization. While they did not record from axons, they speculated that some of the smallest spikes in SbC RGCs may not be transmitted to the brain, complicating the interpretation of the spike activity recorded at the soma.

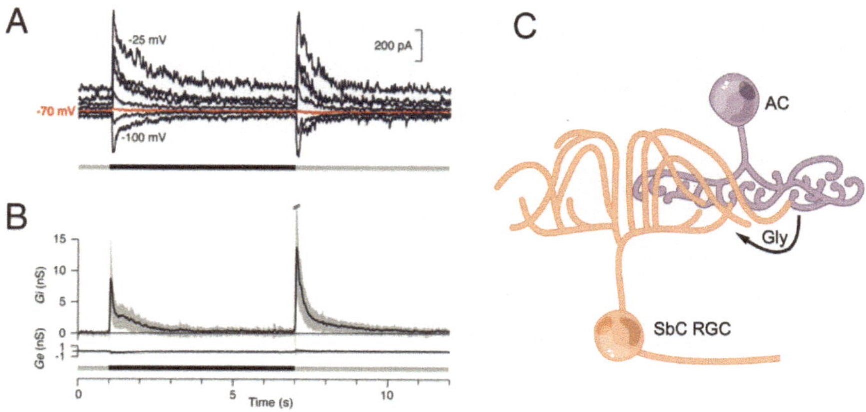

Fig. 6.3 Mechanism of contrast suppression in the rabbit SbC RGC. (A) Voltage-clamp recording of synaptic currents in a rabbit SbC RGC at different holding potentials. Red trace at −70 mV indicates lack of excitatory current at the inhibitory reversal potential. **(B)** Computed inhibitory (Gi) and excitatory (Ge) conductance from **(A)**. **(C)** Minimal circuit diagram for the rabbit SbC RGC. *((A and B) From Sivyer et al. (2010).)*

Transient SbC/ON delayed RGC (mouse)

Functional and morphological characteristics

Tien et al. (2015) were the first to identify an SbC RGC type in mouse retina. They used cholecystokinin (CCK)-Cre mice (Zhu et al., 2014) to target an RGC type with a maintained firing rate around 20 Hz and transient (~0.2–0.5 s) suppression to ON and OFF stimuli (Fig. 6.4A). While these authors named the cell simply SbC RGC, I will add "transient" (t) to distinguish it from the sustained type(s) in the next section. Suppression in these tSbC RGCs had a temporal aspect to its surround. While full-field light stimuli were able to suppress firing completely, the duration of suppression was decreased relative to that for smaller stimuli covering only the RF center (Tien et al., 2016). Morphologically, the tSbC RGCs in these studies resembled the SbC RGCs in rabbit. While they tended to have more densely branched dendrites than the putative rabbit homolog, they were bistratified

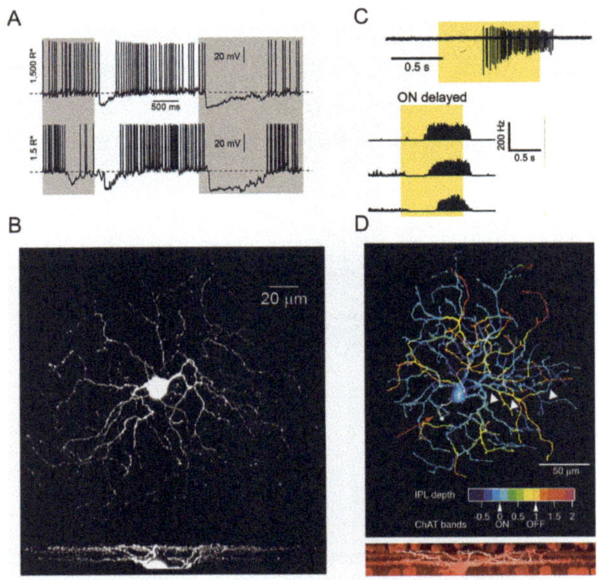

Fig. 6.4 Light responses and morphology of the tSbC/ON delayed RGC in mouse. (**A**) Current-clamp recordings of a tSbC RGC in mouse responding to contrast steps at two different light intensities as indicated. (**B**) Morphology of the tSbC RGC in en face and side views. (**C**) Cell-attached light response and spike responses from three example ON delayed (tSbC) RGCs. (**D**) Morphology of ON delayed RGC in en face and side views. *(From (A) Tien et al. (2015), (B) Tien et al. (2016), and (C, D) Mani and Schwartz (2017).)*

in similar IPL sublaminae, and importantly, they had recursive dendrites (Fig. 6.4B) (Lee et al., 2016; Tien et al., 2016; Tien et al., 2015).

Later, my lab identified an RGC type in WT mice that we named the ON delayed RGC due to the extremely long (~0.3–0.5 s) delay from light onset to its increase in spike rate (Fig. 6.4C) (Mani and Schwartz, 2017). ON delayed RGCs had a maintained firing rate of less than 5 Hz in our hands in darkness. Thus, we classified them as ON RGCs based on their late increase in firing for positive contrast. While differences in recording conditions, including extracellular solutions and light levels, may have contributed to the difference in maintained firing rate, we now have several independent lines of evidence that the ON delayed RGC is, in fact, the same type as the tSbC RGC reported by Tien et al. (2015). First, ON delayed RGCs had a response latency that decreased with spot size, particularly for very large spots extending well beyond their dendrites (Mani and Schwartz, 2017). Second, they were morphologically indistinguishable from the tSbC RGCs and type "73" in the Eyewire museum (Bae et al., 2018), with the same stratification pattern and many recursive dendrites (Fig. 6.4D). Third, we targeted RGCs in the CCK-Cre line and found ON delayed RGCs to be among the labeled cell types (Jacoby and Schwartz, 2018). Finally, as described next, key aspects of the synaptic currents and pharmacology matched between the tSbC and the ON delayed RGC (Jacoby and Schwartz, 2018). Thus, information about mechanism can be combined across these studies, giving us a more complete picture of the circuit.

Mechanism of contrast suppression

At light onset, tSbC RGCs receive strong, transient inhibition and weak, sustained excitation. At light offset, they receive moderate inhibition and a small, sustained decrease in tonic excitation (Fig. 6.5A and D). Both ON and OFF inhibition onto the tSbC RGC use a combination of GABA and glycine receptors (Mani and Schwartz, 2017; Tien et al., 2015), and while there are many elements remaining to be discovered in this circuit, we know of at least one AC type that is involved. A single AC type expresses the gene for the type 3 vesicular glutamate transporter (VGluT3) (Grimes et al., 2011). VGluT3 ACs release glutamate onto RGCs in other retinal circuits (Lee et al., 2016), but in this circuit, they release glycine onto the dendrites of the tSbC RGC (Lee et al., 2016; Tien et al., 2016). Genetic ablation of VGluT3 ACs reduced inhibition to tSbC RGCs, causing a reduction in suppression specifically for small OFF stimuli. Interestingly, suppression for

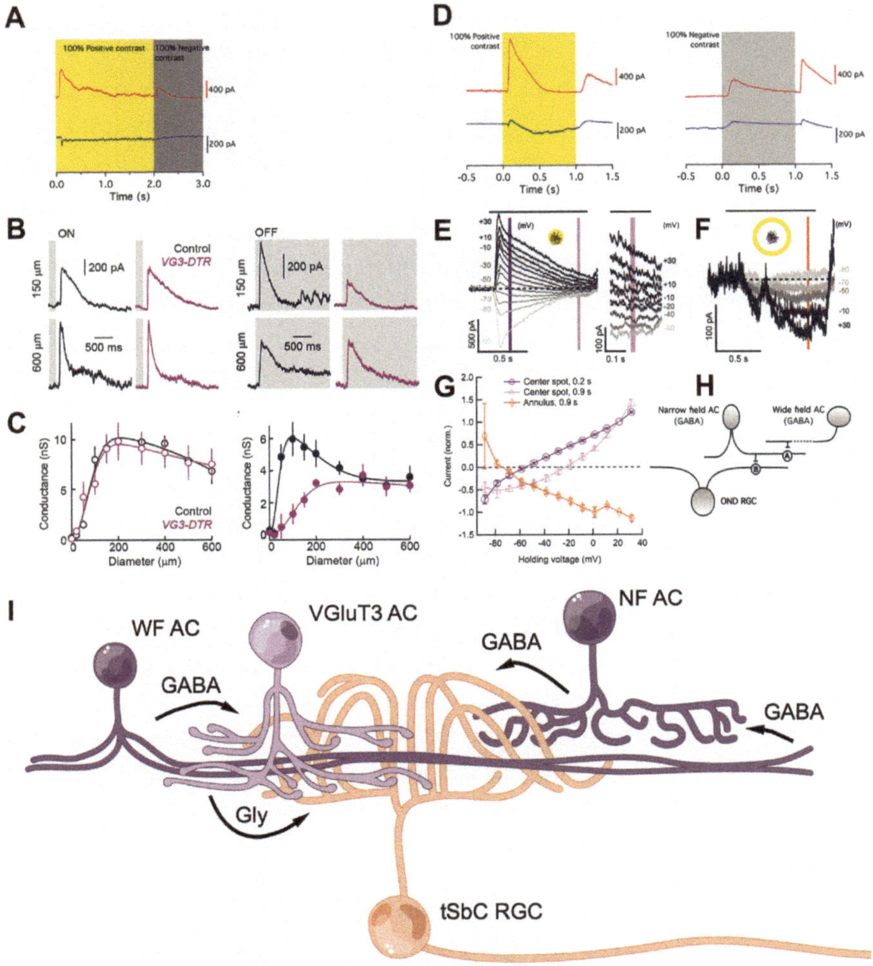

Fig. 6.5 Synaptic circuits upstream of tSbC RGCs. (**A**) Excitatory *(blue)* and inhibitory *(red)* currents in tSbC RGCs. (**B**) Inhibitory current for small and large ON and OFF spots as indicated in WT *(black)* and mice in which VGluT3 ACs are ablated *(magenta)*. (**C**) Summary of inhibitory conductance as a function of spot size in WT *(black)* and VGluT3 ablated mice *(magenta)* for ON *(left)* and OFF *(right)* spots. (**D**) Excitatory *(blue)* and inhibitory *(red)* currents in ON delayed (tSbC) RGCs. (**E**) Voltage-clamp recordings of synaptic currents in ON delayed (tSbC) RGCs for light spots covering the dendritic field. *Shaded regions* indicate time intervals in the IV curves below. (**F**) Same as (**E**) but for an annulus stimulus far beyond the RGC dendrites. (**G**) Current-voltage (IV) curves at the times indicated in (**E** and **F**). (**H**) Schematic of proposed circuit for light-driven disinhibition beyond the RGC dendrites. (**I**) Composite schematic incorporating elements of the circuit upstream of the tSbC RGC from both studies mentioned in the text. *(From (A, D) Jacoby and Schwartz (2018), (B, C) Tien et al. (2016), and (E–H) Mani and Schwartz (2017).)*

large OFF stimuli was unaffected, suggesting that a different AC is responsible for that part of the response (Fig. 6.5B, C, and I). The effect of VGluT3 AC ablation on ON suppression was more complex because excitation was decreased, likely through an indirect pathway, resulting in an increase in suppression (Tien et al., 2016). While we do not know the identity of the ACs responsible for ON suppression in the tSbC RGC, we do know some of their properties. As spot size increases, ON inhibition becomes smaller and more transient. This surround suppression of inhibition is mediated by $GABA_C$ receptors, either on bipolar cells or on the neurites of the ACs that inhibit tSbC RGCs at light onset (Fig. 6.5I) (Mani and Schwartz, 2017).

The tSbC RGC also has a noncanonical pathway that allows it to increase its firing rate above baseline, after the suppression period, for large ON stimuli. We measured the RF center profile of the tSbC RGC, and we noticed that its diameter was almost four times the dendritic field size of the RGC. Annulus stimuli confirmed that light well beyond the dendrites of the RGC could elicit increases in firing. Since excitation onto RGCs typically comes from bipolar cells, which have small receptive fields compared to RGCs, it was a puzzle that tSbC RGCs could be activated for stimuli so far away from their dendrites. Voltage-clamp measurements and pharmacology revealed that this distant activation was not from excitation, but rather from disinhibition. Large spots or distant annuli caused a decrease in a tonic $GABA_B$ conductance (Fig. 6.5E–H) (Mani and Schwartz, 2017). Thus, tSbC RGCs are activated by distant ON stimuli via this serial inhibition circuit (Fig. 6.5I).

Sustained SbC RGC (mouse)

Functional and morphological characteristics

My lab identified the sustained (s)SbC RGCs in mouse (Jacoby et al., 2015), and we admittedly underestimated the diversity of the SbC class, referring to them simply as SbC RGCs. The sSbC had a maintained firing rate of ~15 Hz, and it was typically suppressed completely for the duration of a 1 s light or dark step at 100% contrast. Lower contrast stimuli elicited shorter durations of suppression (Fig. 6.6A). We later showed that suppression in sSbC RGCs can last for up to 20 s for a bright light step from darkness (Fig. 6.6B) (Jacoby and Schwartz, 2018). While our study used light stimuli of 250 μm, around the size of the dendritic field, we have subsequently probed sSbC RGCs with larger light stimuli. These experiments suggested that sSbC RGCs fall into a bimodal distribution with some cells having an RF surround and thus suppressing less for large stimuli and others showing

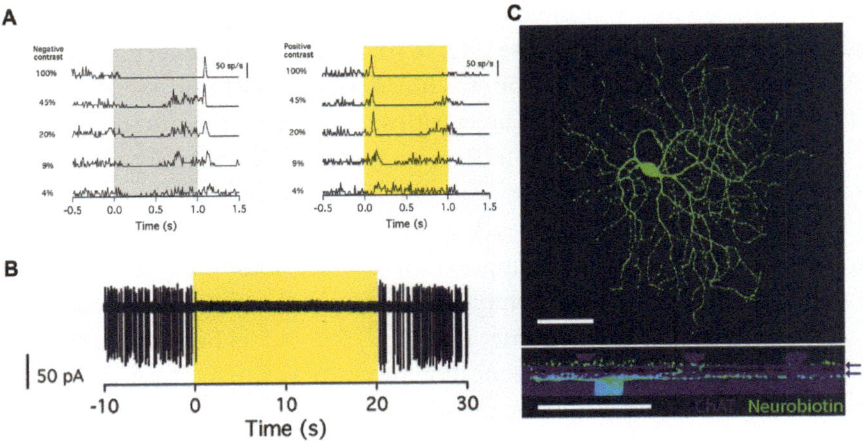

Fig. 6.6 Light responses and morphology of the sSbC RGC in mouse. (**A**) Firing rate of an sSbC RGC to positive and negative contrast steps of various amplitudes as indicated. (**B**) Sustained suppression of an sSbC RGC for a 20-s light step. (**C**) Morphology of an sSbC RGC in en face *(top)* and side *(bottom)* views. Scale bars = 50 μm. *(From (A, C) Jacoby et al. (2015) and (B) Jacoby and Schwartz (2018).)*

no change in suppression for large stimuli. This result and anatomical evidence described next have caused us to provisionally separate sSbC RGCs into two types, named sSbC EW27 and sSbC EW28, by their anatomical equivalents in the Eyewire museum (Bae et al., 2018; Goetz et al., 2021).

Like the SbC in rabbit, the sSbC RGCs we found in mouse were large and bistratified, but they lacked the recursive dendrites that were the hallmark of the rabbit cell (Fig. 6.6C) (Jacoby et al., 2015). Detailed comparisons to reconstructions from electron microscopy (EM) suggest that there are two morphological types that may align with the two function types of sSbC RGCs. Type "27" cells have larger dendritic trees, with fewer bifurcations, whereas type "28" cells have smaller dendritic trees, and dense, highly branched morphology (Bae et al., 2018). The two morphological types also differ slightly in stratification in the IPL, with the type "28" cell having its ON stratum slightly closer to the ganglion cell layer.

Mechanism of contrast suppression

Voltage-clamp recordings in sSbC RGCs revealed a different circuit than the one described in rabbit or the tSbC in mouse (Jacoby et al., 2015). ON steps elicited a large and very sustained inhibitory conductance while

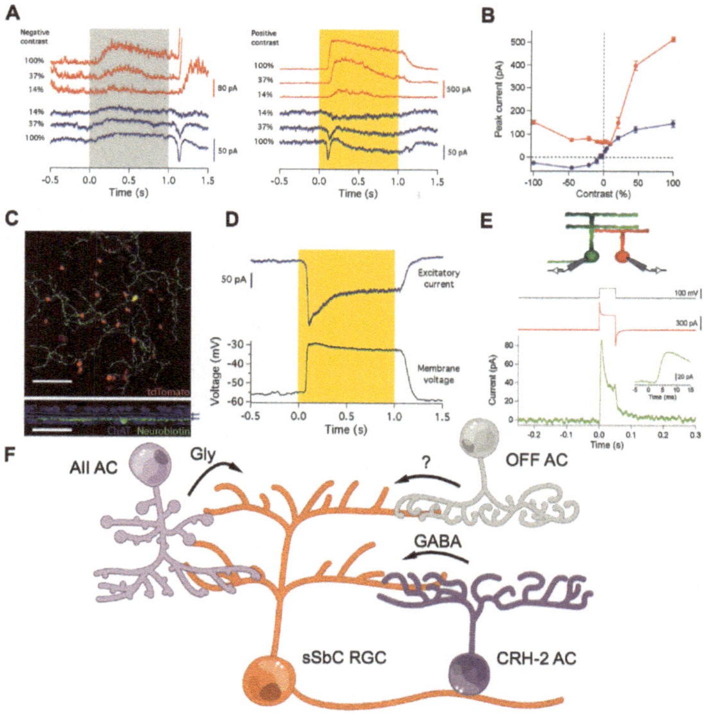

Fig. 6.7 Mechanism of ON contrast suppression in sSbC RGCs in mouse. (**A**) Excitatory *(blue)* and inhibitory *(red)* synaptic currents to sSbC RGCs in mouse for positive and negative contrasts as indicated. (**B**) Summary across cells of excitatory and inhibitory currents as a function of contrast. (**C**) En face *(top)* and side view *(bottom)* of the morphology of the CRH-2 AC. Scale bar = 50 μm. (**D**) Excitatory current *(top)* and membrane voltage in a CRH-2 AC responding to a light step. (**E**) Paired recordings demonstrating direct connectivity between a presynaptic CRH-2 AC *(red)* and a postsynaptic sSbC RGC *(green)*. (**F**) Schematic of the sSbC circuit. *((A–E) Modified from Jacoby et al. (2015).)*

OFF steps elicited much smaller and somewhat more transient inhibition. At light ON, the increase in inhibition was accompanied by a small excitatory conductance with a very transient phase and a more sustained slow, late phase. Tonic excitation was decreased at light OFF (Fig. 6.7A and B). Pharmacology experiments revealed that the large and sustained inhibitory conductance responsible for suppression at light onset contained a combination of GABAergic and glycinergic components. By blocking all ionotropic glutamate receptors, we identified that AII ACs contribute to the glycinergic component. To identify the specific AC responsible for the GABAergic component, we used a transgenic mouse, expressing Cre under the corticotropin-releasing hormone (CRH) promoter (Zhu et al., 2014).

The most prevalent AC labeled in this line was the CRH-2 AC (Zhu et al., 2014), a medium-field, displaced AC with no sodium spikes, and a sustained depolarization to light (Fig. 6.7C and D). Paired recordings demonstrated that CRH-2 ACs release GABA onto sSbC RGCs (Fig. 6.7E) (Jacoby et al., 2015). Ablating the CRH-2 ACs surrounding an sSbC RGC changed its contrast response so that it was no longer suppressed for positive contrasts (Jacoby et al., 2015). Thus, CRH-2 ACs are necessary for the SbC computation in sSbCs RGCs.

Bursty suppressed-by-contrast (bSbC) RGC (mouse)

Functional and morphological characteristics

We have discovered a third (or fourth if you count sSbC as two) type of SbC RGC in mouse that we call the bursty (b)SbC RGC. Functionally, bSbC RGCs are differentiated from the other SbC RGC types in mouse by their higher maintained firing rate (~40 Hz), their highly variable spike amplitudes (Fig. 6.8A), and their lack of an RF surround. Suppression in the bSbC RGC is around the same duration as in the tSbC RGC or a bit longer (0.5 to 1 s), but rather than spot size, it depends on contrast (Fig. 6.8B).

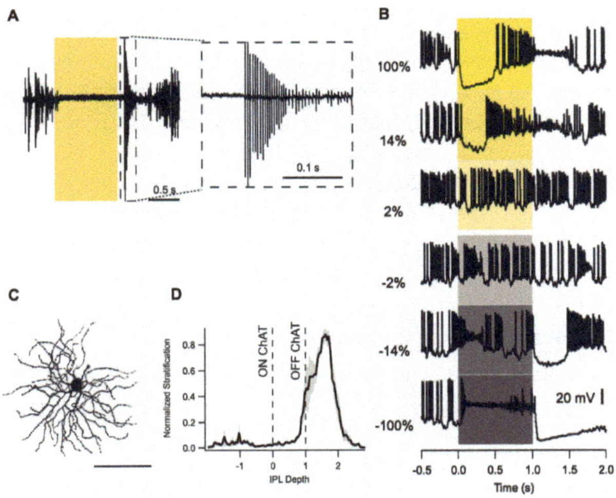

Fig. 6.8 Light responses and morphology of the bSbC RGC in mouse. (**A**) Cell-attached recording of a bSbC RGC in mouse. Inset shows a magnified view of the large spike amplitude adaptation. (**B**) Current-clamp recordings of a bSbC RGC at positive and negative contrast as indicated. (**C**) En face morphology of a bSbC RGC. Scale bar = 100 μm. (**D**) Stratification profile of the bSbC RGC. *Shaded region* is SD across cells. *(Wienbar and Schwartz, manuscript in preparation.)*

Morphologically, bSbC RGCs are very different from the other SbC RGC types. They are monostratified in the OFF layer of the IPL, and they have larger somas than all other RGC types except the three alpha types (Fig. 6.8C and D).

Mechanism of contrast suppression

The mechanism of contrast suppression is also very different in bSbC RGCs than the other SbC types. At light onset, bSbC RGCs receive strong, transient inhibition and a strong decrease in tonic excitation; thus, their spike suppression is a natural result of their conductances. At light offset, the conductance changes are in the opposite direction. Excitation increases while inhibition decreases (Fig. 6.9 A–C, and E). Therefore, OFF suppression would not be the natural result of the synaptic inputs; a spike rate increase would be expected.

The intrinsic properties of the bSbC RGC play a key role in its suppression for OFF stimuli. The synaptic inputs drive the cell into depolarization block, where too high a proportion of the sodium channels are inactivated to generate an action potential (Fig. 6.8C). While the bSbC is clearly morphologically and functionally distinct from the SbC described in rabbit (Sivyer and Vaney, 2010), the two cell types share the feature of sodium channel inactivation (Sivyer et al., 2010). We made paired recordings from the soma and axon of bSbC RGCs to measure spike propagation. These measurements confirmed that small amplitude spikes (with a maximum slope less than 50 V/s) often fail to propagate down the axon (Fig. 6.9D). Thus, depolarization block at light offset is part of the mechanism by which these RGCs transmit an SbC contrast response function to their downstream targets in the brain.

Evidence for SbC RGC projections in the brain

Neurons with SbC response profiles have been found throughout the visual system, especially in the pathway from retina to dLGN to V1 that is most closely associated with visual perception (Niell and Stryker, 2008; Piscopo et al., 2013; Román Rosón et al., 2019; Rompani et al., 2017; Suresh et al., 2016; Tailby et al., 2007). Over the last decade, several different research groups have investigated the functional signals transmitted from the retina to dLGN using a variety of complementary techniques, including extracellular (Piscopo et al., 2013; Román Rosón et al., 2019) and intracellular (Suresh et al., 2016) recordings in dLGN, trans-synaptic viral tracing

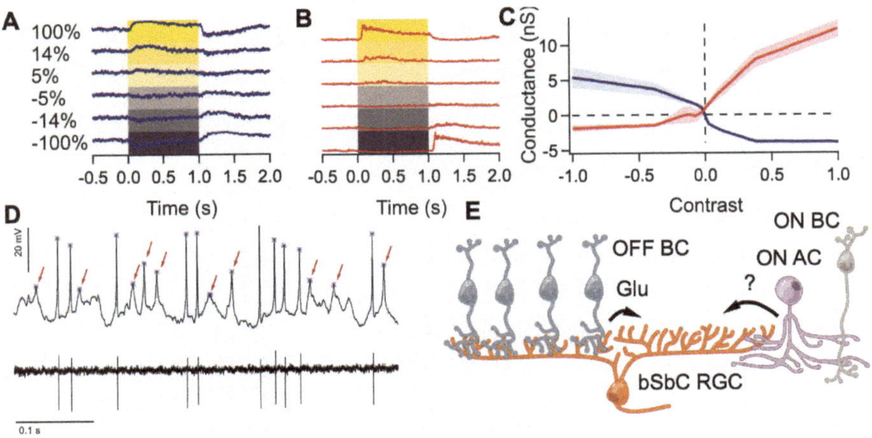

Fig. 6.9 Mechanism of contrast suppression in sSbC RGCs in mouse. (**A**) Excitatory currents in a bSbC RGC to positive and negative contrast steps as indicated. (**B**) Inhibitory currents for the cell in (**A**). (**C**) Excitatory *(blue)* and inhibitory *(red)* conductances in bSbC RGCs as a function of contrast. *Shaded region* is SEM across cells. (**D**) Simultaneous recording of soma *(top)* and axon *(bottom)* of a bSbC RGC demonstrating that small action potentials do not propagate down the optic nerve. (**E**) Schematic of the bSbC circuit. *(Wienber and Schwartz, manuscript in preparation.)*

(Rompani et al., 2017), and calcium imaging of retinal axons in dLGN (Liang et al., 2018). One of the most consistent findings among these studies has been that SbC RGCs make up a substantial fraction of this signal (Fig. 6.10), in the range of 15%–25%. Most studies have grouped SbC-type responses in dLGN, but Román Rosón et al. (2019) found that dLGN spike responses were best explained by including a combination of several different functional groups of SbC RGCs (Fig. 6.10F). It remains to be determined which of the SbC RGC types project to dLGN and what other brain areas they innervate.

Speculation about the role of SbC RGCs in behavior

A number of theories for the role of SbC RGCs in behavior have been introduced, but none have been tested experimentally. A lack of tools for selective manipulation of SbC circuits has prevented a direct, causal experiment linking these circuits to behavior; aided by the mechanistic studies described earlier, such tools are on the horizon in mice. One theory for the role of SbC RGCs involves accommodation, the process by which animals adjust

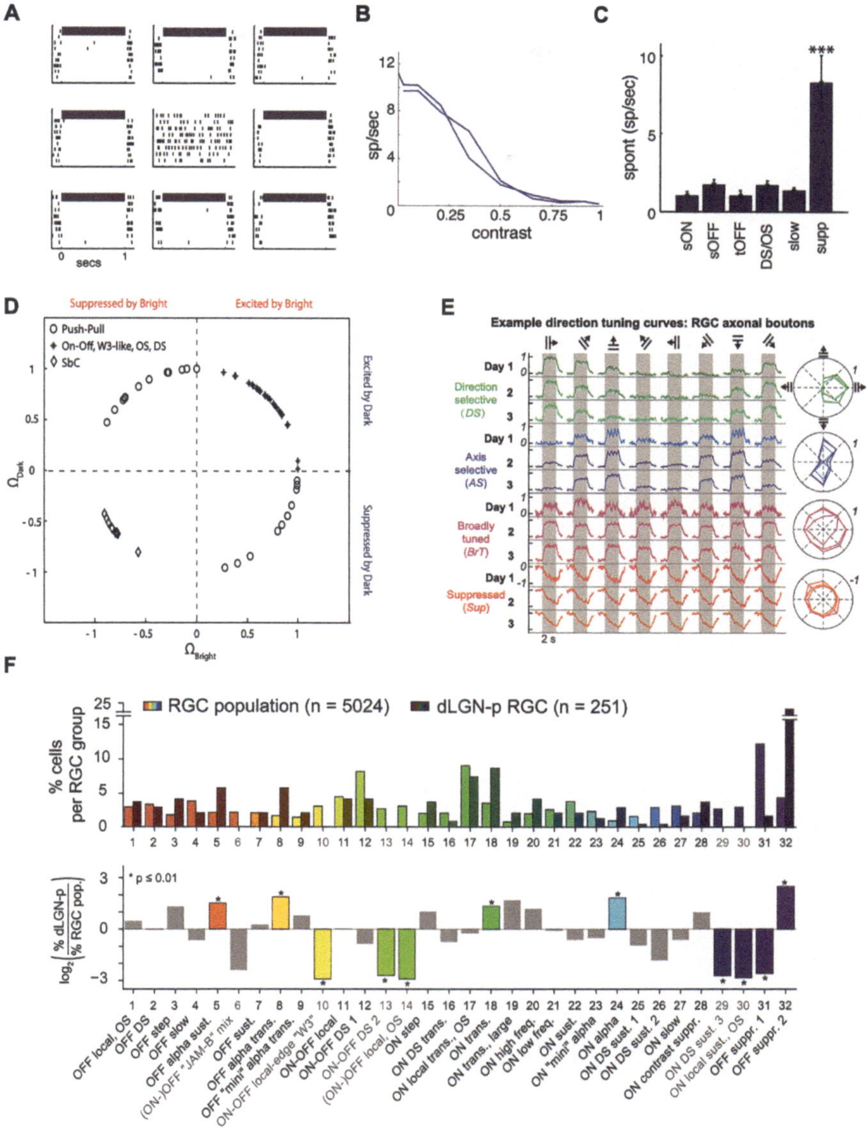

Fig. 6.10 SbC responses in mouse dLGN. (**A**) Spike rasters for a dLGN neuron in response to a grating moved in 8 directions *(time indicated by gray bar)* or for a blank screen *(center raster)*. (**B**) Mean spike rate for two SbC dLGN neurons as a function of grating contrast. (**C**) Spontaneous spike rate of the different neuron types encountered in dLGN. (**D**) Radial scatter plot of contrast profiles of intracellularly recorded cells in mouse dLGN. *Diamond symbols* in the *lower left* quadrant indicate SbC cells that were suppressed by both positive and negative contrasts. (**E**) Calcium responses recorded from RGC axon boutons in dLGN for gratings moving in 8 directions. Suppressed boutons *(red)* showed decreased activation for all stimuli. (**F**) An estimate of the prevalence of RGC connectivity to dLGN based on functional decomposition of dLGN spike responses shows an overrepresentation of the "OFF suppressed 2" RGC group *(rightmost columns in both graphs)*. *(From (A–C) Piscopo et al. (2013), (D) Suresh et al. (2016), (E) Liang et al. (2018), and (F) Román Rosón et al. (2019).)*

the optics of their eyes to alter focal distance. Humans accommodate primarily using our deformable lens (Beatty and Lucero-Wagoner, 2000), but mice depend instead on their pupil to alter depth-of-field (Pinto and Enroth-Cugell, 2000). Regardless of the physical mechanism, the signal about whether an object is in focus must begin in the retina. Rodieck (1967) suggested that SbC RGCs could be part of this process by providing a focus error signal. In-focus images create high contrast edges that would tend to decrease the firing of SbC RGCs, while out-of-focus images create blur on the retina, reducing contrast and increasing the firing rate of SbC RGCs. Accommodation circuits in the brain are mostly described as independent from dLGN (Gamlin, 1999), but the other brain targets of SbC RGCs are unknown, so their participation in accommodation circuits remains an intriguing possibility (Baden et al., 2017; Mani and Schwartz, 2017). Another theory is that SbC RGCs play a role in contrast modulation in the brain (Troy et al., 1989). While there are many mechanisms of contrast adaptation and sensitization within the retina (see Chapter 5), downstream targets of RGCs could engage additional contrast gain modulation by combining SbC RGC responses from those of ON, OFF, and ON-OFF RGCs. Finally, Tien et al. (2015) speculated that SbC RGCs play a role in detecting self-generated visual stimuli, like saccades and blinks. The tSBC RGCs in their study had nonlinear RFs and were thus sensitive to textures across a range of spatial frequencies (Mani and Schwartz, 2017, see Chapter 7). Abrupt changes in texture patterns simulated saccades and abrupt dimming simulated blinks. In both cases, SbC RGCs sharply reduced their firing rate, and a population of SbC RGCs could reliably detect these stimulus events (Tien et al., 2015).

Conclusions and future directions

For more than 30 years, the study of contrast suppression in the retina consisted only of scattered reports of rare cell types. The pioneering work of Sivyer and colleagues in 2010 and then the subsequent discovery of SbC RGCs in mouse began a new era of investigation of the circuit mechanisms of this computation. Remarkably, SbC circuits are now among the most well-studied RGC circuits in the retina (though still a great distance behind the direction-selectivity circuits described in Chapter 11), where we know some of the specific AC types involved and even have genetic access to manipulate them. One of the early conclusions from this work is that multiple SbC mechanisms work in parallel in the retina. Indeed, this is a common theme in retinal computation. Multiple circuit mechanisms work together to support a computation to achieve robustness and flexibility. We are only beginning to understand the constraints that guided the evolution of SbC circuits and how the different SbC RGC types occupy different niches in visual feature space. Perhaps work in the next decade will allow us to link these mechanistic studies in the retina to the targets of SbC RGCs in the brain and their role in behavior.

Landmark paper: Sivyer, B., Taylor, W.R., and Vaney, D.I. (2010). Uniformity detector retinal ganglion cells fire complex spikes and receive only light-evoked inhibition. Proceedings of the National Academy of Sciences of the United States of America *107*(12), 5628–5633.

Discussion questions:

1. Raw spike data and spike rasters from the SbC RGCs are shown in Fig. 1 A and B. What do you notice about these traces that are not captured in the peristimulus time histogram below (panel C)? Pay special attention to the fifth and ninth cells in the spike rasters. Come up with a theory that explains the behavior of these cells.
2. Pharmacological manipulations like those shown in Fig. 2C were also performed in the other mechanistic studies of SbC RGCs mentioned in this chapter. Why are results from these types of experiments hard to interpret? How do the authors in each of these studies approach this problem?
3. What do the results shown in Fig. 3 reveal about the bursting mechanism in these RGCs? Does this influence your answer to question 1?
4. The authors state that excitatory current to SbC RGCs is not significantly modulated by light in their conditions, but the tonic current is reduced by synaptic blockers. Do you think these RGCs receive excitatory synapses from bipolar cells? Do you have alternative interpretations of their results?
5. Explain the relationship between panels E and F in Fig. 4.

Comparison: Tien, N.-W., Pearson, J.T., Heller, C.R., Demas, J., and Kerschensteiner, D. (2015). Genetically identified suppressed-by-contrast retinal ganglion cells reliably signal self-generated visual stimuli. Journal of Neuroscience *35*(30), 10815–10820. vs. Mani, A., and Schwartz, G.W. (2017). Circuit

mechanisms of a retinal ganglion cell with stimulus-dependent response latency and activation beyond its dendrites. Current Biology *27*(4), 1–12.

Discussion questions:

1. I made the argument that these papers are describing the same RGC type with two different names. Argue against me. What is the evidence that they are not the same? Can you come up with ideas for experiments to bring more clarity to this question?
2. Both these studies use bath-applied pharmacology to study the receptor contributions to inhibition in the RGC. How were the results similar and how were they different in the two studies? Can you come up with a reasonable explanation for the discrepancies?

General discussion

How do you think SbC RGC signals are used in the brain? Why are they such a large fraction of the input to dLGN? How would your test your idea?

Programming exercise

To access the programming exercise, please visit https://www.elsevier.com/books-and-journals/book-companion/9780128198964.

SbC RGCs signal contrast with a decrease in firing, so they must maintain a baseline firing rate. But their firing is not periodic; instead they often fire in bursts, likely due to a combination of their synaptic inputs and intrinsic properties. To detect a positive or negative contrast from a population of SbC RGCs, a downstream decoder would need to look for gaps in this ongoing, noisy activity. This exercise examines how the statistics of spike trains within a population of neurons affect the ability to decode a gap in firing.

Load the file "spikeTrains.mat." This file contains two variables: "poissTrains_set" (Poisson spike trains) and "burstTrains_set" (bursty spike trains). Each set is a cell array of 50 trials. Each trial contains a matrix of the spike trains of 100 model neurons (with ones for spikes). There are 2 s of data for each neuron at a sampling rate of 1 kHz. In the first second, each neuron is firing at a mean rate of 20 Hz; in the next second, they all stop firing. This data models the behavior of a population of SbC RGCs to a high contrast stimulus.

1. Use "imagesc" to plot some of the spike data matrices from both data sets. Can you see the differences by eye? Plot the inter-spike interval (ISI) histograms for both data sets. Run a statistical test to prove that they are different even though the mean spike rates are the same.

2. Call the included function "discriminateGap" to simulate a downstream neuron trying to detect the gap in firing. For each neuron, this function keeps a running estimate of the ISI distribution and recognizes pauses that are "std_thres" standard deviations longer than the mean ISI. Start by setting "std_thres" to 2. The other parameters to the function are the "testWindow" in ms (start with 200) and the "responseThres," which is the proportion of neurons detecting a gap required for detection in the downstream neuron. Start with "responseThres" equal to 0.35. The function will return "outliers_poiss_set" and "outliers_burst_set" so you can see the time points at which each neuron detected the gap. The mean down each column will give you the proportion of cells responding at each time point. "RT_poiss" and "RT_burst" are vectors of reaction times in ms for gap detection. NaNs represent failure trials in which either the response threshold was not reached or it was reached erroneously before the time 1 s when the spike rate actually changed. Which spike trains gave shorter detection times in the gap detection task? Is the difference significant? What is your intuition for why this happened?
3. Subsample the neurons to see how performance in the task depends on the number of neurons. Is 0.35 still a reasonable choice for the response threshold? Try different values and run code to optimize the response threshold. Can you show that there is a tradeoff between reaction time and the proportion of correct trials?
4. What about different values for the "std_thres" parameter? Try optimizing this parameter and again exploring the tradeoff between reaction time and proportion correct.
5. Now you can make your own sets of spike trains by using the included function "makeSpikeTrainPair." The inputs "Nseconds," "Ncells," and "Ntrials" are the number of seconds of baseline activity before the firing cessation and the number of cells and trials in the simulation, respectively. "meanRate" is the mean spike rate. "burstRateMultiplier" sets the degree of burstiness by multiplying the mean rate by this value in "high" states and dividing it by this value in "low" states. Start with a value of 4. First, try different mean rates. The reaction time should, of course, decrease as the mean rate increases. What is the shape of this relationship? Is it the same for Poisson and bursty spike trains? At which firing rates are the two models most different in their gap detection performance?

References

Amthor, F.R., Takahashi, E.S., and Oyster, C.W. (1989). Morphologies of rabbit retinal ganglion cells with complex receptive fields. The Journal of Comparative Neurology *280*(1), 97–121.

Baden, T., Schaeffel, F., and Berens, P. (2017). Visual neuroscience: a retinal ganglion cell to report image focus? Current Biology *27*(4), R139–R141.

Bae, J.A., Mu, S., Kim, J.S., Turner, N.L., Tartavull, I., Kemnitz, N., Jordan, C.S., Norton, A.D., Silversmith, W.M., Prentki, R., et al. (2018). Digital museum of retinal ganglion cells with dense anatomy and physiology. Cell *173*(5), 1293–1306.e19.

Beatty, J., and Lucero-Wagoner, B. (2000). The Pupillary System. In Handbook of Psychophysiology (2nd ed.) (Cambridge University Press), pp. 142–162.

Cleland, B.G., and Levick, W.R. (1974). Properties of rarely encountered types of ganglion cells in the cat's retina and on overall classification. The Journal of Physiology *240*(2), 457–492.

De Monasterio, F.M. (1978). Properties of ganglion cells with atypical receptive-field organization in retina of macaques. Journal of Neurophysiology *41*(6), 1435–1449.

Della Santina, L., Kuo, S.P., Yoshimatsu, T., Okawa, H., Suzuki, S.C., Hoon, M., Tsuboyama, K., Rieke, F., and Wong, R.O.L. (2016). Glutamatergic monopolar interneurons provide a novel pathway of excitation in the mouse retina. Current Biology *26*(15), 2070–2077.

Gamlin, P.D.R. (1999). Subcortical neural circuits for ocular accommodation and vergence in primates. Ophthalmic and Physiological Optics *19*(2), 81–89.

Goetz, J., Jessen, Z.F., Jacobi, A., Mani, A., Cooler, S., Greer, D., et al. (2021). Unified classification of mouse retinal ganglion cells using function, morphology, and gene expression. BioRxiv. https://doi.org/10.1101/2021.06.10.447922.

Grimes, W.N., Seal, R.P., Oesch, N., Edwards, R.H., and Diamond, J.S. (2011). Genetic targeting and physiological features of VGLUT3+ amacrine cells. Visual Neuroscience *28*(5), 381–392.

Hoshi, H., Liu, W.L., Massey, S.C., and Mills, S.L. (2009). ON inputs to the OFF layer: bipolar cells that break the stratification rules of the retina. Journal of Neuroscience *29*(28), 8875–8883.

Jacoby, J., and Schwartz, G.W. (2018). Typology and circuitry of suppressed-by-contrast retinal ganglion cells. Frontiers in Cellular Neuroscience *12*, 269.

Jacoby, J., Zhu, Y., DeVries, S.H., and Schwartz, G.W. (2015). An amacrine cell circuit for signaling steady illumination in the retina. Cell Reports *13*(12), 2663–2670.

Lee, S., Zhang, Y., Chen, M., and Zhou, Z.J. (2016). Segregated glycine-glutamate co-transmission from vGluT3 amacrine cells to contrast-suppressed and contrast-enhanced retinal circuits. Neuron *90*(1), 27–34.

Levick, W.R. (1967). Receptive fields and trigger features of ganglion cells in the visual streak of the rabbit's retina. The Journal of Physiology *188*(3), 285–307.

Liang, L., Fratzl, A., Goldey, G., Ramesh, R.N., Sugden, A.U., Morgan, J.L., Chen, C., and Andermann, M.L. (2018). A fine-scale functional logic to convergence from retina to thalamus. Cell *173*(6), 1343–1355.e24.

Mani, A., and Schwartz, G.W. (2017). Circuit mechanisms of a retinal ganglion cell with stimulus-dependent response latency and activation beyond its dendrites. Current Biology *27*(4), 1–12.

Mastronarde, D.N. (1985). Two types of cat retinal ganglion cells that are suppressed by contrast. Vision Research *25*, 1195–1196.

Niell, C.M., and Stryker, M.P. (2008). Highly selective receptive fields in mouse visual cortex. Journal of Neuroscience *28*(30), 7520–7536.

Pinto, L.H., and Enroth-Cugell, C. (2000). Tests of the mouse visual system. Mammalian Genome: Official Journal of the International Mammalian Genome Society *11*(7), 531–536.

Piscopo, D.M., El-Danaf, R.N., Huberman, A.D., and Niell, C.M. (2013). Diverse visual features encoded in mouse lateral geniculate nucleus. Journal of Neuroscience *33*(11), 4642–4656.

Rodieck, R.W. (1967). Receptive fields in the cat retina: a new type. Science *157*(3784), 90–92.

Román Rosón, M., Bauer, Y., Kotkat, A.H., Berens, P., Euler, T., and Busse, L. (2019). Mouse dLGN receives functional input from a diverse population of retinal ganglion cells with limited convergence. Neuron *102*(2), 462–476 e8.

Rompani, S.B., Müllner, F.E., Wanner, A., Zhang, C., Roth, C.N., Yonehara, K., and Roska, B. (2017). Different modes of visual integration in the lateral geniculate nucleus revealed by single-cell-initiated transsynaptic tracing. Neuron *93*(4), 767–776 e6.

Rowe, M.H., and Stone, J. (1976). Properties of ganglion cells in the visual streak of the cat's retina. Journal of Comparative Neurology *169*(1), 99–125.

Sivyer, B., and Vaney, D.I. (2010). Dendritic morphology and tracer-coupling pattern of physiologically identified transient uniformity detector ganglion cells in rabbit retina. Visual Neuroscience *27*(5–6), 159–170.

Sivyer, B., Taylor, W.R., and Vaney, D.I. (2010). Uniformity detector retinal ganglion cells fire complex spikes and receive only light-evoked inhibition. Proceedings of the National Academy of Sciences of the United States of America *107*(12), 5628–5633.

Suresh, V., Çiftçioğlu, U.M., Wang, X., Lala, B.M., Ding, K.R., Smith, W.A., Sommer, F. T., and Hirsch, J.A. (2016). Synaptic contributions to receptive field structure and response properties in the rodent lateral geniculate nucleus of the thalamus. Journal of Neuroscience *36*(43), 10949–10963.

Tailby, C., Solomon, S.G., Dhruv, N.T., Majaj, N.J., Sokol, S.H., and Lennie, P. (2007). A new code for contrast in the primate visual pathway. Journal of Neuroscience *27*(14), 3904–3909.

Tien, N.-W., Pearson, J.T., Heller, C.R., Demas, J., and Kerschensteiner, D. (2015). Genetically identified suppressed-by-contrast retinal ganglion cells reliably signal self-generated visual stimuli. Journal of Neuroscience *35*(30), 10815–10820.

Tien, N.W., Kim, T., and Kerschensteiner, D. (2016). Target-specific glycinergic transmission from VGluT3-expressing amacrine cells shapes suppressive contrast responses in the retina. Cell Reports *15*(7), 1369–1375.

Troy, J.B., Einstein, G., Schuurmans, R.P., Robson, J.G., and Enroth-Cugell, C. (1989). Responses to sinusoidal gratings of two types of very nonlinear retinal ganglion cells of cat. Visual Neuroscience *3*(3), 213–223.

Zhu, Y., Xu, J., Hauswirth, W.W., and DeVries, S.H. (2014). Genetically targeted binary labeling of retinal neurons. Journal of Neuroscience *34*(23), 7845–7861.

PART 3

Spatial features

Texture sensitivity is the ability to perceive fine spatial structure in the world. Sometimes you can look at different kinds of leaves and flowers, like those on this bird of paradise plant, and imagine how they would feel to your fingers.

CHAPTER 7

Texture sensitivity

Gregory William Schwartz[a,b,c]
[a]Departments of Ophthalmology and Physiology, Feinberg School of Medicine, Northwestern University, Chicago, IL, United States
[b]Department of Neurobiology, Weinberg College of Arts and Sciences, Northwestern University, Chicago, IL, United States
[c]Northwestern Interdepartmental Neuroscience Program (NUIN), Northwestern University, Chicago, IL, United States

The concept of a receptive field (RF) is so ubiquitous in sensory neuroscience that one often forgets the assumption of spatial linearity inherent in this construct. Remarkably, Steven Kuffler, who in 1953 first measured RFs of ganglion cells (RGCs) in a mammalian retina, already understood many of the limitations of this simplification of the relationship between a stimulus and a cell's response (Kuffler, 1953). The discussion section of this seminal paper, in particular, is worth a reread as it was decades ahead of its time. Kuffler realized that interactions in retinal circuits upstream of the RGC could elicit responses to stimulation of parts of the RF that could not be predicted by independent sampling over space. He speculated that retinal circuits could be dynamically reconfigured in different stimulus conditions. I use the term "texture sensitivity" to refer to a class of responses to stimuli with spatial structure smaller than that of the RF center—sometimes much smaller—that are not explained by the linear RF. A static RF model does not capture many of the computations we have already discussed because aspects of the cell's response, like luminance or contrast gain, change over time. Texture sensitivity does not require these dynamics; it is not captured by a spatially linear RF model even when the RGC is in steady state.

Perhaps the simplest way to think about texture sensitivity is to consider the pixels of a digital camera. The resolution of a camera is defined by the number of pixels per unit area. If a region of the visual scene containing a dark-light edge falls onto a single pixel such that half the pixel sees black (intensity 0) and half the pixel sees white (intensity 1), it will encode a value of 0.5. Thus, this edge is totally indistinguishable from a patch of uniform gray with an intensity of 0.5. We would say that the edge is below our camera's resolution limit, or equivalently, that its frequency content is beyond the Nyquist frequency of our spatial sampling. Texture sensitivity is the

Retinal Computation
https://doi.org/10.1016/B978-0-12-819896-4.00008-1

property of RGCs that makes them different from pixels in this regard. The majority of RGCs can distinguish a dark-light edge from a uniform patch of gray even when the stimuli are smaller than the RF center. Thus, the spatial resolution of a texture-sensitive RGC is higher than predicted from the size of its RF, and the spatial resolution of a population of RGCs can be higher than predicted from the combination of its RF sizes and packing density.

High resolution in the spatial domain is analogous to other forms of sensitivity that we have described in previous chapters, and it carries the same sorts of clear selective advantages. An animal with higher spatial-resolution vision can recognize threats, mates, or food sources better than its competitors. In humans, acuity is the most important clinical measure of vision. How many letters you can identify on an eye chart determines if you can pilot a plane or drive a car, and it sets the legal definition of blindness. Importantly, however, visual acuity in humans is *not* determined by texture sensitivity. We are among the exceptions in the animal kingdom in this regard because our high-resolution fovea, where single cones drive single RGCs (through single bipolar cells), is indeed more like a pixel encoder (but not quite—see Chapter 9). In the rest of the human retina (99.8% of its area) and the whole retina of most other species, spatial resolution is set not by the linear RF size of each RGC but by their texture sensitivity. However, our own reliance on the fovea for high-resolution vision has influenced the history of vision science research emphasizing the linear, pixel-like aspects of how RGCs encode space. This chapter will cover measurements, functional consequences, circuit mechanisms, and models of texture sensitivity in RGCs.

Nonlinear spatial integration

Texture sensitivity is my term, and I chose it to emphasize function over mechanism, but the same computation is traditionally called "nonlinear spatial integration." RFs exhibiting nonlinear spatial integration are said to contain "subunits." The history of the study of nonlinear spatial integration in the mammalian retina begins with Kuffler in 1953, but a breakthrough came in 1966 when Christina Enroth-Cugell and John Robson developed a test to classify a cell as spatially linear, which they called "X," or nonlinear, "Y" (Enroth-Cugell and Robson, 1966). Their test was both elegant and powerful. The authors used a split-field grating, which modulated sinusoidally over time so that different parts of the monitor showed bright and dark regions. The stimulus is depicted in Fig. 7.1A. The investigators varied

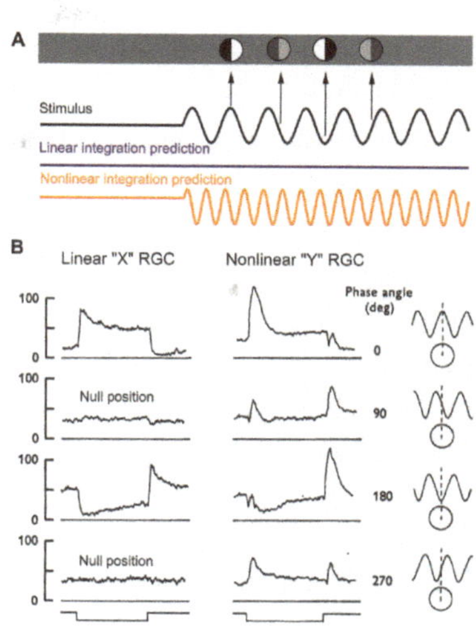

Fig. 7.1 A test of nonlinear spatial integration in retinal ganglion cells. (A) Schematic of a contrast reversing grating centered on the RF of an RGC. The predicted response from a spatially linear RF is zero *(purple)* while the predicted response from a (rectified) nonlinear RF is at twice the frequency of the stimulus *(orange)*. (**B**) The classic experiment first demonstrating the difference between spatially linear (X) and nonlinear (Y) RGCs. Firing rate is plotted in response to gratings at four different positions (phases) with respect to the RF center of each cell. *(Adapted from (A) Grimes et al. (2014) and (B) Enroth-Cugell and Robson (1966).)*

the stimulus position so that the edge fell at different locations within the RF of a recorded RGC. For a linear (X) cell, there existed a position at which the dark and light portions of the grating perfectly canceled within the RF, so the RGC had a "null" position at which the modulation of the grating yielded no response (Fig. 7.1B, left). Nonlinear (Y) cells had no such null position for the grating. Instead, when the dark and light portions covered the same-sized areas of the RF, the RGC responded at twice the modulation frequency of the grating; that is, it increased its firing rate for the preferred stimulus (dark or light, depending on the cell type) on either half of the RF (Fig. 7.1B, right).

It is important to point out one strength and one weakness of this test for spatial linearity. A strength is that it makes no assumptions about the shape of the RF. While it is typically diagrammed for circular or Gaussian RFs, *any* shape has a position in which it can be cut into halves of equal area in any orientation. It is worth the exercise to convince yourself of this fact by taking a piece of paper and drawing the most irregular RF shape you can imagine and then sliding a book across it from left to right on the page until the book covers exactly half the RF area. Changing the orientation of the page relative to the book makes no difference. The weakness of Enroth-Cugell and Robson's test is that it is only valid for ON or OFF cells. As described in Chapter 4, ON and OFF are only two of the four possible contrast tuning profiles for RGCs. ON-OFF and suppressed-by-contrast RGCs will always be classified as nonlinear by this test regardless of the spatial structure of their

RFs because their responses to both ON and OFF polarity make cancellation by such a stimulus impossible. Enroth-Cugell and Robson (1966) went on to introduce another version of their test with a drifting grating. While experimentally convenient because it avoids the need to search different locations for the possible null point, it adds the additional complication of possible nonlinear responses to motion, which are prevalent in retinal circuits (see "Motion" section). Despite these issues, the drifting gratings version of the linearity test is often used in visual areas downstream of the retina (dorsal lateral geniculate nucleus and cortex) to extend the X vs. Y dichotomy.

Receptive field subunits and their functional consequences

So far, both my analogy about a nonlinear pixel and Enoth-Cugell and Robson's X vs. Y test have described texture sensitivity as the ability to signal a dark-light edge that bisects the RF center. This essentially increases the spatial resolution of the RF by a factor of two. In fact, the subunits of an RGC's RF can be considerably smaller than half the RF center size. When probed with gratings at different spatial frequencies, Y cells in the cat responded for frequencies up to 10 times higher than that of the full RF center (Hochstein and Shapley, 1976; Victor and Shapley, 1979). Two-dimensional texture patterns in mouse RGCs also revealed sensitivity at spatial scales 10 times smaller than that of the RF center (Schwartz et al., 2012).

So does an RGC with subunits one-tenth the size of its RF center size encode space the same way as a population of 100 linear RGCs, each with one-tenth the RF diameter? No. The subunits of an RGC's RF do not cancel each other's responses, but they are not independent because their information is ultimately integrated in the RGC and transmitted as a single spike train. While the theoretical computational power of a nonlinear RF is enormous since it is equivalent to a two-layer neural network (Funahashi, 1989; Hornik et al., 1989), several biological constraints, including on the shapes and diversity of nonlinear functions, the dynamic range of neural elements, and encoding into discrete spike trains, have prevented the field from developing a unified theory of texture sensitivity in RGCs.

Several studies, however, have illustrated examples of the kinds of information that can be encoded in an RGC with nonlinear spatial integration (reviewed by Gollisch, 2013). Demb et al. (2001b) showed that spatially nonlinear RGCs respond to "second-order" motion in which the contrast changes across the RF center as a function of time but the luminance does not (Fig. 7.2A). Humans readily perceive second-order motion, and the authors argued that this computation might begin in the retina instead of in the cortex, as previously assumed. As suggested by the term "texture sensitivity," spatially nonlinear RFs also endow an RGC with sensitivity to spatial structure within the RF center (Fig. 7.2B). A texture-sensitive RGC can respond differently based on the spatial contrast within its RF, and its firing rate can be modulated by the spatial frequency content or level of blur in an image (Fig. 7.2C) (Mani and Schwartz, 2017). Texture sensitivity also has important implications for motion, as covered in later chapters (reviewed

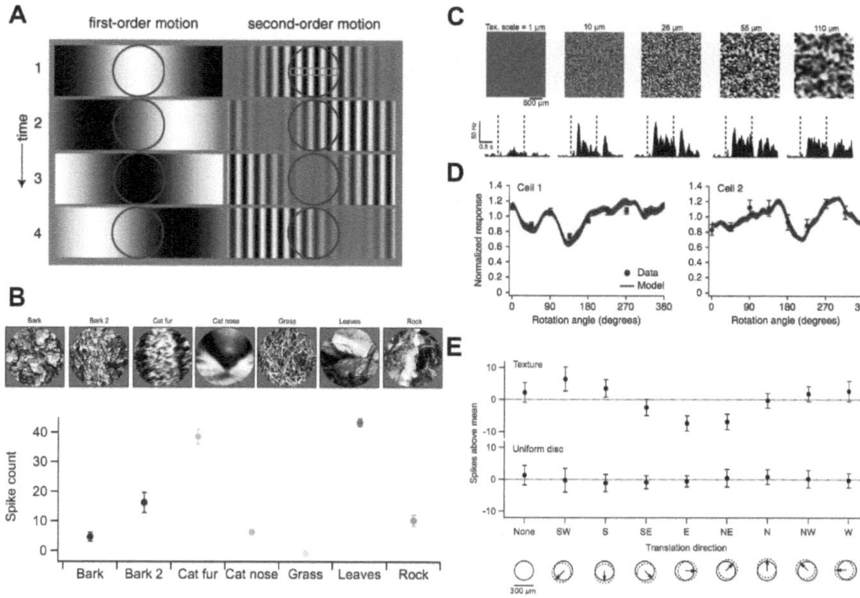

Fig. 7.2 Functions of texture sensitivity. (**A**) Diagram of first-order (luminance changing) and second-order (contrast changing) motion stimuli. (**B**) *Top,* natural images normalized to zero mean contrast and presented to the RF center of ON alpha RGCs. *Bottom,* spike response to each of the stimuli above. (**C**) *Top,* texture stimuli at different spatial scales. *Bottom,* responses of an ON delayed RGC to each of the stimuli above. *Dashed lines* indicate onset and offset of texture presentation from a uniform gray background. (**D**) Responses of two ON alpha RGCs to the same circular texture presented at different angles *(black points)* and predictions from the bottom-up anatomical subunit model (*purple,* see Fig. 7.5A for model construction). (**E**) Responses to the same texture *(top)* or uniform disc *(bottom)* presented at nine different positions relative to the RF center. Scale of the stimulus translation relative to the RF center is depicted below. *((A) From Demb et al. (2001b). Adapted from (B, D, and E) Schwartz et al. (2012) and (C) Mani and Schwartz (2017).)*

by Gollisch, 2013; Gollisch and Meister, 2010). Additional computational benefits may arise, somewhat counter-intuitively, from inhomogeneities in the locations of the RF subunits. Deviations from a Gaussian profile in the placement of subunits allow an RGC to transmit information about small translations or rotations of a texture pattern within its RF (Fig. 7.2D and E) (Schwartz et al., 2012), though whether the brain uses such information remains an open question.

Defocus detection

Another possible function of texture sensitivity is to encode the global level of defocus of an image. High acuity vision requires that the eye's optics focus an image precisely onto the outer segments of the photoreceptors (Fig. 7.3A). Humans can adjust their focus in a limited range by altering lens curvature, but optical problems outside this range, called refractive errors in the clinical literature, are widespread and require corrections like glasses, contact lenses, or surgery.

But how do we know when an image is in focus? This seems like a high-level computation that takes place in the cortex, but remarkably, animal studies have shown that the degree of image defocus is already computed at the level of the retina (Baden et al., 2017; Mani and Schwartz, 2017) (reviewed by Carr and Stell, 1995; Wallman and Winawer, 2004). Not only can this signal be used to adjust focus dynamically, but the retinal defocus signal is critical for regulating eye growth, a process called emmetropization (Fig. 7.3B) (Rabin et al., 1981). Even with the optic nerve severed, animals exposed to blurred visual environments during development have excessive eye growth causing myopia (Fig. 7.3C) (Wallman et al., 1987). The worldwide epidemic of developmental myopia has been attributed to environmental factors in the modern visual world that disrupt the retinal defocus detection circuit and hinder emmetropization (Dolgin, 2015).

Not only can the retina detect the degree of defocus, but amazingly it can also detect the *direction* of defocus. Images focused behind the retina (hyperopia) versus in front of the retina (myopia) require different corrections, increases or decreases in eye growth rate, respectively, and experiments have argued against a trial-and-error strategy (Aleman et al., 2018). There has been considerable speculation about the mechanism by which the retina detects the direction of defocus, with theories

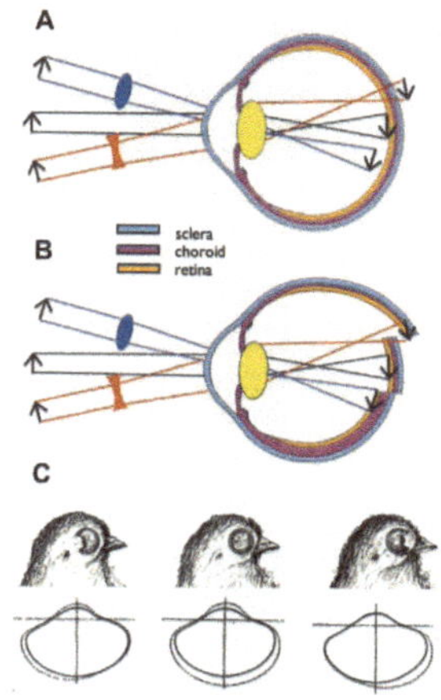

Fig. 7.3 Defocus detection and its role in regulating eye growth. (**A**) Schematic of the eye and the effects of a positive lens *(blue)* and a negative lens *(red)* on the position of the image relative to the retina. The positive lens causes myopic defocus and the negative lens causes hyperopic defocus. (**B**) Similar schematic to (**A**) showing the compensatory changes in eye size and choroid thickness after long-term exposure to each positive or negative defocus. (**C**) *Top,* drawings of pigeons with different lenses that caused hyperopic defocus on the nasal side of the retina *(left)*, the entire retina *(middle)*, or the temporal retina *(right)*. *Bottom,* corresponding changes in eye growth *(dotted line)* relative to the control eye *(black line)* show elongation of the region of the eye in which the image was defocused. *(From (A and B) Wallman and Winawer (2004) and (C) Wallman et al. (1987).)*

including the use of chromatic aberration, shadows of the retinal vasculature, or even the wavefront of the light (reviewed by Wallman and Winawer, 2004).

Is texture sensitivity important for natural scenes?

Texture sensitivity is often measured with artificial stimuli, but it is essential to consider whether it plays a substantial role in RGC responses to more natural visual stimuli. While some investigators have argued that spatially linear models are successful in encoding natural scenes (Bomash et al., 2013; Nirenberg and Pandarinath, 2012), there is a growing body of evidence on substantial and specific failures of these models in salamander (Liu et al., 2017), mouse (Schwartz et al., 2012), and primate retina (Heitman et al., 2016; Turner and Rieke, 2016). In particular, Turner and Rieke (2016) showed that OFF parasol RGCs in primate retina respond to spatial structure within their RF centers for images taken from a database of natural scenes taking into account human eye movements (Fig. 7.4A–D) and that these nonlinear responses accentuated edges in the scenes (Fig. 7.4E). Interestingly, ON parasol RGCs responded more linearly (Fig. 7.4B and D), largely due to the cancellation of nonlinear excitation by nonlinear inhibition, and the degree of texture sensitivity in both parasol RGC types was modulated by the contrast of the surround in natural scenes (Turner et al., 2018). A systematic investigation of the role of texture sensitivity across RGC types and species is an avenue for future research.

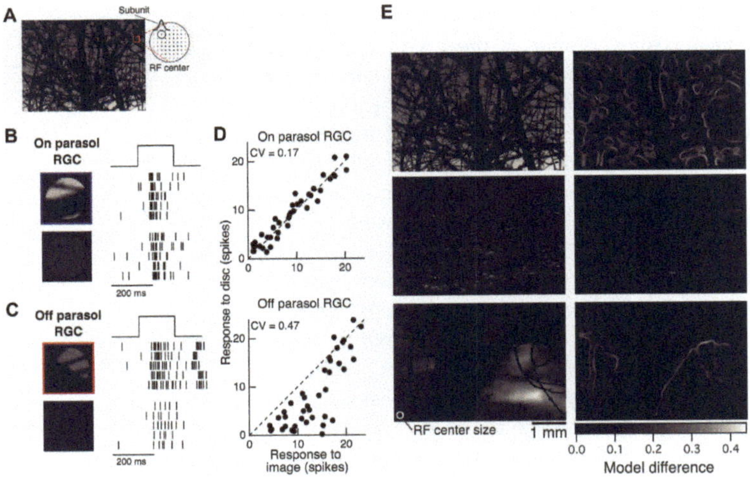

Fig. 7.4 Texture sensitivity impacts RGC responses to natural scenes. (**A**) Natural scene showing the size of a parasol RGC's RF and of the nonlinear subunits. (**B**) ON parasol RGC responding to the presentation of a natural image patch *(top)* and a linear equivalent disc *(bottom)* with the same total contrast but no spatial structure. (**C**) Same as (**B**) for an OFF parasol RGC. (**D**) Scatter plots of responses to many different image patches and linear equivalent discs in ON *(top)* and OFF *(bottom)* parasol RGCs. (**E**) Three natural scene images *(left)* and corresponding heatmaps of the difference between how the scenes are represented by spatially linear versus nonlinear models *(right)*. *(Modified from Turner and Rieke (2016).)*

The biological substrate of receptive field subunits

The "vertical" excitatory pathway consists of two levels of convergence: photoreceptor synapses onto bipolar cells (BCs) and BC synapses onto RGCs. From the perspective of synaptic physiology, the fact that the RGC, a third-order neuron, integrates its inputs nonlinearly is not surprising. Indeed, BCs form the RF subunits of RGCs. This fact has been demonstrated by voltage-clamp recordings showing spatially nonlinear excitatory input to RGCs (Demb et al., 2001a) and even the alignment of nonlinear subunits to morphologically identified BCs (Liu et al., 2017; Schwartz et al., 2012) (see Fig. 7.5). It is instructive to think about what it requires for spatial integration in an RGC to be linear (reviewed in Schwartz and Rieke, 2011). Both sets of synapses upstream of the RGC would have to maintain high tonic rates of glutamate release such that positive and negative contrast steps yielded changes in release that were equal and opposite. Linearity would also have to be maintained postsynaptically in the glutamate

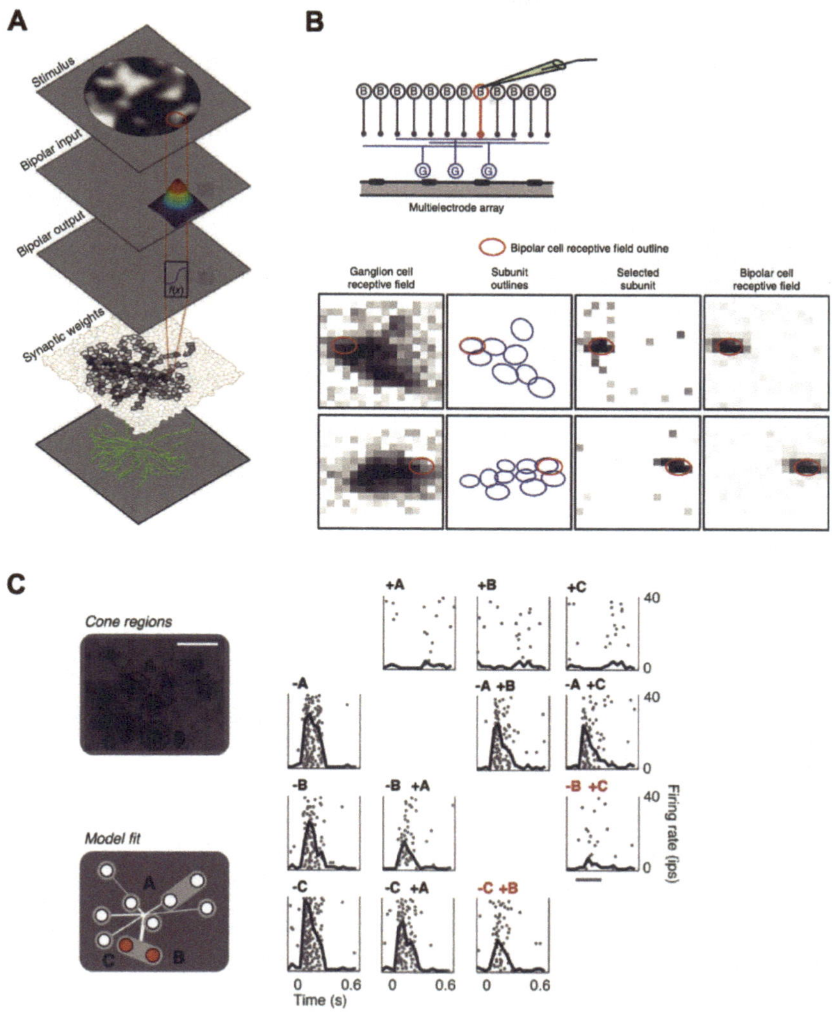

Fig. 7.5 Nonlinear RGC models that incorporate the positions of RF subunits. (**A**) Diagram of a bottom-up spatially nonlinear RF model for an ON alpha RGC. Subunit weights are based on an anatomical model of synaptic connectivity with bipolar cells. (**B**) *Top*, schematic of the experimental paradigm in which a single bipolar cell is recorded with a sharp electrode while spikes from many RGCs around it are recorded on a multi-electrode array. *Bottom*, RGC RFs and computed component subunit RF ellipses *(blue)*. The recorded bipolar cell's RF is well aligned with one of the computed subunits in both examples. (**C**) *Top left*, islands of sensitivity in the RF of a peripheral midget RGC correspond to individual cone locations. *Right*, stimulating pairs of cones with opposite contrast stimuli tests for cancellation, which is seen for cones B and C *(highlighted in red)*. *Bottom left*, model of subunit structure of the midget RGC showing the cones B and C share a bipolar cell subunit, but cone A provides the only input to its bipolar cell subunit. *(Modified from (A) Schwartz et al. (2012), (B) Liu et al. (2017), and (C) Freeman et al. (2015).)*

receptors and the electrical cable properties of the BCs and RGC. While photoreceptors indeed have high tonic glutamate release and special mechanisms to linearize their output, like a nearly linear relationship between calcium concentration and release rate (Shapley, 2009; Thoreson et al., 2004), most BCs have nonlinear outputs (Franke et al., 2017; Gollisch and Meister, 2010). In fact, spatial linearity at the level of RGCs is often achieved by nonlinear inhibition from the ON pathway canceling nonlinear excitation from the OFF pathway, or vice versa, in a push-pull circuit as we discussed in Chapter 4 (Molnar et al., 2009). Even photoreceptor synapses are nonlinear under certain conditions (Burkhardt et al., 1998), and recent work has suggested that this can lead to texture sensitivity in BCs themselves, creating subunits at the RGC up to the spatial scale of photoreceptors (Schreyer and Gollisch, 2021).

Spatially nonlinear RF models

To achieve a model for a spatially nonlinear RGC with better predictive power than the standard linear RF model, one needs to know (1) the size of the subunits, (2) the shape of the nonlinearity, and (3) the spatial location of each subunit (Enroth-Cugell and Freeman, 1987). There has been great success in measuring the first and second of these properties through a combination of experiments with different stimuli, like gratings at various spatial frequencies, and computational studies, for example, to estimate the shape of the subunit nonlinearity that best fits the data (Freeman et al., 2015; Shah et al., 2020). The problem of estimating the locations of the subunits is considerably more difficult. An anatomical approach using an image of the RGC's dendrites worked in mouse (Fig. 7.5A) (Schwartz et al., 2012), and more high-throughput analysis approaches that rely only on the spike data have shown great promise for RGCs in salamander (Fig. 7.5B) (Liu et al., 2017) and primate (Fig. 7.5C) (Freeman et al., 2015; Shah et al., 2020). Despite this progress on modeling texture sensitivity in RGCs, many studies continue to rely on spatially linear RF models of RGCs, like the linear-nonlinear (LN) model.

Which RGCs are X and Y cells?

There has been considerable confusion in the literature about the identities of X and Y RGCs. Enroth-Cugell and Robson's study and many subsequent investigations of spatial integration in RGCs throughout the 1970s and 1980s were performed in cat. X cells became synonymous with "brisk sustained" or "beta" RGCs, which come in ON and OFF varieties and represent true anatomical and functional types in cat (Wassle et al., 1981). Similarly, Y cells became synonymous with "brisk transient" or "alpha" RGCs, also comprised of ON and OFF types. Other RGC types in cat received either Greek or English letter names, with varying degrees of certainty as to whether they represented a single type. In primate, midget and parasol RGCs have been identified as the homologs of the cat X and Y cells, respectively (Shapley and Hugh Perry, 1986). Homology in mouse is not straightforward as no "X-like" RGC has been identified, and there are three "alpha" types that have important functional and morphological differences from their namesakes in the cat (Pang et al., 2003).

However, there are more fundamental problems with assigning linear or nonlinear spatial integration as the sole defining characteristic of an RGC type. First, we now know that there are more than 30 distinct RGC types in most mammalian species and likely even more types in other vertebrates (Baden et al., 2020). A single dichotomy of X vs. Y does not map onto this functional diversity. Furthermore, spatial integration in RGCs is one of many RF properties that can change dynamically, for example, in different lighting conditions (Wienbar and Schwartz, 2018). The same RGC type in mouse was shown to be spatially linear in scotopic conditions and nonlinear in mesopic and photopic conditions (Grimes et al., 2014), and the spatial nonlinearity of some RGCs in primate depends on the contrast in the RF surround (Turner et al., 2018). As described earlier, our understanding of texture sensitivity's biological basis enables us to view this property from a new perspective. Nonlinear spatial integration comes in a broad range of flavors and degrees among RGC types, and it should be thought of as the rule rather than the exception. The example of the OFF delta/OFF sustained alpha RGC in Chapter 4 (Contrast sensitivity) describes some of the constraints involved in making a cell respond linearly to contrast. A similar push-pull mechanism is at work at the level of the synaptic input to midget RGCs in primate retina so that nonlinear excitation can effectively cancel nonlinear inhibition to linearize the RGC's spike response (Cafaro and Rieke, 2013).

Landmark paper: Kuffler, S.W. (1953). Discharge patterns and functional organization of mammalian retina. Journal of Neurophysiology *16*(1), 37–68.

Discussion questions:

1. Describe how the receptive field in Fig. 3 was constructed. Kuffler could have instead presented spots of the same intensity at different locations and plotted the number of spikes recorded for each stimulus location. Are the two methods equivalent? Why or why not?
2. On pages 44–45, Kuffler contrasts the previous operational definition of a receptive field from Hartline with the "functional" receptive field. What is the difference? How do the results of Figs. 4, 5, and 7 demonstrate "fluidity" in the receptive field? Is there is a single type of experiment that could reveal the full extent of a functional RGC receptive field?
3. In the Discussion section "Psychophysical aspects," Kuffler makes two connections that have each become entire fields of research. First, explain how the results in this paper could be related to the perceptions of contrast and acuity and how eye movements might play a role. Second, summarize his "field from the trees" argument about the population code of the retina. Neuroscientists increasingly record from many neurons simultaneously these days. In your reading of the modern literature, do experimenters still generally search for "a 1:1 agreement between discharge patterns and information"?

Landmark paper: Enroth-Cugell, C., and Robson, J.G. (1966). The contrast sensitivity of retinal ganglion cells of the cat. The Journal of Physiology *187*(3), 517–552.

Discussion questions:

1. What is the recording method in this paper? Contrast it with that of Kuffler. Do you think this could have affected the sampling of RGC types?
2. What is the relative proportion of X vs. Y cells the authors measured?
3. In the second paragraph of the Discussion, the authors make the distinction between linearity of spatial summation and linearity of the RGC response. What is the difference? Can they be measured independently?
4. In the Discussion section "Psychophysical correlates," the authors relate their results to human sensitivity to grating patterns. Other than the species difference, do you see major flaws in their logic?

Landmark paper: Schwartz, G.W., Okawa, H., Dunn, F.A., Morgan, J.L., Kerschensteiner, D., Wong, R.O., and Rieke, F. (2012). The spatial structure of a nonlinear receptive field. Nature Neuroscience *15*(11), 1572–1580.

Discussion questions:

1. How are the measures of subunit size different in Fig. 1 vs. Fig. 2? Comment on the strengths and weaknesses of these different approaches.
2. Explain the experiment in Fig. 5. What does the nonlinearity index represent? How does this experiment demonstrate the biological substrate of the subunits?
3. How was the nonlinearity between the BCs and the RGC measured in this study? Where is this measurement shown? How can one understand this function from the perspective of synaptic physiology?
4. Do you think that the information in the RGC response about texture seed, rotation, translation (Fig. 1), or location (Supplemental Fig. 6) is used by the brain? What difficulties are associated with reading out such a signal?
5. RGCs of the same type usually have considerable overlap in their RFs. How do you think texture sensitivity within each RGC of a given type interacts with the RGC population code?

Programming exercise

To access the programming exercise, please visit https://www.elsevier.com/books-and-journals/book-companion/9780128198964.

The package "nonlinear_subunits_model.zip" contains the "guts" of a model of a nonlinear RGC RF built from bipolar cell subunits. In this exercise, you will generate texture stimuli and see how parameters of the model affect how the RGC responds. You can run the script "subunit_model.m" to go through the steps of the model. Parameters of the simulation are in the first block.

1. Use only a single blur factor for your textures (10 μm) but try four different random seeds. For each stimulus, use "subplot" to display the stimulus and next to it the activation pattern of the bipolar cells. Plot each bipolar cell at the location of its RF center and use either the size of the marker or its color to demonstrate its linear activation.
2. Repeat the previous step, but now add a plot for the glutamate release of each bipolar cell, following the nonlinearity.
3. Vary the Gaussian blur of the textures and plot how the RGC responds as a function of this parameter.

4. For each Gaussian blur factor, try 10 different random seeds. How discriminable are the different seeds relative to the different blur factors? Does the discriminability across seeds vary with blur factor? Explain your intuition for the results.
5. How do the RF size of the bipolar cell subunits and their spacing relate to the scale of texture sensitivity in the RGC? Make figures to back up your argument.
6. The parameter "position_noise" determines the regularity of the bipolar cell grid. Plot the locations of the bipolar cells ("bpX vs. bpY") across a range of this parameter to see the effect. Does changing this parameter change the spatial scale of texture sensitivity in the RGC? Does it change the discriminability of different random seeds? Why or why not?
7. Now explore the bipolar cell nonlinearity. It is implemented as a cumulative Gaussian, so it has a parameter for the inflection point (exc_NL_offset) and one for the slope (exc_NL_sigma). As you vary the inflection point from 0.05 to 0.5, what happens to the spatial scale of texture sensitivity in the RGC? Why? What happens if you make the inflection point 0? Why? Now vary the slope from 0.1 to 0.6. How does this affect texture sensitivity? Why?

References

Aleman, A.C., Wang, M., and Schaeffel, F. (2018). Reading and myopia: contrast polarity matters. Scientific Reports *8*(1).

Baden, T., Schaeffel, F., and Berens, P. (2017). Visual neuroscience: a retinal ganglion cell to report image focus? Current Biology *27*(4), R139–R141.

Baden, T., Euler, T., and Berens, P. (2020). Understanding the retinal basis of vision across species. Nature Reviews Neuroscience *21*(1), 5–20.

Bomash, I., Roudi, Y., and Nirenberg, S. (2013). A virtual retina for studying population coding. PLoS ONE *8*(1), e53363.

Burkhardt, D.A., Fahey, P.K., and Sikora, M. (1998). Responses of ganglion cells to contrast steps in the light-adapted retina of the tiger salamander. Visual Neuroscience *15*(2), 219–229.

Cafaro, J., and Rieke, F. (2013). Regulation of spatial selectivity by crossover inhibition. Journal of Neuroscience *33*(15), 6310–6320.

Carr, B.J., and Stell, W.K. (1995). The Science Behind Myopia. In Webvision: The Organization of the Retina and Visual System. http://www.ncbi.nlm.nih.gov/pubmed/29266913.

Demb, J.B., Zaghloul, K.A., Haarsma, L., and Sterling, P. (2001a). Bipolar cells contribute to nonlinear spatial summation in the brisk-transient (Y) ganglion cell in mammalian retina. Journal of Neuroscience *21*(19), 7447–7454.

Demb, J.B., Zaghloul, K., and Sterling, P. (2001b). Cellular basis for the response to second-order motion cues in Y retinal ganglion cells. Neuron *32*(4), 711–721.

Dolgin, E. (2015). The myopia boom. Nature *519*(7543), 276–278.

Enroth-Cugell, C., and Freeman, A.W. (1987). The receptive-field spatial structure of cat retinal Y cells. The Journal of Physiology *384*(1), 49–79.

Enroth-Cugell, C., and Robson, J.G. (1966). The contrast sensitivity of retinal ganglion cells of the cat. The Journal of Physiology *187*(3), 517–552.

Franke, K., Berens, P., Schubert, T., Bethge, M., Euler, T., and Baden, T. (2017). Inhibition decorrelates visual feature representations in the inner retina. Nature *542*(7642), 439–444.

Freeman, J., Field, G.D., Li, P.H., Greschner, M., Gunning, D.E., Mathieson, K., Sher, A., Litke, A.M., Paninski, L., Simoncelli, E.P., et al. (2015). Mapping nonlinear receptive field structure in primate retina at single cone resolution. eLife *4*, e05241.

Funahashi, K.-I. (1989). On the approximate realization of continuous mappings by neural networks. Neural Networks *2*(3), 183–192.

Gollisch, T. (2013). Features and functions of nonlinear spatial integration by retinal ganglion cells. Journal of Physiology, Paris *107*(5), 338–348.

Gollisch, T., and Meister, M. (2010). Eye smarter than scientists believed: neural computations in circuits of the retina. Neuron *65*(2), 150–164.

Grimes, W.N., Schwartz, G.W., and Rieke, F. (2014). The synaptic and circuit mechanisms underlying a change in spatial encoding in the retina. Neuron *82*, 460–473.

Heitman, A., Brackbill, N., Greschner, M., Sher, A., Litke, A.M., and Chichilnisky, E.J. (2016). Testing pseudo-linear models of responses to natural scenes in primate retina. BioRxiv, 045336.

Hochstein, S., and Shapley, R.M. (1976). Linear and nonlinear spatial subunits in Y cat retinal ganglion cells. The Journal of Physiology *262*(2), 265–284.

Hornik, K., Stinchcombe, M., and White, H. (1989). Multilayer feedforward networks are universal approximators. Neural Networks *2*(5), 359–366.

Kuffler, S.W. (1953). Discharge patterns and functional organization of mammalian retina. Journal of Neurophysiology *16*(1), 37–68.

Liu, J.K., Schreyer, H.M., Onken, A., Rozenblit, F., Khani, M.H., Krishnamoorthy, V., Panzeri, S., and Gollisch, T. (2017). Inference of neuronal functional circuitry with spike-triggered non-negative matrix factorization. Nature Communications *8*(1), 149.

Mani, A., and Schwartz, G.W. (2017). Circuit mechanisms of a retinal ganglion cell with stimulus-dependent response latency and activation beyond its dendrites. Current Biology *27*(4), 1–12.

Molnar, A., Hsueh, H.-A., Roska, B., and Werblin, F.S. (2009). Crossover inhibition in the retina: circuitry that compensates for nonlinear rectifying synaptic transmission. Journal of Computational Neuroscience *27*(3), 569–590.

Nirenberg, S., and Pandarinath, C. (2012). Retinal prosthetic strategy with the capacity to restore normal vision. Proceedings of the National Academy of Sciences of the United States of America *109*(37), 15012–15017.

Pang, J.-J., Gao, F., and Wu, S.M. (2003). Light-evoked excitatory and inhibitory synaptic inputs to ON and OFF alpha ganglion cells in the mouse retina. Journal of Neuroscience *23*(14), 6063–6073.

Rabin, J., Van Sluyters, R.C., and Malach, R. (1981). Emmetropization: a vision-dependent phenomenon. Investigative Ophthalmology & Visual Science *20*(4), 561–564.

Schreyer, H.M., and Gollisch, T. (2021). Nonlinear spatial integration in retinal bipolar cells shapes the encoding of artificial and natural stimuli. Neuron *109*, 1692–1706.e8. https://doi.org/10.1016/j.neuron.2021.03.015.

Schwartz, G., and Rieke, F. (2011). Nonlinear spatial encoding by retinal ganglion cells: when $1 + 1 \neq 2$. Journal of General Physiology *138*(3), 283–290.

Schwartz, G.W., Okawa, H., Dunn, F.A., Morgan, J.L., Kerschensteiner, D., Wong, R.O., and Rieke, F. (2012). The spatial structure of a nonlinear receptive field. Nature Neuroscience *15*(11), 1572–1580.

Shah, N.P., Brackbill, N., Rhoades, C., Kling, A., Goetz, G., Litke, A.M., Sher, A., Simoncelli, E., and Chichilnisky, E.J. (2020). Inference of nonlinear receptive field subunits with spike-triggered clustering. eLife *9*, e45743.

Shapley, R. (2009). Linear and nonlinear systems analysis of the visual system: why does it seem so linear? A review dedicated to the memory of Henk Spekreijse. Vision Research *49*(9), 907–921.

Shapley, R., and Hugh Perry, V. (1986). Cat and monkey retinal ganglion cells and their visual functional roles. Trends in Neurosciences *9*(C), 229–235.

Thoreson, W.B., Rabl, K., Townes-Anderson, E., and Heidelberger, R. (2004). A highly Ca2+-sensitive pool of vesicles contributes to linearity at the rod photoreceptor ribbon synapse. Neuron *42*(4), 595–605.

Turner, M.H., and Rieke, F. (2016). Synaptic rectification controls nonlinear spatial integration of natural visual inputs. Neuron *90*(6), 1257–1271.

Turner, M.H., Schwartz, G.W., and Rieke, F. (2018). Receptive field center-surround interactions mediate context-dependent spatial contrast encoding in the retina. eLife *7*, e38841.

Victor, J.D., and Shapley, R.M. (1979). The nonlinear pathway of Y ganglion cells in the cat retina. The Journal of General Physiology *74*(6), 671–689.

Wallman, J., and Winawer, J. (2004). Homeostasis of eye growth and the question of myopia. Neuron *43*(4), 447–468.

Wallman, J., Gottlieb, M.D., Rajaram, V., and Fugate-Wentzek, L.A. (1987). Local retinal regions control local eye growth and myopia. Science *237*(4810), 73–77.

Wassle, H., Boycott, B.B., and Illing, R.B. (1981). Morphology and mosaic of on- and off-beta cells in the cat retina and some functional considerations. Proceedings of the Royal Society of London - Biological Sciences *212*(1187), 177–195.

Wienbar, S., and Schwartz, G.W. (2018). The dynamic receptive fields of retinal ganglion cells. Progress in Retinal and Eye Research *67*, 102–117.

Blending in with one's surroundings is a common defensive strategy in nature that exploits the computation of surround suppression. This insect has mastered the skill. While it is often described in terms of average luminance, other aspects of the surroundings of an image, like contrast, texture, and color, can also provide context to shape the response to stimuli in the receptive field center.

CHAPTER 8

Surround suppression

Gregory William Schwartz[a,b,c]
[a]Departments of Ophthalmology and Physiology, Feinberg School of Medicine, Northwestern University, Chicago, IL, United States
[b]Department of Neurobiology, Weinberg College of Arts and Sciences, Northwestern University, Chicago, IL, United States
[c]Northwestern Interdepartmental Neuroscience Program (NUIN), Northwestern University, Chicago, IL, United States

Surround suppression is perhaps the most well-known retinal computation. Often called lateral inhibition, it is ubiquitous in the central nervous system. Defined broadly, surround suppression in a visual neuron is the collection of mechanisms that create a spatially extended region, the surround, which opposes the receptive field (RF) center. Early accounts of surround suppression in the retina focused on the opposition of response polarity, like the opposition of an ON RF center by an OFF surround. This limited definition has persisted in the literature and textbooks. I will describe a more inclusive view of the RF surround as a collection of circuits specialized for different kinds of suppression in different RGCs. The RF surround of a visual neuron may oppose, for example, the color, motion, direction, or orientation selectivity of the RF center instead of, or in addition to, its polarity.

Perceptual consequences of surround suppression are well documented (Foley, 2019; Gheorghiu et al., 2014; Hallum and Movshon, 2014), and they are the source of countless visual illusions (Spillmann, 1994). However, because of this computation's ubiquity in the visual system, it can be difficult to attribute particular perceptual surround phenomena to the retina. In this chapter, I will cover the aspects of surround suppression known to originate in retinal circuits.

Mechanisms of surround suppression in the retina can be divided into those occurring in the outer retina via horizontal cells and those occurring in the inner retina via amacrine cells. Much of the current research focus on surround suppression is at different levels of analysis in the outer versus the inner retina. In the outer retina, horizontal cells impact the center-surround organization of receptive fields beginning at the photoreceptor output synapse, though recent work has suggested additional functional roles for horizontal cells (Drinnenberg et al., 2018; Kemmler et al., 2014). The synaptic

Retinal Computation
https://doi.org/10.1016/B978-0-12-819896-4.00009-3

mechanisms of horizontal cell signaling, however, remain controversial. In the inner retina, the diversity of amacrine cell types and their synaptic circuits—onto bipolar cells, onto retinal ganglion cells, and onto other amacrine cells—have led to a different research focus. Rather than computing average luminance in the surround, these amacrine circuits often compute a different feature of the surround, like global motion, to suppress the RF center. Thus, the research focus in the inner retina is often on exploring the functional consequences of amacrine cell circuits for surround suppression rather than the details of their synaptic mechanisms.

Outer retinal mechanisms: Horizontal cells

Surround suppression at the level of photoreceptors was discovered in the retina of the horseshoe crab by Hartline and colleagues in the 1950s and was part of the body of work that led to his Nobel Prize in 1967 (Hartline et al., 1956; Hartline and Ratliff, 1957). The neurons responsible for lateral interactions between photoreceptors are called horizontal cells (HCs) because of their horizontally elongated morphology within the plane of the retina. HCs are remarkably well conserved across vertebrate species from fish to humans, where they comprise 3–4 main classes (Fig. 8.1) (reviewed by Boije et al., 2016). H1 HCs (also called B-type) have separate axonal and dendritic compartments which contact rods and cones, respectively. H2 HCs (A-type) have no axon and typically make only cone contacts. H3 and H4 HCs (C type) have a more "candelabra" shape than H2 HCs, with H4 HCs having the larger dendritic field. Some species have additional subtypes within these major classes that are generally differentiated based on the cone types they contact. Thus, the diversity of HC types in the vertebrate retina is closely related to the color-processing requirements of vision in each species, a topic I will return to in Chapter 16 (Color processing). Some of the prominent model species in

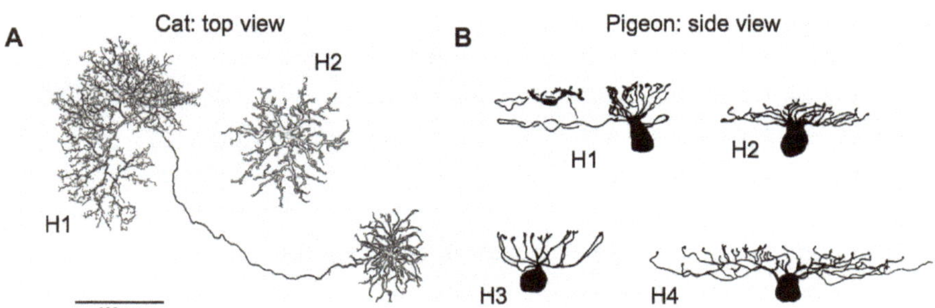

Fig. 8.1 Morphologies of horizontal cells. (**A**) Top view of H1 and H2 horizontal cell types from cat retina. (**B**) Side view of HC types 1–4 from pigeon retina. *((A) Modified from Diamond (2017) and originally from Boycott et al. (1978), and (B) modified from Mariani (1987).)*

retinal research represent this range of HC types. Zebrafish and chickens, both of which have tetrachromatic vision, have 4 HCs (H1–H4) (Connaughton et al., 2004; Fischer et al., 2007; Gallego et al., 1975). Salamanders (Zhang et al., 2006), guinea pigs (Gonzälez-soriano, 1994), cats (Boycott et al., 1978), and rabbits (Bloomfield and Miller, 1982) have H1 and H2 types, as do many primates, including macaques (Wässle et al., 2000), marmosets (Chan et al., 1997), and humans (Kolb et al., 1992). Mice and rats, both nocturnal species, have only H1 HCs (Gonzälez-soriano, 1994).

HCs embody a fascinating intersection between local and global processing. Their fine dendritic tips form invaginating contacts within cone synapses where they can both sense and affect changes in glutamate, protons, and potassium ions in this nanometer-scale micro-environment (Chapot et al., 2017a,b). On the other hand, HCs are very strongly electrically coupled by a network of gap junctions (using connexin-57 in mice) enabling them to average signals over thousands of photoreceptors and large regions of visual space (Hombach et al., 2004). The nature of the interplay between local and global signaling in HCs has been a topic of research for decades, and electron-microscopic reconstructions and modeling continue to offer new insights (Chapot et al., 2017a,b).

Three mechanisms have been proposed to explain the suppression of cone output by horizontal cells (Fig. 8.2) (reviewed by Chapot et al., 2017a,b; Kramer and Davenport, 2015; Thoreson and Mangel, 2012). The GABA mechanism is the most straightforward. Like other inhibitory neurons, horizontal cells may synthesize and release GABA, though whether they do in vivo is controversial (Deniz et al., 2011; Hirano et al., 2018). There have been reports of GABA receptor expression in cones (Deniz et al., 2019) and GABA-mediated hyperpolarization of cones (Deniz

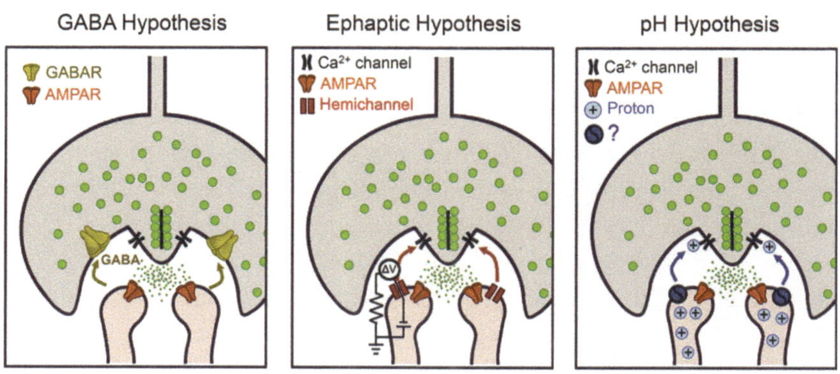

Fig. 8.2 Three possible mechanisms for horizontal cell feedback onto cones. *(From Kramer and Davenport (2015).)*

et al., 2011). Another proposed mechanism involves the release of protons by horizontal cells into the synaptic cleft, thus lowering pH to modulate the Ca^{2+} channels in the cones that control glutamate release (Beckwith-Cohen et al., 2019; Davenport et al., 2008). Finally, the ephaptic mechanism suggests that depolarization of horizontal cells causes an increase in the concentration of K^{+} ions in the synaptic cleft, thereby diminishing the voltage gradient used by the voltage-gated Ca^{2+} channels in the cone pedicle (Fahrenfort et al., 2009; Vroman et al., 2013). As in many cases in biology, these mechanisms are not mutually exclusive, and their actions may even be intertwined (Chapot et al., 2017a,b; Fahrenfort et al., 2009; Kramer and Davenport, 2015; Thoreson et al., 2008).

Inner retinal mechanisms

Surround suppression in the inner retina depends on amacrine cells (ACs) transmitting inhibitory signals onto the axon terminals of bipolar cells (BCs), other ACs, and/or the dendrites of RGCs. Wide-field ACs, which tend to be GABAergic, are particularly well suited for this role because they collect signals from a large area of visual space. With a few notable exceptions, the particular types of wide-field ACs responsible for surround suppression in different retinal circuits remain unknown.

Inhibition onto bipolar cells and other ACs

Excitation to BCs is spatially restricted to their narrow dendritic fields where most BC types receive input from <8 cones within a diameter of ~20 μm (in mice; Wassle et al., 2009). Conversely, inhibition arrives mostly at the BC axon terminal, via ACs that can cover huge regions of space up to millimeters. Thus, BCs are poised to be a major site of surround suppression.

Voltage-clamp recordings in RGCs have long implicated BCs in surround suppression. Isolated excitatory synaptic input to RGCs has shown strong surround suppression in some cases, suggesting a source upstream of the RGC. Importantly, the degree of surround suppression in excitatory input varies widely across RGC types, suggesting a site downstream of horizontal cell mechanisms in the outer retina shared by all cones. While several studies have used electrophysiology to measure the receptive fields of BCs (Baccus et al., 2008; Berntson and Taylor, 2000; Dacey et al., 2000; Schwartz et al., 2012), difficulties in accessing these cells in the middle of the retina have prevented widespread adoption of this technique. The advent of the fluorescent glutamate sensor, iGluSnFR (Marvin et al., 2013), enabled a new set of optical measurements of BC output that has dramatically increased the throughput of studies of BC receptive fields (Fig. 8.3).

Borghuis et al. (2013) performed one of the first iGluSnFR studies in the retina. These investigators used spots of different diameters and measured robust surround suppression in both ON and OFF BCs (Fig. 8.3A and B). Subsequently, Franke et al. (2017) used iGluSnFR to measure the outputs of more than 13,000 BCs. They clustered the BC terminals into functional

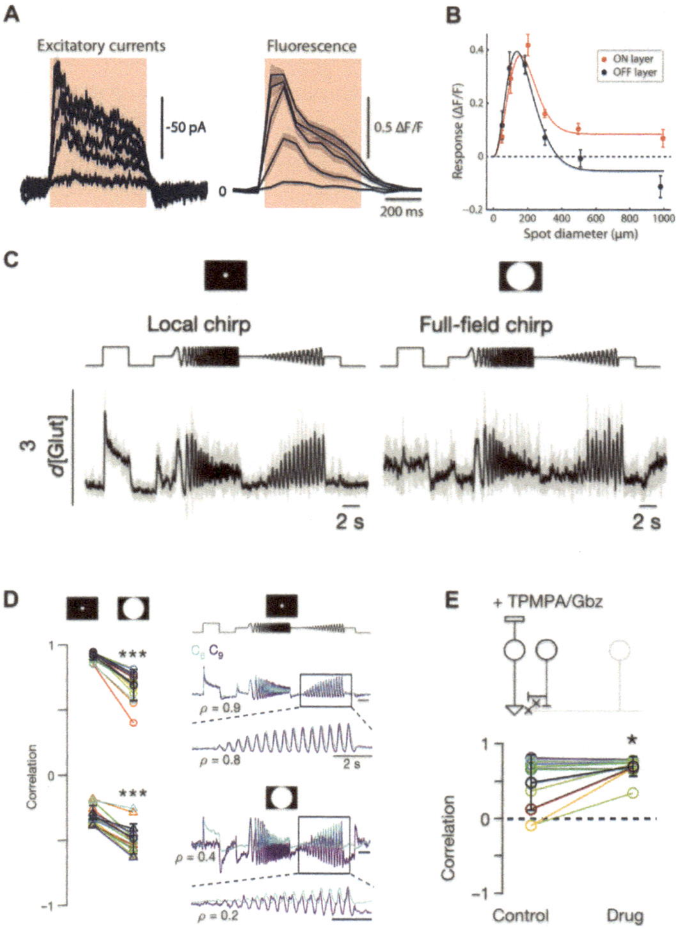

Fig. 8.3 iGluSnFR measurements of surround suppression in bipolar cells. (**A**) *Left*, excitatory currents (inverted) to a series of light steps of different contrast. *Right*, corresponding iGluSnFR fluorescence measurements. (**B**) iGluSnFR responses in ON and OFF layers of the IPL to spots of different sizes. (**C**) *Left*, mean iGluSnFR response of a group of BC terminals to a local chirp stimulus (shown above). *Right*, response of the same group of terminals to a global chirp. (**D**) *Left*, correlation coefficients for all pairs of BC types for the local chirp *(left)* and the global chirp *(right)*. Positive correlations are for BC types of the same polarity; negative for those of opposite polarity. *Right*, examples of clusters 8 and 9 showing high correlation for the local chirp but lower correlation for the global chirp. (**E**) Block of GABAergic inhibition increases BC glutamate correlations in response to a global chirp. *(Modified from (A, B) Borghuis et al. (2013) and (C–E) Franke et al. (2017).)*

types corresponding to the known anatomical BC types in mice. Using both small and large spots and pharmacological manipulations of inhibition, they showed that BC types are differentially suppressed by ACs to create their receptive field surrounds. The different effects of inhibition on each BC type had the functional consequence of decorrelating the responses of the BC channels (Fig. 8.3C–E).

In the rod bipolar pathway, both rod bipolar cells (RBCs) and AII ACs are potential sites of surround suppression. A recent study combined electrophysiology, optogenetic circuit tracing, and connectomics to demonstrate that a wide-field spiking AC, called the nNOS-1 AC, is a major source of inhibition in the rod bipolar pathway, making inhibitory contacts with both AII ACs and RBCs (Park et al., 2020).

Since many ACs have complex response properties, circuits in the inner retina can use visual features of the surround other than mean luminance to suppress the receptive field center. As a field, we are only beginning to understand the multilayered AC circuits that converge onto BC terminals, but one general feature of the surround we know is encoded in these circuits is motion. Both spiking ACs (Zhang et al., 2012) and a nonspiking GABAergic cell called the TH2 AC (Kim and Kerschensteiner, 2017) have been implicated in a circuit that suppresses RGC responses to global motion, making the RGC selective for local motion (see Chapter 12). Inhibition appears to target multiple locations in this circuit, including GABA synapses onto an excitatory, glutamatergic AC (Kim and Kerschensteiner, 2017) as well as BCs and the RGC dendrites (Zhang et al., 2012). The next section of the book will cover motion circuits in greater detail.

Inhibition onto ganglion cells

Inhibition directly onto the dendrites of RGCs can also contribute to surround suppression (Farrow et al., 2013), and many RGCs use both "presynaptic inhibition" (onto BCs) and "postsynaptic inhibition" (on the RGC dendrites) to create the RF surround (Fig. 8.4) (Jacoby and Schwartz, 2017). The classical center-surround model of RGC receptive fields includes excitation and inhibition centered at the same location. The inhibition has a smaller peak than the excitation but a much wider extent, leading to the familiar difference-of-Gaussians profile. As mentioned earlier, the term lateral inhibition is used as a synonym for surround suppression, but the concentric spatial arrangement of center and surround in the difference-of-Gaussians model is not really "lateral" in the geometric sense of the word. A recently identified RGC type named the PixON RGC has a "truly lateral" inhibitory surround (Johnson et al., 2018). The receptive field center of the PixON RGC receives

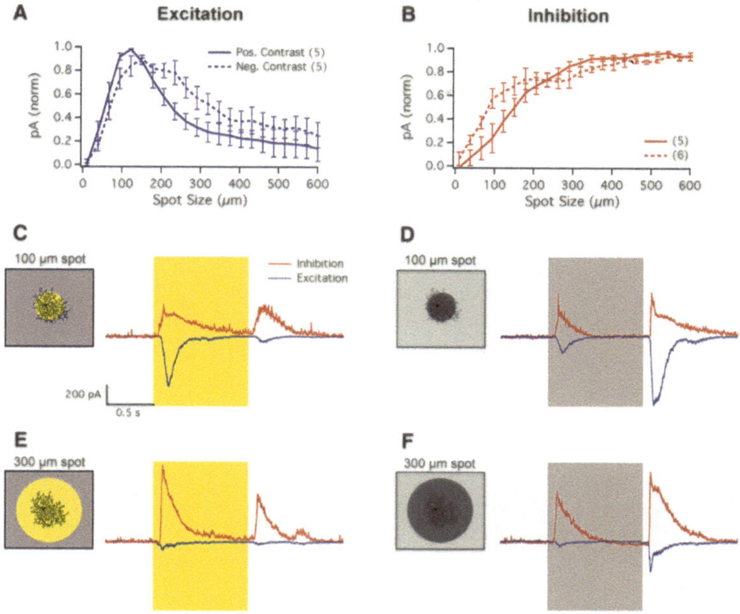

Fig. 8.4 HD1 RGCs use both presynaptic and postsynaptic inhibition to support surround suppression in their spike responses. (**A**) Normalized excitatory current in HD1 RGCs for positive *(solid line)* or negative *(dashed line)* contrast spots of different sizes. Error bars are SEM across cells. (**B**) Same as (**A**) for inhibitory currents. (**C–F**) Examples of synaptic currents from an HD1 RGC for small and large spots of positive and negative contrast. *(From Jacoby and Schwartz (2017).)*

only excitation, and the surround receives only inhibition (Fig. 8.5). Surround suppression in this circuit involves spiking ACs that synapse directly onto the PixON RGC's dendrites. These ACs collectively have a receptive field that forms an annulus around the receptive field center of the RGC. The upstream mechanisms leading to this "truly lateral" inhibition remain unknown.

Nonlinear surround suppression

Another consequence of surround suppression in the inner retina being carried by ACs is that spatial integration in the surround can be nonlinear. Unlike in the outer retina, where the coupled network of horizontal cells averages luminance fairly linearly across space, the receptive fields of most ACs are made up of subunits from BCs. Therefore, patterned stimuli can drive surround suppression even when their mean luminance is the same as the center. In other words, the spatial frequency selectivity of a suppressive surround generated in the inner retina is higher than that generated in the outer retina by horizontal cells.

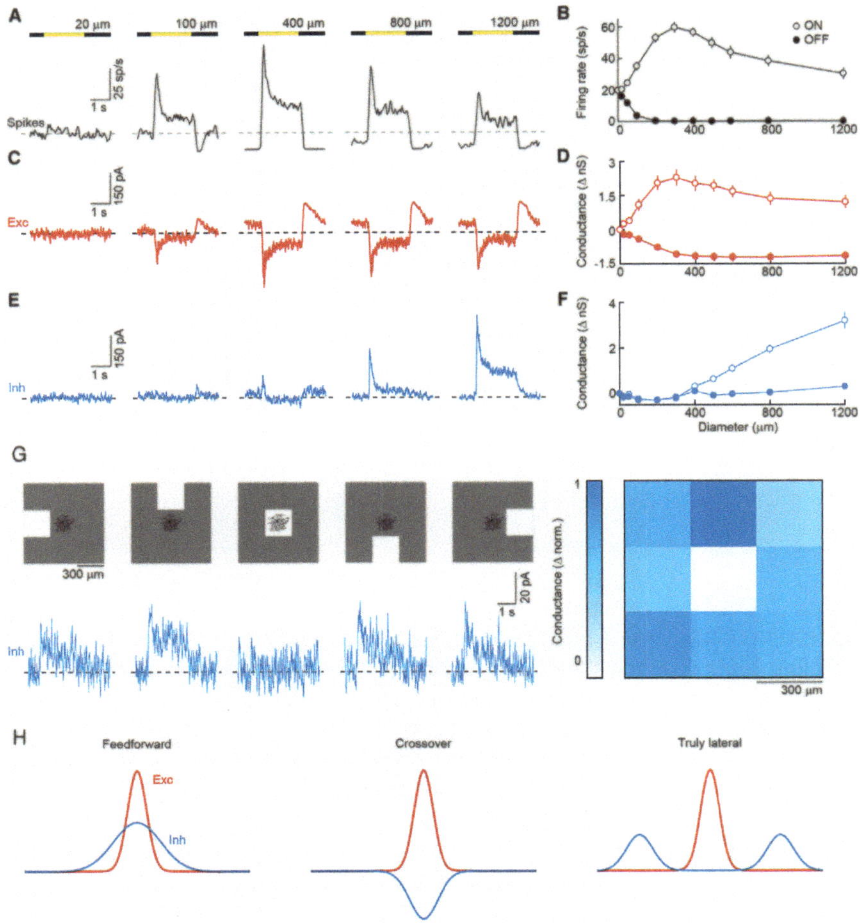

Fig. 8.5 Truly lateral inhibition onto the dendrites of the PixON RGC. (**A**) Example spike responses from a PixON RGC for spots of different sizes as indicated above traces. (**B**) Mean firing rate across cells for ON and OFF spots as a function of spot size. (**C** and **D**) Same as (**A** and **B**) for excitatory currents. (**E** and **F**) Same as (**A** and **B**) for inhibitory currents. (**G**) Left, 3 × 3 grid stimulus where the central square covers the dendritic field of the PixON RGC. Inhibitory current trace is shown below each square. *Right*, mean inhibitory conductance elicited by each of the 9 squares of the grid stimulus. (**H**) Different possible spatial arrangements of excitation and inhibition in the RF of an RGC. Inhibition in the PixON RGC is "truly lateral." *(Modified from Johnson et al. (2018).)*

This property was tested directly in a study that measured responses of both HCs and surround suppression in RGCs of the guinea pig retina (Fig. 8.6A–C) (Zaghloul et al., 2007). The authors first determined the spatial frequency sensitivity of HCs. Then, they recorded from RGCs and stimulated the RF surround with a grating with a spatial frequency beyond the

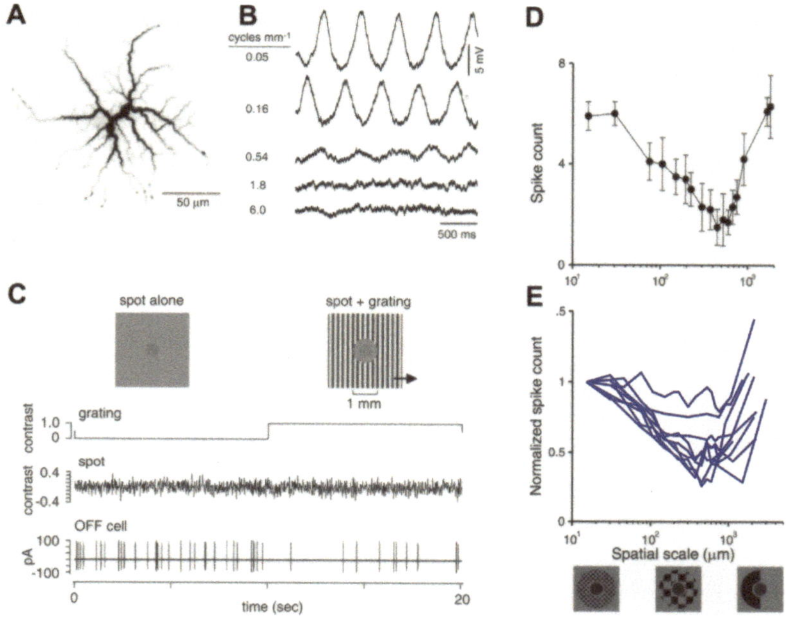

Fig. 8.6 Receptive field surrounds are comprised of nonlinear subunits. (**A**) Image of an HC recorded in whole-mount guinea pig retina. (**B**) HC responses to drifting gratings at the spatial frequencies indicated. (**C**) Stimulus paradigm in which a central spot varies randomly in the RF center either in the presence or in the absence of a high spatial frequency drifting grating in the surround. Spike responses from an OFF RGC *(bottom)* show suppression from the grating. (**D**) Mean spike response of a salamander RGC stimulated with a fixed spot in the RF center and a surround consisting of squares of different sizes (x-axis, stimuli shown below). (**E**) Same as (**D**) for a population of RGCs. Each line is a different cell. *(From (A–C) Zaghloul et al. (2007) and (D, E) Takeshita and Gollisch (2014).)*

limits of HCs. The high-frequency grating elicited robust surround suppression in both ON and OFF RGCs, and voltage-clamp measurements showed that the dominant mechanism was inhibition onto BC terminals, with a smaller contribution from direct inhibition onto RGC dendrites in OFF RGCs. Nonlinear spatial integration in the RF surround was also investigated in the salamander retina using isoresponse contours (Fig. 8.6D and E) (Takeshita and Gollisch, 2014). As we will see in Chapter 12, nonlinear surround suppression plays a key role in the circuits for object motion sensitivity.

Surround influences beyond suppression

ACs inhibit not only BCs and RGC dendrites but also each other. Sign reversals are possible via this AC to AC inhibition motif known as "serial inhibition." Thus, instead of suppressing the receptive field center, surround activation can increase the RF center's response through disinhibition. Early in the history of retinal circuit research, James McIlwain (1964) discovered such a phenomenon. He presented a moving dark spot in the periphery away from the RF center of an ON RGC, and ON responses in the center were enhanced. Enhancement of the center response by movement in the far surround became known as the McIlwain effect. It occurs for some RGCs but not for others, and the circuit mechanisms remain unresolved (Barlow et al., 1977; McIlwain, 1966). My lab identified a disinhibitory circuit in the ON delayed RGC that causes the cell to respond with increases in spike rate for stimuli well beyond its dendritic field (Mani and Schwartz, 2017) (see Chapter 6). The effects of surround stimulation cannot always be neatly characterized as either enhancing or suppressing the RF center response. Turner et al. (2018) presented parasol RGCs in primate retina with natural images either restricted to the RF center or covering both center and surround. Surround activation could either increase or decrease the center response, but it consistently caused spatial integration to be more linear in the RF center.

Conclusions and future directions

As one of the first retinal computations identified, researchers have been studying the mechanisms of surround suppression for over 7 decades. Some aspects of this work have become archetypes for research outside the retina. Horizontal cells are among the best-characterized interneurons in the central nervous system in terms of biophysics, synaptic physiology, and electrical coupling. Comparative work on horizontal cells has extended to dozens of species. Conversely, biology's complexity and diversity have frustrated efforts at a comprehensive understanding of surround suppression in bipolar and ganglion cells. Over 60 AC types (Yan et al., 2020) interact in multilayered circuits converging on both BC terminals and RGC dendrites. Early concepts of surround suppression, like the linear difference-of-Gaussians RF, have proven insufficient to capture the functional diversity of interactions between the RF center and surround in the networks of the inner retina. While work on surround suppression will undoubtedly continue to yield new insights into retinal computation, perhaps our terminology and our goals have to change a bit. Rather than a single computation, surround suppression (and activation through disinhibition) really encompasses a whole class of computations that compare visual stimuli at different locations in space. Instead of a unified model of surround suppression at the level of RGCs, the more likely outcome of the coming decades of research will be a collection of discoveries about how retinal circuits interact across visual space to represent objects, colors, contours, local and global motion, and countless other features of the visual scene.

Landmark paper: Takeshita, D., and Gollisch, T. (2014). Nonlinear spatial integration in the receptive field surround of retinal ganglion cells. The Journal of Neuroscience *34*(22), 7548–7561.

Discussion questions:

1. All the measurements in this paper are spikes recorded on a multi-electrode array, yet the authors claim in the introduction that they are studying spatial nonlinearity independent of the nonlinearity in spike generation. How is this accomplished? Explain in words and with a sketch or a plot.
2. What are the differences that define "quadratic" and "homogeneity detector" RGCs in this study? How is surround suppression similar and how is it different between these two classes of RGCs?
3. Figs. 7 and 8 investigate the spatial scale of nonlinear spatial integration in the surround. How do you interpret the nonmonotonic relationships between checker size and various measures of surround nonlinearity and surround suppression? How does this contrast with the classical view of the RF surround?
4. What do the results of this study mean for typical measurements of RF center and surround with spatiotemporal white noise (checkerboard) stimuli?

Programming exercise

To access the programming exercise, please visit https://www.elsevier.com/books-and-journals/book-companion/9780128198964.

The file RF_model.zip contains the code for this exercise. The main script is called "RF_center_surround_model.m." The purpose of this simulation is to explore how the RF center and surround interact and how measurements of both components can depend on details of the measurement technique.

The model has an RF center modeled as a 2D Gaussian in space with an SD of 50 μm and an RF surround with an SD of 200 μm. Center and surround share the same temporal filter with opposite sign, and they are added linearly to generate a voltage activation signal in the RGC. A spike generation nonlinearity is modeled as a cumulative Gaussian rectified at zero.

A checkerboard white noise stimulus is presented to this model, and both the spatial and temporal filters are applied to create the voltage activation signal. Spikes are generated probabilistically from the spike nonlinearity. From this spike train, the spike-triggered average (STA) is computed with the "computeSTA" function, which returns the full 3D STA as well as separate spatial and temporal (RF center) components. These represent the spatial and temporal RF measurements from the stimulated experiment.

The "setup" section of the script contains the parameters you will vary to explore the effects on the measured RF.

1. Run the stimulation with the default parameters and use "imagesc" to plot the resulting spatial RF, "STA_space." Can you make out the RF center and surround or only the center? Try looking at a Gaussian smoothed version of the RF, "STA_space_smooth," and explore some different values of the smoothing parameter. Note that the surround weight was twice that of the center in the model. Why then does the center appear so much stronger?
2. Does the RF center SD estimated from the 2D Gaussian fit match the SD that went into the model? Try decreasing the surround weight to 0.25. Try increasing the surround weight to 5. How did these changes alter the estimated size of the RF center? Explain why this happened. Can RF center and RF surround be measured independently with this kind of experiment?
3. Set the surround strength back to 2, and now explore a range of values for "checkerSize." This parameter changes the size of the checkers in the test stimulus but does not change the model of the RGC's RF. How does the RF vary as a function of checkerSize? Are the effects similar or different in the RF center vs. the RF surround?
4. Create an index to measure center:surround ratio. How does this index vary as a function of checker size? Explain why you got this result and why it is a problem for interpretations of center:surround ratio using STAs computed from checkerboard stimuli. Can you think of an experiment that mitigates this problem?
5. Vary "NL_halfMax_point." How does the spike count in the model depend on this parameter? For a fixed experiment time ("stimTime" parameter), does making the model give you more spikes improve the signal-to-noise ratio of your RF measurement? At what point is increasing the spike rate with this parameter giving you diminishing returns? What speculations about RF measurements in high vs. low firing-rate RGCs can you make based on these results?

Advanced

6. Alter the model so both the center and surround are comprised of nonlinear subunits. You can use the same cumulative Gaussian nonlinearity for the center and surround subunits and then make the spike generation nonlinearity a simple rectification. Start with center subunits of 20 μm SD and surround subunits of 40 μm SD. Do the trends observed earlier change? Is the subunit model generally more or less sensitive to different stimulus parameters, like checker size?
7. What if the center and surround have subunits of the same size? Is the model now robust to checker size? Why or why not? Now make the nonlinearity different in the center and surround subunits while keeping their size the same. How does the nonlinearity effect the RF measurement? Does center:surround ratio depend on checker size in this model?

References

Baccus, S.A.A., Olveczky, B.P.P., Manu, M., and Meister, M. (2008). A retinal circuit that computes object motion. Journal of Neuroscience *28*(27), 6807–6817.

Barlow, H.B., Derrington, A.M., Harris, L.R., and Lennie, P. (1977). The effects of remote retinal stimulation on the responses of cat retinal ganglion cells. The Journal of Physiology *269*(1), 177–194.

Beckwith-Cohen, B., Holzhausen, L.C., Wang, T.M., Rajappa, R., and Kramer, R.H. (2019). Localizing proton-mediated inhibitory feedback at the retinal horizontal cell–cone synapse with genetically-encoded pH probes. Journal of Neuroscience *39*(4), 651–662.

Berntson, A., and Taylor, W.R. (2000). Response characteristics and receptive field widths of on-bipolar cells in the mouse retina. The Journal of Physiology *524*(3), 879–889.

Bloomfield, S.A., and Miller, R.F. (1982). A physiological and morphological study of the horizontal cell types of the rabbit retina. Journal of Comparative Neurology *208*(3), 288–303.

Boije, H., Fard, S.S., Edqvist, P.H., and Hallböök, F. (2016). Horizontal cells, the odd ones out in the retina, give insights into development and disease. Frontiers in Neuroanatomy *10*(July), 77.

Borghuis, B.G., Marvin, J.S., Looger, L.L., and Demb, J.B. (2013). Two-photon imaging of nonlinear glutamate release dynamics at bipolar cell synapses in the mouse retina. Journal of Neuroscience *33*(27), 10972–10985.

Boycott, B.B., Peichl, L., and Wassle, H. (1978). Morphological types of horizontal cell in the retina of the domestic cat. Proceedings of the Royal Society of London: Biological Sciences *203*(1152), 229–245.

Chan, T.L., Goodchild, A.K., and Martin, P.R. (1997). The morphology and distribution of horizontal cells in the retina of a New World monkey, the marmoset *Callithrix jacchus*: a comparison with macaque monkey. Visual Neuroscience *14*(1), 125–140.

Chapot, C.A., Behrens, C., Rogerson, L.E., Baden, T., Pop, S., Berens, P., Euler, T., and Schubert, T. (2017a). Local signals in mouse horizontal cell dendrites. Current Biology *27*(23), 3603–3615.e5.

Chapot, C.A., Euler, T., and Schubert, T. (2017b). How do horizontal cells 'talk' to cone photoreceptors? Different levels of complexity at the cone–horizontal cell synapse. Journal of Physiology *595*(16), 5495–5506.

Connaughton, V.P., Graham, D., and Nelson, R. (2004). Identification and morphological classification of horizontal, bipolar, and amacrine cells within the zebrafish retina. Journal of Comparative Neurology *477*(4), 371–385.

Dacey, D., Packer, O.S., Diller, L., Brainard, D., Peterson, B., and Lee, B. (2000). Center surround receptive field structure of cone bipolar cells in primate retina. Vision Research *40*(14), 1801–1811.

Davenport, C.M., Detwiler, P.B., and Dacey, D.M. (2008). Effects of pH buffering on horizontal and ganglion cell light responses in primate retina: evidence for the proton hypothesis of surround formation. Journal of Neuroscience *28*(2), 456–464.

Deniz, S., Wersinger, E., Schwab, Y., Mura, C., Erdelyi, F., Szabó, G., Rendon, A., Sahel, J. A., Picaud, S., and Roux, M.J. (2011). Mammalian retinal horizontal cells are unconventional GABAergic neurons. Journal of Neurochemistry *116*(3), 350–362.

Deniz, S., Wersinger, E., Picaud, S., and Roux, M.J. (2019). Evidence for functional GABAA but not GABAC receptors in mouse cone photoreceptors. Visual Neuroscience *36*, E005.

Drinnenberg, A., Franke, F., Morikawa, R.K., Juttner, J., Hillier, D., Hantz, P., Hierlemann, A., Azeredo da Silveira, R., and Roska, B. (2018). How diverse retinal functions arise from feedback at the first visual synapse. Neuron *99*(1), 117–134.e11.

Diamond, J.S. (2017). Inhibitory interneurons in the retina: types, circuitry, and function. Annual Review of Vision Science *3*,1–24.

Fahrenfort, I., Steijaert, M., Sjoerdsma, T., Vickers, E., Ripps, H., van Asselt, J., Endeman, D., Klooster, J., Numan, R., ten Eikelder, H., et al. (2009). Hemichannel-mediated and pH-based feedback from horizontal cells to cones in the vertebrate retina. PLoS ONE *4*(6), e6090.

Farrow, K., Teixeira, M., Szikra, T., Viney, T.J., Balint, K., Yonehara, K., and Roska, B. (2013). Ambient illumination toggles a neuronal circuit switch in the retina and visual perception at cone threshold. Neuron *78*(2), 325–338.

Fischer, A.J., Stanke, J.J., Aloisio, G., Hoy, H., and Stell, W.K. (2007). Heterogeneity of horizontal cells in the chicken retina. Journal of Comparative Neurology *500*(6), 1154–1171.

Foley, J.M. (2019). Lateral effects in pattern vision. Journal of Vision *19*(9), 8.

Franke, K., Berens, P., Schubert, T., Bethge, M., Euler, T., and Baden, T. (2017). Inhibition decorrelates visual feature representations in the inner retina. Nature *542*(7642), 439–444.

Gallego, A., Baron, M., and Gayoso, M. (1975). Horizontal cells of the avian retina. Vision Research *15*(8–9), 1029–1030.

Gheorghiu, E., Kingdom, F.A.A., and Petkov, N. (2014). Contextual modulation as de-texturizer. Vision Research *104*, 12–23.

Gonzälez-soriano, J. (1994). Morphological types of horizontal cell in rodent retinae: a comparison of rat, mouse, gerbil, and guinea pig. Visual Neuroscience *11*(3), 501–517.

Hallum, L.E., and Movshon, J.A. (2014). Surround suppression supports second-order feature encoding by macaque V1 and V2 neurons. Vision Research *104*, 24–35.

Hartline, H.K., and Ratliff, F. (1957). Inhibitory interaction of receptor units in the eye of Limulus. The Journal of General Physiology *40*(3), 357–376.

Hartline, H.K., Wagner, H.G., and Ratliff, F. (1956). Inhibition in the eye of Limulus. The Journal of General Physiology *39*(5), 651–673.

Hirano, A.A., Liu, X., Brecha, N.C., and Barnes, S. (2018). Analysis of feedback signaling from horizontal cells to photoreceptors in mice. Methods in Molecular Biology *1753*, 179–189.

Hombach, S., Janssen-Bienhold, U., Söhl, G., Schubert, T., Büssow, H., Ott, T., Weiler, R., and Willecke, K. (2004). Functional expression of connexin57 in horizontal cells of the mouse retina. European Journal of Neuroscience *19*(10), 2633–2640.

Jacoby, J., and Schwartz, G.W. (2017). Three small-receptive-field ganglion cells in the mouse retina are distinctly tuned to size, speed, and object motion. The Journal of Neuroscience *37*(3), 610–625.

Johnson, K.P., Zhao, L., and Kerschensteiner, D. (2018). A pixel-encoder retinal ganglion cell with spatially offset excitatory and inhibitory receptive fields. Cell Reports *22*(6), 1462–1472.

Kemmler, R., Schultz, K., Dedek, K., Euler, T., and Schubert, T. (2014). Differential regulation of cone calcium signals by different horizontal cell feedback mechanisms in the mouse retina. Journal of Neuroscience *34*(35), 11826–11843.

Kim, T., and Kerschensteiner, D. (2017). Inhibitory control of feature selectivity in an object motion sensitive circuit of the retina. Cell Reports *19*(7), 1343–1350.

Kolb, H., Linberg, K.A., and Fisher, S.K. (1992). Neurons of the human retina: a Golgi study. Journal of Comparative Neurology *318*(2), 147–187.

Kramer, R.H., and Davenport, C.M. (2015). Lateral inhibition in the vertebrate retina: the case of the missing neurotransmitter. PLoS Biology *13*(12), e1002322.

Mani, A., and Schwartz, G.W. (2017). Circuit mechanisms of a retinal ganglion cell with stimulus-dependent response latency and activation beyond its dendrites. Current Biology *27*(4), 1–12.

Mariani, A.P. (1987). Neuronal and synaptic organization of the outer plexiform layer of the pigeon retina. The American Journal of Anatomy *179*(1), 25–39.

Marvin, J.S., Borghuis, B.G., Tian, L., Cichon, J., Harnett, M.T., Akerboom, J., Gordus, A., Renninger, S.L., Chen, T.W., Bargmann, C.I., et al. (2013). An optimized fluorescent probe for visualizing glutamate neurotransmission. Nature Methods *10*(2), 162–170.

McIlwain, J.T. (1964). Receptive fields of optic tract axons and lateral geniculate cells: peripheral extent and barbituate sensitivity. Journal of Neurophysiology *27*, 1154–1173.

McIlwain, J.T. (1966). Some evidence concerning the physiological basis of the periphery effect in the cat's retina. Experimental Brain Research *1*(3), 265–271.

Park, S.J.H., Lieberman, E.E., Ke, J.B., Rho, N., Ghorbani, P., Rahmani, P., Jun, N.Y., Lee, H.L., Kim, I.J., Briggman, K.L., et al. (2020). Connectomic analysis reveals an interneuron with an integral role in the retinal circuit for night vision. eLife *9*, e56077.

Schwartz, G.W., Okawa, H., Dunn, F.A., Morgan, J.L., Kerschensteiner, D., Wong, R.O., and Rieke, F. (2012). The spatial structure of a nonlinear receptive field. Nature Neuroscience *15*(11), 1572–1580.

Spillmann, L. (1994). The Hermann grid illusion: a tool for studying human perspective field organization. Perception *23*(6), 691–708.

Takeshita, D., and Gollisch, T. (2014). Nonlinear spatial integration in the receptive field surround of retinal ganglion cells. The Journal of Neuroscience *34*(22), 7548–7561.

Thoreson, W.B., and Mangel, S.C. (2012). Lateral interactions in the outer retina. Progress in Retinal and Eye Research *31*(5), 407–441.

Thoreson, W.B., Babai, N., and Bartoletti, T.M. (2008). Feedback from horizontal cells to rod photoreceptors in vertebrate retina. Journal of Neuroscience *28*(22), 5691–5695.

Turner, M.H., Schwartz, G.W., and Rieke, F. (2018). Receptive field center-surround interactions mediate context-dependent spatial contrast encoding in the retina. eLife 7, e38841.

Vroman, R., Klaassen, L.J., and Kamermans, M. (2013). Ephaptic communication in the vertebrate retina. Frontiers in Human Neuroscience 7, 612.

Wässle, H., Dacey, D.M., Haun, T., Haverkamp, S., Grünert, U., and Boycott, B.B. (2000). The mosaic of horizontal cells in the macaque monkey retina: with a comment on biplexiform ganglion cells. Visual Neuroscience *17*(4), 591–608.

Wassle, H., Puller, C., Muller, F., and Haverkamp, S. (2009). Cone contacts, mosaics, and territories of bipolar cells in the mouse retina. Journal of Neuroscience *29*(1), 106–117.
Yan, W., Laboulaye, M.A., Tran, N.M., Whitney, I.E., Benhar, I., and Sanes, J.R. (2020). Mouse retinal cell atlas: molecular identification of over sixty amacrine cell types. The Journal of Neuroscience *40*(27), 5177–5195.
Zaghloul, K.A., Manookin, M.B., Borghuis, B.G., Boahen, K., and Demb, J.B. (2007). Functional circuitry for peripheral suppression in mammalian Y-type retinal ganglion cells. Journal of Neurophysiology *97*(6), 4327–4340.
Zhang, J., Zhang, A.J., and Wu, S.M. (2006). Immunocytochemical analysis of GABA-positive and calretinin-positive horizontal cells in the tiger salamander retina. Journal of Comparative Neurology *499*(3), 432–441.
Zhang, Y., Kim, I.J., Sanes, J.R., and Meister, M. (2012). The most numerous ganglion cell type of the mouse retina is a selective feature detector. Proceedings of the National Academy of Sciences of the United States of America *109*(36), E2391–E2398.

Although much of retinal computation is concerned with the identification of what objects are present in a visual scene, object localization is instead about *where* something is located. For hunting animals, like this chameleon, there is clear selective pressure on tracking small prey objects as accurately and precisely as possible.

CHAPTER 9

Object localization

Gregory William Schwartz[a,b,c] and Devon Greer[c]
[a]Departments of Ophthalmology and Physiology, Feinberg School of Medicine, Northwestern University, Chicago, IL, United States
[b]Department of Neurobiology, Weinberg College of Arts and Sciences, Northwestern University, Chicago, IL, United States
[c]Northwestern Interdepartmental Neuroscience Program (NUIN), Northwestern University, Chicago, IL, United States

Most of the computations we have considered so far have focused broadly on the identification of visual stimuli. Photon detection (Chapter 1), Absolute luminance detection (Chapter 3), Contrast sensitivity (Chapter 4), and Texture sensitivity (Chapter 7) all inform the brain about what is present in the visual scene, but none of these computations explicitly encode *where* information. In vision, *where* is just as important as *what*. Precision in localizing prey objects can determine whether an animal catches a meal or whether it goes hungry. Motion plays a major role in the localization of prey and other objects in the environment, and subsequent chapters will address the roles of self-motion (Chapter 12, Object motion sensitivity) and trajectory estimation (Chapter 13, Motion anticipation). Here, we consider the localization of static objects, a necessary foundation for these motion computations.

The limits of behavioral performance in objection localization across the animal kingdom are remarkable. Archerfish can shoot a stream of water to strike a 3-mm insect up to 60 cm away (20 times their body length). At this distance, the target subtends ~0.1 degrees of visual angle[1] and 1–2 photoreceptors (Ben-Simon et al., 2012; Segev et al., 2007; Timmermans, 2009). Larval zebrafish detect UV-bright paramecia <100 μm in size at a distance of 3.25 mm. This target subtends a visual angle of 1.8 degrees, less than the spacing between UV cones, so it is likely detected from the activation of a single cone (Yoshimatsu et al., 2020). When the fish makes its strike, it

[1] While in other chapters I have generally specified distances in units of microns on the retina, here I will use degrees of visual angle because most of the measurements come from psychophysics, where this is the natural unit. The conversion factor for these units depends on eye size: 140 μm/° in humans, 90 μm/° in archerfish and 30 μm/° in mice. Degrees of visual angle are subdivided into "minutes" and "seconds" (also sometimes called "arcmin" and "arcsec" and here used as units of distance, not time) with 60 seconds per minute and 60 minutes per degree. Single (′) and double (″) primes are used to denote minutes and seconds, respectively.

Retinal Computation
https://doi.org/10.1016/B978-0-12-819896-4.00010-X

is still up to 1 mm away from the target, so position information accurate enough for capture must be present in ~5 cones (Patterson et al., 2013). Chameleons track moving insects with a precision down to 1 degree before striking with their tongues at distances up to 25 cm (1.5 times body length) (Flanders, 1985; Harkness, 1977). Eagles can identify and track a rabbit from 1.5 km away (Richards and Short, 1998), and their visual acuity for grating patterns reaches 140 cycles per degree (cpd), more than twice that of humans (Reymond, 1985). The vernier acuity threshold in humans (discussed in the following section), at which they can detect spatial displacement, is as small as 5″ of visual angle, less than one-tenth of the minimum separation between cones in the central fovea (Westheimer and McKee, 1977).

Some of the most elegant and quantitative experiments in sensory neuroscience come from systems in which the limits of behavior can be closely matched to the performance of individual neurons or neural populations. The most compelling example in vision science of this level of understanding between signal and noise properties of neurons and behavior was covered in Chapter 1, but object localization, particularly in the context of prey detection, has similar potential for matching neurons to behavior. Just as there is an evident selective pressure to respond to fewer photons than your competitors, there is pressure to localize prey objects more precisely than competitors. Unlike with photon detection, our understanding of the neural basis for spatial acuity lags far behind the measurements from psychophysics. As we will see, spatial acuity is often an emergent property of a population of cells, so measurements of single cells give us an incomplete picture.

Spatial acuity and hyperacuity

We are all probably anecdotally aware of spatial acuity measurements from visits to an optometrist or ophthalmologist (or even a vision exam for a driver's license) in which we must identify letters down to the smallest size possible. The most difficult of these discriminations are between letters like C and O or F and E that differ by only a single arc or line segment. The best human observers can distinguish these letters reliably when the size of the arc or line segment is about 1′ of visual angle, equal to the spacing of the cone mosaic in the central fovea (Westheimer, 1981). This resolution limit is called simply "acuity" or Snellen acuity, referencing the Snellen eye chart (Westheimer, 1981). Another way to measure acuity is by presenting sinusoidal gratings at finer and finer spatial scales until the observer cannot distinguish the grating from a uniform gray. Human acuity measured in this paradigm is also about 1′ of visual angle (60 cpd) (Westheimer, 2012). Across

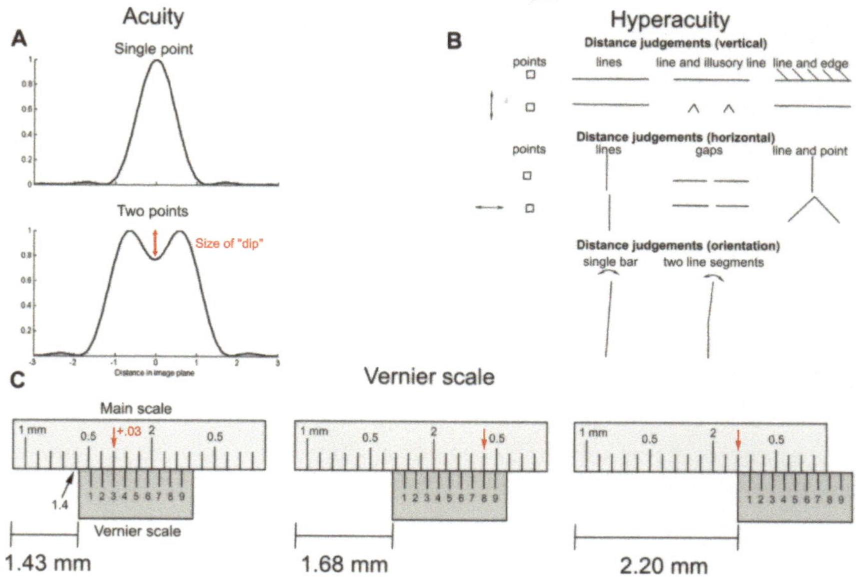

Fig. 9.1 Acuity and hyperacuity. (**A**) *Top*, point-spread function of any optical system, like the eye. *Bottom*, the superposition of two nearby spots. The ability to distinguish the two points from a single point depends on the size of the central "dip" *(red)* relative to noise. The Rayleigh criterion states that the separation must be greater than the half-width of the point-spread function to be resolvable. (**B**) Various kinds of psychophysical tasks have been used to measure the hyperacuity threshold. *Top row*, vertical distance judgments; *middle row*, horizontal distance judgments; *bottom row*, orientation judgments. (**C**) Diagrams of a vernier scale showing measurements of 1.43 mm, 1.68 mm, and 2.20 mm. Measurement precision for this device relies on hyperacuity. *(Modified from (A) Westheimer (2012) and (B) Westheimer (1981).)*

positions in the retina and across species, acuity measured by grating discrimination is generally in good agreement with photoreceptor density (Fig. 9.1A) (Ben-Simon et al., 2012; Blake et al., 1974; Campbell and Green, 1965; Gaffney and Hodos, 2003; Miller and Murphy, 1995; Over and Moore, 1981; Reymond, 1985; Timney and Keil, 1992; Van Hof, 1967; Wiesenfeld and Branchek, 1976).

This would be a short chapter indeed if the story ended there. Just like a digital camera, we could calculate the precision with which an animal could localize an object by knowing the spacing of its photoreceptor "pixels." However, to quote Gerald Westheimer (1975), the leading expert on spatial acuity for over four decades, "The spoilers are not just in the wings, but have been actually on the stage for a long while." In the 1850s, Alfred Volkmann measured the just-noticeable visual distance in tasks where subjects had to decide whether bars were closer together or further apart than a fixed standard. Remarkably, he obtained a value of 12″ of visual angle, one-fifth of the minimum cone spacing in the fovea. Ewald Hering is generally credited with recognizing this discrepancy between different spatial acuity measures in an 1899 speech in which he argued with Helmholtz on the topic. This history and English translations of both Hering's speech and Volkmann's earlier work, which he cites, are provided in a captivating account by Strasburger et al. (2018).

Hering's students went on to study the limits of spatial discrimination in a variety of different paradigms (Fig. 9.1B). The most well-known of these experiments is called the "vernier" task, and the corresponding discrimination threshold is called "vernier acuity." In this task, the observer judges whether two or more sets of lines are continuous or offset along the perpendicular axis (Best, 1900). The term traces its roots to the vernier caliper, a device invented in 1631 by Pierre Vernier, in which measurement precision is improved because the operator uses line segment alignment to estimate increments between the smallest divisions (Fig. 9.1C) (Barrow, 1750).

Gerald Westheimer (1975) coined the term "hyperacuity" in his famous editorial "Spatial Acuity and Hyperacuity" to refer to thresholds in spatial tasks that are substantially smaller than Snellen acuity or photoreceptor spacing. Westheimer and his contemporaries performed psychophysical measurements to test various aspects of hyperacuity and constrain its mechanisms. Several comprehensive review articles by Westheimer (1981, 2009, 2012) cover the details of these experiments much more thoroughly than I will here, but I will list several important conclusions from these studies that are especially informative about the underlying computation.

More than they have pointed to a specific mechanism, the psychophysics experiments have served to rule out various mechanisms involving other retinal computations. Hyperacuity is not simply a manifestation of luminance or contrast sensitivity because stimuli with equal total luminance or equal flux can nonetheless be distinguished by width down to $6''$. Thresholds do not depend on whether the stimuli are lines, edges, bright, or dark (Fig. 9.2A). In fact, the stimuli need not be lines at all. Similar thresholds are measured for distances between points or even gaps in continuous lines. The orientation of real lines or the illusory line connecting two spots does not affect threshold (Westheimer and McKee, 1977). Spatiotemporal sampling with fixational eye movements (as suggested by Hering) (Strasburger et al., 2018) does not seem to play a role because presentations as brief as 10 ms are sufficient to reach the lowest threshold (Westheimer and Pettet, 1990) (but see Jiang et al., 2017).

As hyperacuity tasks ask subjects to judge alignment, distance, or width, they must involve a comparison between two objects. What is the optimal

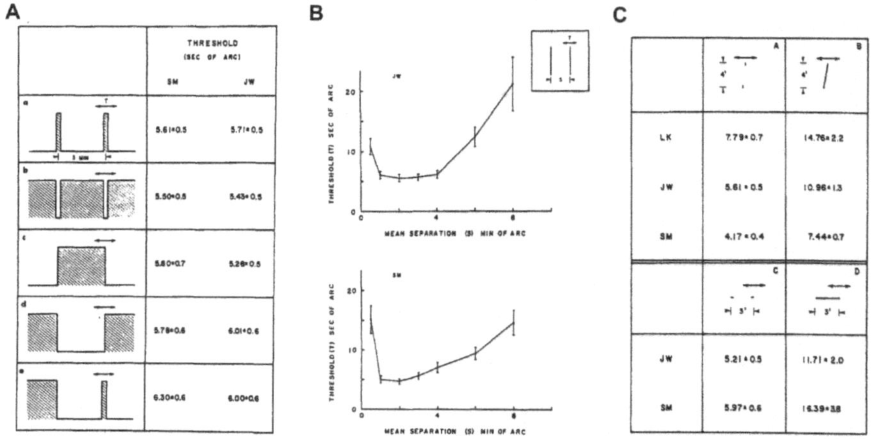

Fig. 9.2 The effects of different stimulus configurations on vernier acuity. (A) Diagrams of different stimulus configurations and corresponding vernier acuity thresholds for two subjects. *Top to bottom*: Distance judgments between two bright lines, or two dark lines, width judgments for a bright bar or a dark bar, the distance between a bright edge and a line. (**B**) Vernier acuity thresholds for two subjects for the two-line separation task as a function of mean separation between the lines. (**C**) *Top*, horizontal displacement vernier acuity thresholds for small stimuli separated by 4′ *(left)* or connected by a line *(right)*. *Bottom*, similar experiment with the spots horizontally aligned. *(Modified from Westheimer and McKee (1977).)*

separation in space and time of the objects being compared? Westheimer and colleagues performed a series of experiments to answer this question and concluded that thresholds are lowest (most sensitive) when the objects are presented within ~20 ms and between 2′ and 5′ apart (Westheimer, 2012). While the dependence of threshold on temporal contiguity is monotonic (simultaneous presentation is best), subjects can learn to compare distances to an internal standard (i.e., to determine that lines separated by 3′06″ are further apart than the average value of 3′00″ across many trials). Interestingly, threshold dependence on spatial separation between objects is nonmonotonic (Fig. 9.2B). If the objects are closer together than 2′, they begin to merge into a single object because of the Snellen acuity limit imposed by the photoreceptor mosaic, and hyperacuity thresholds suffer. This was most elegantly demonstrated in an experiment showing that the hyperacuity threshold for the horizontal separation of two points separated by 4′ is the standard 5–6″, but it gets worse by a factor of two when the same two points are connected by a line segment (Fig. 9.2C) (Westheimer and McKee, 1977). Beyond 5′ separation, hyperacuity thresholds also suffer, suggesting that the comparison mechanism integrates over approximately this area.

Hyperacuity thresholds are also lower in conditions in which stimuli are dim, low contrast, or presented outside the fovea (Westheimer, 2012).

The generalizability of hyperacuity to very different kinds of objects and its dependence on even higher-order abstractions like visual "crowding" (Levi, 2008) suggest that the neural "comparator" that estimates distance is likely beyond the retina, probably in the cortex. Nonetheless, this comparator must rely on precise representations of object position from the retina. Thus, at the level of retinal computation, the goal is not to find a neuron that fires differentially for lines $3'00''$ and $3'06''$ apart, but rather to estimate the precision and signal-to-noise ratio (SNR) with which retinal neurons convey information about object position (Geisler, 1984). Next, we consider what is known about the precision with which retinal ganglion cells (RGCs), the output neurons of the retina, convey location information.

Object localization outside the fovea

Much of the work on hyperacuity has focused on the most central part of the fovea, the specialized region where single cones, through midget bipolar cells (BCs), provide the RF center of single midget RGCs. But 99.8% of the retinal surface (and the visual field) in humans is outside the fovea (Kolb, 1995). While many species have a region of the retina (or several) with increased receptor density, most lack a region with 1:1:1 cone:BC: RGC connectivity (Baden et al., 2020). Object localization is critical in the periphery for humans and other foveal species because we rely on foveating eye movements for high acuity vision. Accurately localizing an object outside our central view allows us to place it on the fovea with a single saccade (Schall, 1995). Species without a fovea, like zebrafish and archerfish, still perform remarkably precise visual behaviors, especially in the context of prey detection, so there must be mechanisms for this computation that do not rely solely on the metabolically demanding dense packing of retinal neurons. When multiple photoreceptors converge at the level of the RGC, is spatial resolution degraded by a factor equal to the convergence? Are certain RGC types specialized for spatial acuity because of low convergence or other circuit mechanisms?

"Spot-detector" RGCs

Lettvin and colleagues' seminal paper "What the frog's eye tells the frog's brain" first introduced the "bug detector" RGC, a cell that responded well to small, dark moving objects but failed to respond to large objects or full-field edges (Lettvin et al., 1959). A functionally similar RGC type that responded to small

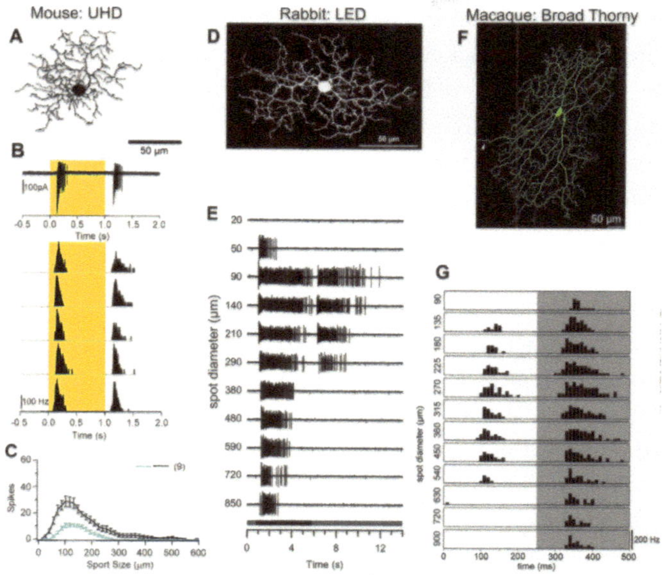

Fig. 9.3 Spot-detector RGCs in three mammalian species. (**A**) Morphology of an HD1 RGC from mouse retina. (**B**) Example spike responses from HD1 RGCs for small spots centered on the RF. (**C**) Spike count at light ON *(cyan)* and OFF *(black)* as a function of spot diameter for HD1 RGCs. (**D**) Morphology of a rabbit LED RGC. (**E**) Spike traces from a rabbit LED responding to spots of different sizes. (**F**) Morphology of a broad thorny RGC in macaque. (**G**) Spike rates for a broad thorny RGC responding to spots of different sizes. *(Modified from (A–C) Jacoby and Schwartz (2017), (D, E) van Wyk et al. (2006), and (F, G) Puller et al. (2015).)*

dark or light edges within its receptive field but not to wide-field edges was identified in the rabbit retina and named the "local edge detector" (LED) (Levick, 1967). Similar "spot-detector" RGCs have been found in mouse (Zhang et al., 2012), cat (Berson et al., 1998), and guinea pig (Xu et al., 2005), all species that rely on vision for prey detection despite lacking a fovea (Fig. 9.3). The macaque, which has a fovea similar to that of humans, also has one or more spot-detector RGC types, located mostly in the periphery (Calkins and Sterling, 2007; Dacey et al., 2003), and one of these types, the "broad thorny" RGC (Fig. 9.3F and G), has been suggested to play a role in encoding an error signal for smooth pursuit eye movements (Puller et al., 2015).

Spot-detector RGCs share common morphological and physiological features across species, but recent work has shown that subtle differences, both between species and between different spot-detector RGC types in the same species, may underlie functional specialization. Spot-detector RGCs tend to have small, dense dendritic fields in either a single stratum or two strata near the middle of the inner plexiform layer (IPL). Because their dendritic fields are small, mosaic coverage requires that these RGC types be more densely packed than their larger neighbors. Thus, they are among the most numerous RGC types in several species (Zhang et al., 2012) (though midget RGCs far outnumber them in primates). Spot detectors also tend to respond transiently to ON or OFF contrasts, but there are some notable exceptions. The original bug detector RGC discovered in frogs was an OFF cell. And while transient ON-OFF spot detectors exist in both rabbit and mouse (Sivyer et al., 2011; Zhang et al., 2012), both of these species also have a much more sustained, small-RF RGC type. This was the LED first described in rabbits, and the homologous cell has retained that name in mice (Famiglietti, 2005; Russell and Werblin, 2010; van Wyk et al., 2006; Zeck et al., 2005). Three additional spot-detector RGC types were found in mouse and named the "high-definition" RGCs (HD1, HD2, and UHD). While all three are transient ON-OFF cells, they differ in ON: OFF ratio, size, and speed tuning (Jacoby and Schwartz, 2017). In zebrafish, RGC axon terminals with spot-detector responses have been recorded, and several morphological RGC types have been suggested as possible sources (Semmelhack et al., 2014).

Mechanistically, spot-detector RGCs achieve their sensitivity profile from strong surround suppression (see Chapter 8). Measurements of surround suppression in these cells have typically revealed both presynaptic (at the level of BC terminals) and postsynaptic (inhibition to RGC dendrites) components (Jacoby and Schwartz, 2017; Puller et al., 2015; Sivyer et al., 2011; van Wyk et al., 2006; Zhang et al., 2012). Importantly, however, the small linear RF centers of these RGCs are not sufficient to explain the hyperacuity thresholds measured behaviorally, even outside the fovea. The following sections explore possible solutions to this apparent paradox, including sub-RF-center spatial resolution at the single RGC level and hyperacuity at the RGC population level.

While the functional properties and evolutionary conservation of spot-detector RGCs, especially given their high density, suggest that they play a role in object localization, for the most part, connections to behavior have been circumstantial. A notable exception is the work by Semmelhack et al. (2014), who laser-ablated a projection field of spot-detector RGCs in larval zebrafish and demonstrated a marked reduction in hunting behavior (Fig. 9.4). In mice, many neurons in the superior colliculus (SC) have robust spot detection responses (Gale and Murphy, 2014; Wang et al., 2010), and chemogenetic inactivation of subpopulations of cells in SC has been shown to alter prey-capture behavior (Hoy et al., 2019). Since ~90% of RGC types

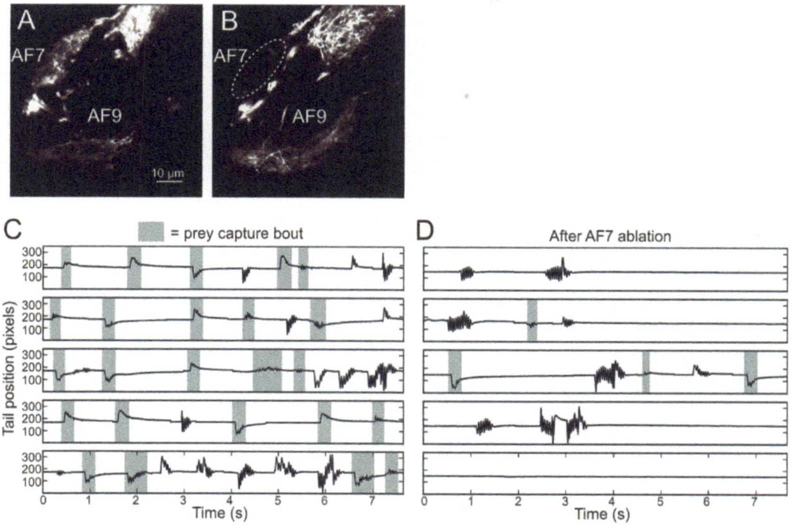

Fig. 9.4 Ablation of a subset of RGC terminals disrupts hunting behavior in zebrafish. (**A**) In vivo image of a plane of the larval zebrafish brain containing RGC arborization fields (AFs) 7 and 9. (**B**) Same region after the laser ablation of AF7. (**C**) Plots of tail position over time with prey detection bouts highlighted in blue. (**D**) Same animal as in (**C**) after AF7 ablation. *(From Semmelhack (2014).)*

project to the SC in mice (Ellis et al., 2016), it is unclear which specific types are involved in objection localization for hunting.

Spatial information in RGCs at smaller scales than the linear RF center

The spatial scale of information transmission from a Gaussian RF is not simply equal to the standard deviation of the Gaussian. While nonlinear spatial integration plays an important role in this regard (see Chapter 7 and the following section), even linear RFs can respond to stimulus displacements much smaller than their standard deviation. Shapley and Victor (1986) measured hyperacuity thresholds in spatially linear (X) RGCs of the cat retina and worked out the theoretical basis of hyperacuity in a linear RF (Fig. 9.5). They showed that for a sine-wave grating at the optimal spatial frequency (which can also be calculated), the displacement threshold was equal to $\frac{r\sqrt{e}}{Z\sqrt{2}}$, where r is the radius of the linear RF, and Z is the SNR. For a linear RF with an SNR of 12, the displacement threshold is 10 times smaller than the RF radius.

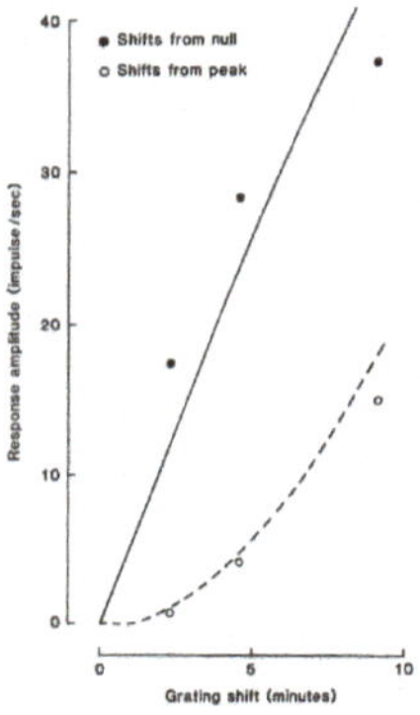

Fig. 9.5 Responses in linear cat RGCs to displacements much smaller than the RF diameter. Recording from a cat X RGC with a linear RF diameter of 1.25 degrees of visual angle. A split field grating (0.8 cpd) was displaced by different amounts (x-axis) centered on either the null (bisecting) point of the RF *(filled symbols)* or the peak of the RF *(open symbols)*. Spike responses reveal sensitivity for very small displacements compared with the RF diameter. *(From Shapley and Victor (1986).)*

RGCs with nonlinear spatial integration carry information near the spatial scale of their subunits rather than that of the linear RF center (Schwartz et al., 2012) (see Chapter 7). But in both the linear and nonlinear cases, the information is not exactly a position signal. Rather, the computation is defined in terms of a binary choice: "Did the grating move?" or "Is the stimulus a fine texture or a uniform gray?"

Consider the example of the displacement threshold for a grating. If the RF were circularly symmetric, as for a two-dimensional Gaussian, the cell would carry no information about the displacement angle. Equivalently, a circularly symmetric RF, linear or nonlinear, can encode the radius of displacement of a small spot from the RF center but not its angle. At the single-cell level, this ambiguity can only be overcome by introducing asymmetries in the RF. Deviations from perfect circular symmetry are common in RF measurements (Gauthier et al., 2009), and they are more pronounced at the level of nonlinear subunits (Schwartz et al., 2012). While decoding information from asymmetric RFs is not trivial (the downstream neuron would need a representation of the asymmetry, possibly acquired by learning), the gain in information about spot position has been demonstrated in a nonlinear RF model (Fig. 9.6) (Schwartz et al., 2012). RF asymmetries among a population of RGCs are considered in the following section.

Spatial information in RGC populations

So far, we have considered hyperacuity at the level of single RGCs, but in the central fovea, all hyperacuity paradigms must involve RGC populations because the arguments earlier about hyperacuity in single RGCs are invalidated by the zero divergence midget pathway (Fig. 9.7). For a midget RGC

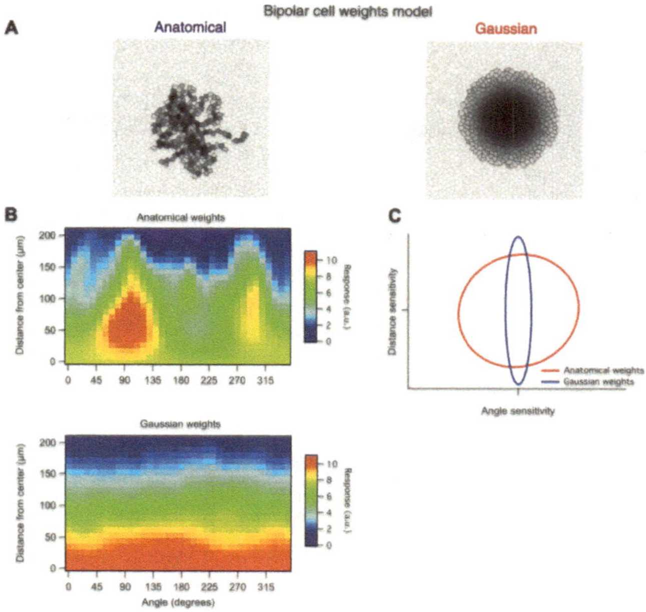

Fig. 9.6 Inhomogeneous RGC RFs provide more spatial information than Gaussian RFs. (**A**) Bipolar mosaic with weights assigned based on the morphology of an ON alpha RGC *(left)* or a Gaussian approximation of its RF *(right)*. (**B**) *Top*, heatmap of anatomical model response for different stimulus positions. Y-axis is the distance of the spot from the model's center of mass; x-axis is the angle. *Bottom*, same plot for the Gaussian model. (**C**) Variance of the model responses across angles (x-axis) and distances (y-axis) for the anatomical model *(red)* and the Gaussian model *(blue)*. The area inside each circle represents the total information about spot position present in the response distribution. *(Modified from Schwartz et al. (2012).)*

with a single cone contributing to the RF center, there cannot be nonlinear subunits, and the RF can no longer be modeled as a Gaussian. Photons captured in different parts of a single cone outer segment contribute equally to the cone's photoresponse.

Homogeneous populations

Each RGC type forms an ordered mosaic to ensure nearly uniform sampling of visual space (at least locally). Importantly, however, RGC mosaics include extensive overlap as measured both in RFs (Baden et al., 2016) and in the two-dimensional overlap of dendritic fields within each RGC type (Bae et al., 2018) (Fig. 9.8).

Because of the overrepresentation of space in RGC mosaics, even the smallest "point source" stimuli are encoded by the spikes of ~3–4 cells,

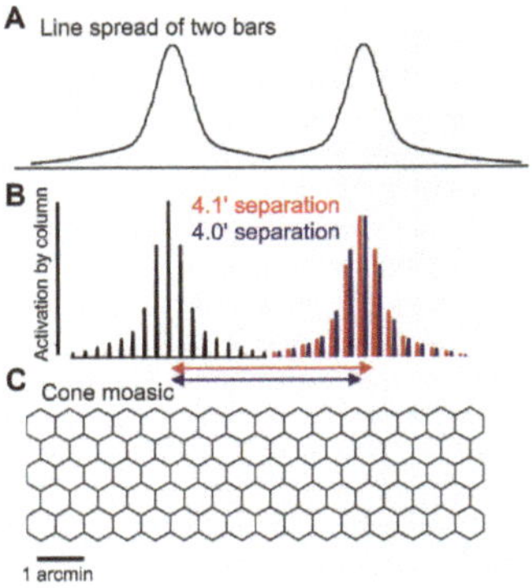

Fig. 9.7 Schematic of the two-line hyperacuity task as seen by the cone mosaic in the central human fovea. (**A**) The line spread function of two bars separated by 4′. (**B**) Representation of cone activation for the vernier acuity task near threshold. The left *(black)* distribution is the cone activity for each column for the fixed left bar. The right distributions are the cone activity for the right bar at a separation of 4.0′ *(blue)* and 4.1′ *(red)*. (**C**) Representation of the central fovea cone mosaic to scale with (**A** and **B**). *(Modified from Westheimer (2012).)*

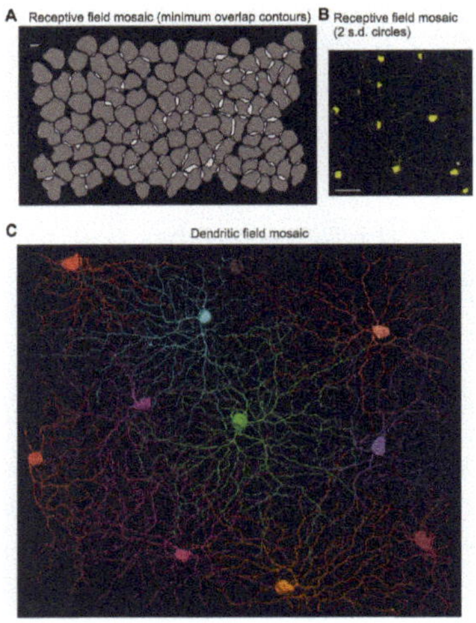

Fig. 9.8 Mosaic organization of RGC RFs and dendritic fields. (**A**) Mosaic of OFF parasol RGC RFs from peripheral macaque retina. Contours are drawn to minimize overlap to reveal positioning, not to represent the full extent of each RF. (**B**) Mosaic of OFF transient alpha RGCs from mouse *(yellow somas)* along with their 2 SD RF boundaries. (**C**) Mosaic of dendrites of a single morphological type of RGC (type "4on") in mouse retina. *(Modified from (A) Gauthier et al. (2009) and (B) Baden et al. (2016). (C) From museum.eyewire.org.)*

and larger RGC populations represent the larger stimuli that are used in typical hyperacuity experiments in psychophysics. Lee et al. (1995) measured displacement thresholds in single midget and parasol RGCs of the nonhuman primate and then constructed models of populations of each RGC type to predict thresholds in the two-edge displacement task. For low-contrast stimuli, their estimates of displacement thresholds from parasol RGCs agreed well with those measured behaviorally in humans; for high (>20%) contrast stimuli, both midget and parasol RGC populations were consistent with human performance (Fig. 9.9).

Parasol and midget RGCs comprise four types, with separate ON and OFF varieties of each, so different populations must represent positive versus

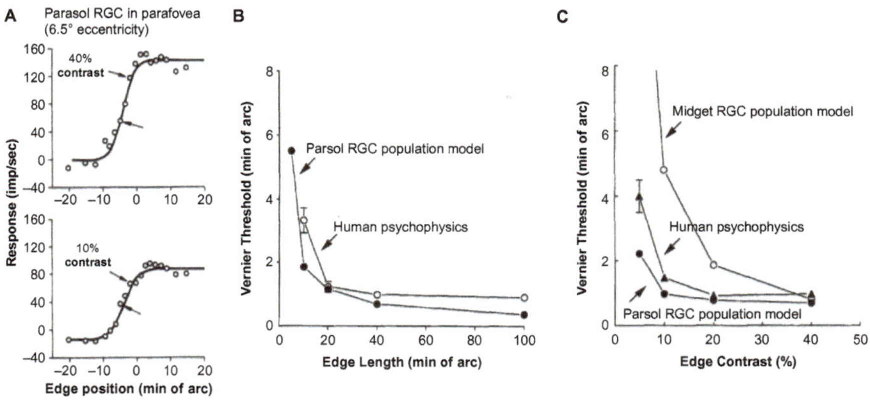

Fig. 9.9 Spatial precision of responses in primate RGCs and population models. (A) *Top,* response of a primate parasol RGC to 40% contrast edges at different positions relative to the RF center. *Arrows* indicate the region of highest slope used to compute the spatial sensitivity. *Bottom,* responses of the same RGC for a 10% contrast bar. (**B**) Vernier threshold for a model of the parasol RGC population constructed from the sensitivity measurements in (**A**) and known mosaic spacing *(filled symbols)* compared with human performance at the same eccentricity *(open symbols).* (**C**) Parasol and midget RGC population models of vernier thresholds as a function of stimulus contrast in comparison with human performance. *(Modified from Lee et al. (1995).)*

negative contrast stimuli. But ON-OFF RGCs (the class to which most spot-detector RGCs belong) could, in principle, represent the position of both dark and light edges. Most ON-OFF RGCs, including spot-detector types, have well-aligned ON and OFF RFs. One type in mouse, however, called the F-mini-ON RGC, has a systematic shift such that its OFF RF is ~25 μm ventral of its ON RF (Cooler and Schwartz, 2020). In a model of the F-mini-ON population, the ON-OFF offset RFs were shown to improve the accuracy with which the cells represent the position of a dark-light edge (particularly a horizontal edge) relative to aligned ON-OFF RFs (Fig. 9.10). The F-mini-ON population model could represent bar position down to 0.6 degrees, less than 12% of the RF diameter of a single F-mini-ON RGC.

Heterogeneous populations

The representation of space need not involve only one RGC type at a time. Estimates of the number of RGC types in vertebrates typically range from 20 to 60. The mouse, where they have been most thoroughly characterized, has ~45 RGC types (Baden et al., 2016; Bae et al., 2018). Since most RGC

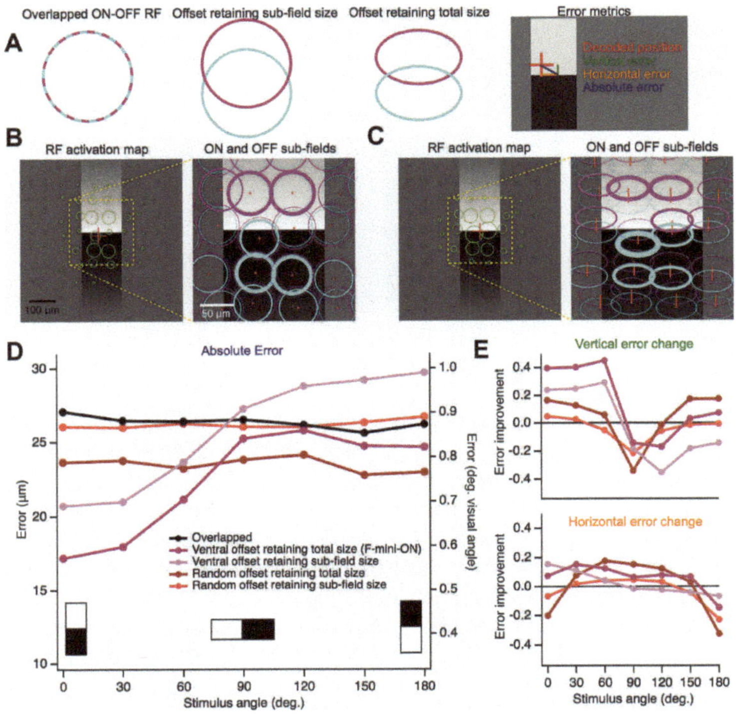

Fig. 9.10 A multicell model of object localization shows an advantage of offset ON-OFF RFs. (**A**) Schematics of the different RF structures and error metrics. (**B**) RF activation map *(left)* for a single instantiation of the overlapped RF model. The stimulus is the black-white edge. *Green circles* show the positions of each RGC with the radius of the circle proportional to its response. *Red cross* indicates the center of mass of the RGC responses. Magnified portion *(right)* shows the ON *(magenta)* and OFF *(cyan)* subfields of each RF. Line thickness is proportional to activation of each subfield. (**C**) RF activation map for the same positions and stimulus as in (**B**) but for the offset RF model retaining total size. Orange lines in the magnified view *(right)* connect ON and OFF subfields from the same RGC. (**D**) Absolute error in decoded edge position for each model. Points are means of 500 iterations of each model with different RGC and stimulus positions. Error bars for SEM are obscured by symbols. (**E**) Vertical *(top)* and horizontal *(bottom)* components of the error in (**D**) shown as a change ratio relative to the overlapped RF result. *(Modified from Cooler and Schwartz (2020).)*

mosaics are independent (but see Roy et al., 2021), the coverage factor in a heterogeneous RGC population can be enormous (up to ~150). While it is unlikely that all RGC types contribute to spatial discrimination tasks, it is informative to consider how position encoding precision scales with the size of a heterogeneous RGC population with highly overlapping RFs.

This was the nature of the task studied by Soo et al. (2011) in the salamander retina (Fig. 9.11). The authors stimulated single cones, including

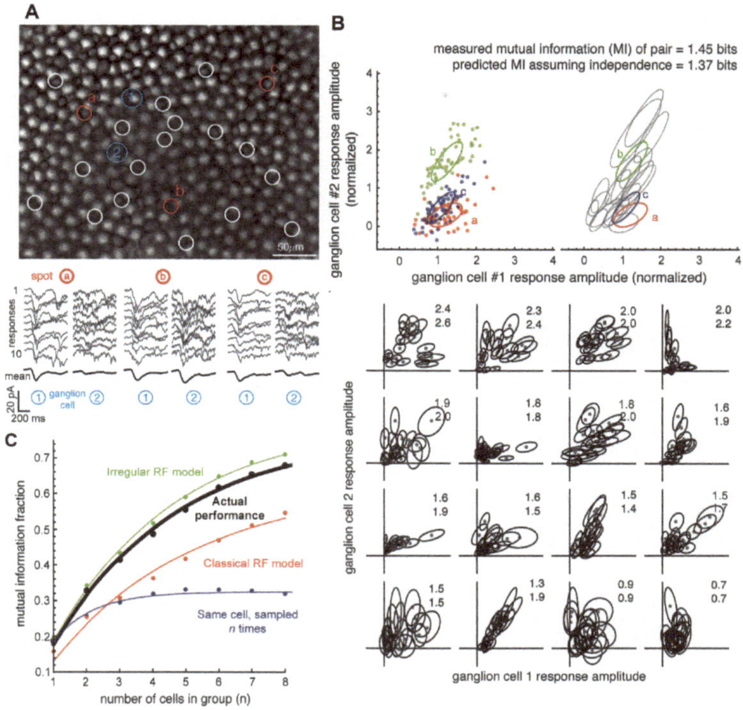

Fig. 9.11 RF inhomogeneities aid in object localization in a heterogeneous, overlapping population of salamander RGCs. (**A**) *Top*, image of the photoreceptor array in a whole-mount salamander retina. Small white and red circles indicate the positions of stimulated cones. Large blue circles labeled 1 and 2 indicate positions of simultaneously recorded RGCs. Bottom, 10 examples of responses for RGCs 1 and 2 to spots a, b, and c. (**B**) *Top left*, scatter plot of the amplitudes of the responses of the two RGCs for spots a, b, and c. Matching colored ellipses are 2D Gaussian fits to the response clouds. *Top right*, response ellipses for all 20 spots. Mutual information between the two-RGC response patterns and spot position is listed along with the mutual information assuming independence between the responses of the two RGCs. Total information possible is $\log_2(20) = 4.32$ bits. *Bottom*, similar plots for 16 more RGC pairs ordered according to their mutual information. (**C**) Fraction of mutual information about spot position computed by synthetically combining responses from populations of up to eight RGCs *(black line)*. Repeating sampling of the same cell reduces noise but rapidly saturates at low information content *(blue line)*. Gaussian RF models fail to reach the performance of the population *(red line)*. Irregular RFs measured with fine-scale white noise *(green line)* match the performance level of the recorded responses. *(Modified from Soo et al. (2011).)*

adjacent ones, with small spots of light and measured the information about spot location contained in the excitatory synaptic inputs to pairs of RGCs (of any type) with highly overlapping RFs. The most informative RGC pair encoded 55% of the possible information about spot location, and model populations of five RGCs reached 74%. Like the single-cell result described earlier, irregularities in the RFs were beneficial for position encoding in the population. Gaussian approximations of the RFs encoded less spatial information in the population than the detailed RFs measured with small spots. Since they were different between cells, irregularities tended to decorrelate the responses from RGCs with highly overlapping RFs, thus increasing the information content over the more correlated Gaussian model.

Conclusions and future directions

Object localization in the retina is certainly more complex than it is in a pixel camera: a fact that many vision scientists do not fully appreciate—including myself before I began researching this chapter! Detailed hyperacuity measurements in human psychophysics and prey detection in animal models have established a strong foundation on which to connect the activity of neurons to the sensitivity limits of behavior. The types of experiments needed to make this connection quantitative have been rare; they require measurements of both the signal and noise distributions of populations of retinal neurons. Thus, there remains a host of important discoveries to be made in this very fundamental line of research about how the retina represents location.

Landmark paper: Westheimer, G., and McKee, S.P. (1977). Spatial configurations for visual hyperacuity. Vision Research *17*(8), 941–947.

Discussion questions:

1. What is the stimulus configuration in Experiment 2 (Fig. 2)? How do the authors explain the shape of the hyperacuity threshold curve in this experiment and the differences between conditions A, B, and C?
2. If the cells making distance comparisons for vernier acuity are located in primary visual cortex, it would stand to reason that line orientation, a component of the responses of cells in that region, would play a role. What data in this paper argue for or against a role for orientation selectivity in the computation?
3. Explain the stimulus configuration for the data shown in Fig. 8 and what this reveals about the nature of the computation.
4. What additional spatial or spatiotemporal configurations would you test to reveal constraints on the neural mechanisms underlying hyperacuity?

Landmark paper: Lee, B.B., Wehrhahn, C., Westheimer, G., and Kremers, J. (1995). The spatial precision of macaque ganglion cell responses in relation to vernier acuity of human observers. Vision Research *35*(19), 2743–2758.

Discussion questions:

1. What is the range of retinal eccentricities for the RGCs recorded in this study? What are the RF center sizes of parasol and midget RGCs in macaque at this eccentricity?
2. What do the arrows indicate in Fig. 2? How do the authors compute the position sensitivity of RGCs in a single number?
3. How do the curves for parasols (MC pathway) and midgets (PC pathway) differ in Fig. 4? How do the authors interpret these differences?
4. What is ROC analysis? How is it used in Figs. 5, 6, and 7?
5. What assumptions do the authors use to convert their single-cell measurements into a population model? How do they support each of these assumptions?
6. What are the main conclusions from the population models, and how are they supported by Fig. 11?

Landmark paper: Soo, F.S., Schwartz, G.W., Sadeghi, K., and Berry, M.J. (2011). Fine spatial information represented in a population of retinal ganglion cells. The Journal of Neuroscience *31*(6), 2145–2155.

Discussion questions:

1. Why do the authors record excitatory synaptic inputs from the RGCs rather than their spikes? What are the advantages and disadvantages of this approach for studying the computation of object localization?
2. How are individual responses quantified? What data in Fig. 2 justify this choice?
3. How do the smallest discriminable location differences compare to the RF sizes of these RGCs?
4. What assumptions go into the information calculation shown in Fig. 3? What do the numbers in panel b tell you about the relative independence of responses from the two simultaneously recorded RGCs? Why is this result surprising?
5. What is the difference between the black and gray points in Fig. 4? What does this result suggest about the role of noise correlations in object localization? Is two simultaneously recorded cells enough to reach a strong conclusion about noise correlations? Why or why not?

6. Which data in Figs. 6–8 support the conclusion that RF irregularities enhance object localization in RGC populations? What do you think is the physiological basis of these irregularities? Are you surprised that they are uncorrelated between nearby RGCs? Why or why not?

Programming exercise

To access the programming exercise, please visit https://www.elsevier.com/books-and-journals/book-companion/9780128198964.

The exercise explores how an array of cones in the human fovea might encode the position of bar stimuli in a vernier acuity task. The script "vernierAcuity_twoBar.m" sets up the simulation parameters in the first block of code. Subsequent code blocks create the array of cone positions, compute their voronoi areas to create the hexagonal grid, model the activation across the array for the two bars, and then use a sum of two Gaussians to fit the activation profile and estimate the distance between the bars. To run the simulation a number of times for different bar separations and collect the distance estimates, you can use the function "testVernierThreshold.m," which returns the percentage of the time that the model detects a distance between bars that exceeds the average value. This is the same task as used by Westheimer and McKee (1977).

1. Run the simulation script and make a plot with the voronoi domain of each cone colored by its level of activation. Now try different bar separations and different amounts of optical blur to see how the activation pattern changes.
2. Obtain the psychometric function for the model for different values of optical blur ranging from 0 to 2.5 arcmin (note the pixel-to-arcmin conversion factor in the script). Compute the vernier acuity threshold from these curves. Is threshold monotonically related to optical blur? Why or why not?
3. Now test different ranges of bar separation. In particular, make the bars on average only 1 arcmin apart and repeat part 2. Do the results change? Does this match what Westheimer and McKee (1977) found?
4. Does the vernier acuity threshold for very small or zero optical blur change if you introduce noise into the cone mosaic with the "hexArrayNoise" parameter? Explain this result.
5. How would you summarize any counterintuitive effects of noise that you observed in this model? How does this break a common dogma about blur and optical resolution?

References

Baden, T., Berens, P., Franke, K., Román Rosón, M., Bethge, M., and Euler, T. (2016). The functional diversity of retinal ganglion cells in the mouse. Nature *529*(7586), 345–350.

Baden, T., Euler, T., and Berens, P. (2020). Understanding the retinal basis of vision across species. Nature Reviews Neuroscience *21*(1), 5–20.

Bae, J.A., Mu, S., Kim, J.S., Turner, N.L., Tartavull, I., Kemnitz, N., Jordan, C.S., Norton, A.D., Silversmith, W.M., Prentki, R., et al. (2018). Digital museum of retinal ganglion cells with dense anatomy and physiology. Cell *173*(5), 1293–1306.e19.

Barrow, J. (1750). Navigatio Britannica, or, A Complete System of Navigation.

Ben-Simon, A., Ben-Shahar, O., Vasserman, G., Ben-Tov, M., and Segev, R. (2012). Visual acuity in the archerfish: behavior, anatomy, and neurophysiology. Journal of Vision *12*(12), 18.

Berson, D.M., Pu, M., and Famiglietti, E.V. (1998). The zeta cell: a new ganglion cell type in cat retina. Journal of Comparative Neurology *399*(2), 269–288.

Best, F. (1900). Ueber die Grenze der Erkennbarkeit von Lagenunterschieden. Albrecht von Græfe's Archiv Für Ophthalmologie *51*, 453–460.

Blake, R., Cool, S.J., and Crawford, M.L.J. (1974). Visual resolution in the cat. Vision Research *14*(11), 1211–1217.

Calkins, D.J., and Sterling, P. (2007). Microcircuitry for two types of achromatic ganglion cell in primate fovea. Journal of Neuroscience *27*(10), 2646–2653.

Campbell, F.W., and Green, D.G. (1965). Optical and retinal factors affecting visual resolution. The Journal of Physiology *181*(3), 576–593.

Cooler, S., and Schwartz, G.W. (2020). An offset ON-OFF receptive field is created by gap junctions between distinct types of retinal ganglion cells. Nature Neuroscience *24*, 105–115.

Dacey, D.M., Peterson, B.B., Robinson, F.R., and Gamlin, P.D. (2003). Fireworks in the primate retina: in vitro photodynamics reveals diverse LGN-projecting ganglion cell types. Neuron *37*(1), 15–27.

Ellis, E.M., Gauvain, G., Sivyer, B., and Murphy, G.J. (2016). Shared and distinct retinal input to the mouse superior colliculus and dorsal lateral geniculate nucleus. Journal of Neurophysiology *116*(2), 602–610.

Famiglietti, E.V. (2005). Synaptic organization of complex ganglion cells in rabbit retina: type and arrangement of inputs to directionally selective and local-edge-detector cells. Journal of Comparative Neurology *484*(4), 357–391.

Flanders, M. (1985). Visually guided head movement in the African chameleon. Vision Research *25*(7), 935–942.

Gaffney, M.F., and Hodos, W. (2003). The visual acuity and refractive state of the American kestrel (*Falco sparverius*). Vision Research *43*(19), 2053–2059.

Gale, S.D., and Murphy, G.J. (2014). Distinct representation and distribution of visual information by specific cell types in mouse superficial superior colliculus. The Journal of Neuroscience *34*(40), 13458–13471.

Gauthier, J.L., Field, G.D., Sher, A., Greschner, M., Shlens, J., Litke, A.M., and Chichilnisky, E.J. (2009). Receptive fields in primate retina are coordinated to sample visual space more uniformly. PLoS Biology 7(4), 0747–0755.

Geisler, W.S. (1984). Physical limits of acuity and hyperacuity. Journal of the Optical Society of America A *1*(7), 775.

Harkness, L. (1977). Chameleons use accommodation cues to judge distance. Nature *267* (5609), 346–349.

Hoy, J.L., Bishop, H.I., and Niell, C.M. (2019). Defined cell types in superior colliculus make distinct contributions to prey capture behavior in the mouse. Current Biology *29*(23), 4130–4138.e5.

Jacoby, J., and Schwartz, G.W. (2017). Three small-receptive-field ganglion cells in the mouse retina are distinctly tuned to size, speed, and object motion. The Journal of Neuroscience *37*(3), 610–625.

Jiang, H., Cottaris, N., Golden, J., Brainard, D., Farrell, J.E., and Wandell, B.A. (2017). Simulating Retinal Encoding: Factors Influencing Vernier Acuity. In IS and T International Symposium on Electronic Imaging Science and Technology, pp. 177–181.

Kolb, H. (1995). Photoreceptors. In Webvision: The Organization of the Retina and Visual System..

Lee, B.B., Wehrhahn, C., Westheimer, G., and Kremers, J. (1995). The spatial precision of macaque ganglion cell responses in relation to vernier acuity of human observers. Vision Research *35*(19), 2743–2758.

Lettvin, J., Maturana, H., McCulloch, W., and Pitts, W. (1959). What the frog's eye tells the frog's brain. Proceedings of the IRE *47*(11), 1940–1951.

Levi, D.M. (2008). Crowding—an essential bottleneck for object recognition: a mini-review. Vision Research *48*(5), 635–654.

Levick, W.R. (1967). Receptive fields and trigger features of ganglion cells in the visual streak of the rabbit's retina. The Journal of Physiology *188*(3), 285–307.

Miller, P.E., and Murphy, C.J. (1995). Vision in dogs. Journal of the American Veterinary Medical Association *207*(12), 1623–1634.

Over, R., and Moore, D. (1981). Spatial acuity of the chicken. Brain Research *211*(2), 424–426.

Patterson, B.W., Abraham, A.O., MacIver, M.A., and McLean, D.L. (2013). Visually guided gradation of prey capture movements in larval zebrafish. Journal of Experimental Biology *216*(16), 3071–3083.

Puller, C., Manookin, M.B., Neitz, J., Rieke, F., and Neitz, M. (2015). Broad thorny ganglion cells: a candidate for visual pursuit error signaling in the primate retina. Journal of Neuroscience *35*(13), 5397–5408.

Reymond, L. (1985). Spatial visual acuity of the eagle *Aquila audax*: a behavioural, optical and anatomical investigation. Vision Research *25*(10), 1477–1491.

Richards, J.D., and Short, J. (1998). Wedge-tailed eagle *Aquila audax* predation on endangered mammals and rabbits at Shark Bay, Western Australia. Emu *98*(1), 23–31.

Roy, S., Jun, N.Y., Davis, E.L., Pearson, J., and Field, G.D. (2021). Inter-mosaic coordination of retinal receptive fields. Nature *592*, 409–413. https://doi.org/10.1038/s41586-021-03317-5.

Russell, T.L., and Werblin, F.S. (2010). Retinal synaptic pathways underlying the response of the rabbit local edge detector. Journal of Neurophysiology *103*(5), 2757–2769.

Schall, J.D. (1995). Neural basis of saccade target selection. Reviews in the Neurosciences *6* (1), 63–85.

Schwartz, G.W., Okawa, H., Dunn, F.A., Morgan, J.L., Kerschensteiner, D., Wong, R.O., and Rieke, F. (2012). The spatial structure of a nonlinear receptive field. Nature Neuroscience *15*(11), 1572–1580.

Segev, R., Schneidman, E., Goodhouse, J., and Berry, M.J. (2007). Role of eye movements in the retinal code for a size discrimination task. Journal of Neurophysiology *98*(3), 1380–1391.

Semmelhack, J.L., Donovan, J.C., Thiele, T.R., Kuehn, E., Laurell, E., and Baier, H. (2014). A dedicated visual pathway for prey detection in larval zebrafish. eLife *3*,e04878.

Shapley, R., and Victor, J. (1986). Hyperacuity in cat retinal ganglion cells. Science *231*(4741), 999–1002.

Sivyer, B., Venkataramani, S., Taylor, W.R., and Vaney, D.I. (2011). A novel type of complex ganglion cell in rabbit retina. Journal of Comparative Neurology *519*(16), 3128–3138.

Soo, F.S., Schwartz, G.W., Sadeghi, K., and Berry, M.J. (2011). Fine spatial information represented in a population of retinal ganglion cells. Journal of Neuroscience *31*(6), 2145–2155.

Strasburger, H., Huber, J., and Rose, D. (2018). Ewald Hering's (1899) on the limits of visual acuity: a translation and commentary: with a supplement on Alfred Volkmann's (1863) physiological investigations in the field of optics. I-Perception *9*(3). 10.1177/2041669518763675.

Timmermans, P.J.A. (2009). Prey catching in the archer fish: marksmanship, and endurance of squirting at an aerial target. Netherlands Journal of Zoology *50*(4), 411–423.

Timney, B., and Keil, K. (1992). Visual acuity in the horse. Vision Research *32*(12), 2289–2293.

Van Hof, M.W. (1967). Visual acuity in the rabbit. Vision Research 7(9–10), 749–751.

van Wyk, M., Taylor, W.R., and Vaney, D.I. (2006). Local edge detectors: a substrate for fine spatial vision at low temporal frequencies in rabbit retina. Journal of Neuroscience *26*(51), 13250–13263.

Wang, L., Sarnaik, R., Rangarajan, K., Liu, X., and Cang, J. (2010). Visual receptive field properties of neurons in the superficial superior colliculus of the mouse. Journal of Neuroscience *30*(49), 16573–16584.

Westheimer, G. (1975). Visual acuity and hyperacuity. Investigative Ophthalmology *14*(8), 570–572.

Westheimer, G. (1981). Visual Hyperacuity. In Progress in Sensory Physiology (Springer), pp. 1–30.

Westheimer, G. (2009). Hyperacuity. In Encyclopedia of Neuroscience (Elsevier), pp. 45–50.

Westheimer, G. (2012). Optical superresolution and visual hyperacuity. Progress in Retinal and Eye Research *31*(5), 467–480.

Westheimer, G., and McKee, S.P. (1977). Spatial configurations for visual hyperacuity. Vision Research *17*(8), 941–947.

Westheimer, G., and Pettet, M.W. (1990). Contrast and duration of exposure differentially affect vernier and stereoscopic acuity. Proceedings of the Royal Society B: Biological Sciences *241*(1300), 42–46.

Wiesenfeld, Z., and Branchek, T. (1976). Refractive state and visual acuity in the hooded rat. Vision Research *16*(8), 823–827.

Xu, Y., Dhingra, N.K., Smith, R.G., and Sterling, P. (2005). Sluggish and brisk ganglion cells detect contrast with similar sensitivity. Journal of Neurophysiology *93*(5), 2388–2395.

Yoshimatsu, T., Schröder, C., Nevala, N.E., Berens, P., and Baden, T. (2020). Fovea-like photoreceptor specializations underlie single UV cone driven prey-capture behavior in zebrafish. Neuron *107*(2), 320–337.e6.

Zeck, G.M., Xiao, Q., and Masland, R.H. (2005). The spatial filtering properties of local edge detectors and brisk-sustained retinal ganglion cells. European Journal of Neuroscience *22*(8), 2016–2026.

Zhang, Y., Kim, I.J., Sanes, J.R., and Meister, M. (2012). The most numerous ganglion cell type of the mouse retina is a selective feature detector. Proceedings of the National Academy of Sciences of the United States of America *109*(36), E2391–E2398.

Orientated edges are important cues for navigation and object perception. While orientation selectivity was made famous by early work in the visual cortex, the computation begins in the retina. Separate populations of retinal neurons represent horizontal and vertical orientations, perhaps to guide animals as they navigate through environments like forests where these orientations are abundantly represented by linear objects such as trees.

CHAPTER 10

Orientation selectivity

Gregory William Schwartz[a,b,c]
[a]Departments of Ophthalmology and Physiology, Feinberg School of Medicine, Northwestern University, Chicago, IL, United States
[b]Department of Neurobiology, Weinberg College of Arts and Sciences, Northwestern University, Chicago, IL, United States
[c]Northwestern Interdepartmental Neuroscience Program (NUIN), Northwestern University, Chicago, IL, United States

Orientation selectivity (OS) is one of the best-known computations in vision science, largely because of the Nobel Prize-winning work of Hubel and Wiesel in cat visual cortex (Hubel and Wiesel, 1962; Priebe, 2016). OS in the retina has received much less attention even though its discovery was contemporaneous with Hubel and Wiesel's work (Levick, 1967; Maturana and Frenk, 1963). For nearly five decades after the first papers describing OS retinal ganglion cells (RGCs), there was only fragmentary data about the mechanism of retinal OS. The past decade has seen a resurgence of interest in retinal OS, yet central questions remain both about circuit mechanisms within the retina and the readout of OS RGCs in the brain. An excellent recent review summarized progress in the identification of OS retinal circuits across vertebrates (Antinucci and Hindges, 2018).

OS retinal ganglion cells

Retinal ganglion cells responding selectively to horizontally oriented stimuli were first discovered in the retina of the pigeon (Fig. 10.1A) (Maturana and Frenk, 1963). In the rabbit retina, Levick (1967) reported both horizontally and vertically selective OS RGCs (Fig. 10.1B). Both OS RGC types in rabbits were confirmed in studies from different groups (Amthor et al., 1989; Bloomfield, 1994; Caldwell and Daw, 1978b). Unlike the pinwheels in the visual cortex that represent all orientations (Bonhoeffer and Grinvald, 1991), most reports of OS RGCs in mammals have identified only cell types preferring cardinal (horizontal or vertical) orientations (Fig. 10.2). Zebrafish retina, however, appears to contain OS RGCs with oblique orientation preference as well as the cardinal types (Antinucci et al., 2016).

Retinal Computation
https://doi.org/10.1016/B978-0-12-819896-4.00011-1

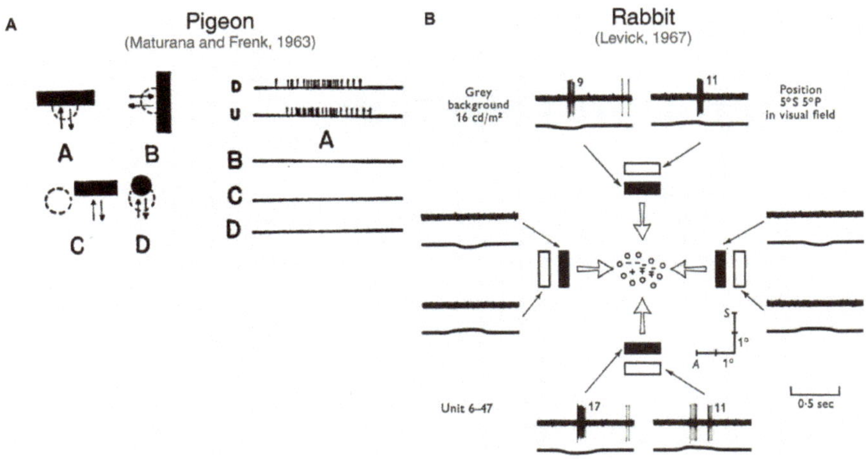

Fig. 10.1 The first reports of OS RGCs in vertebrates. Responses from OS RGCs in pigeon (**A**) and rabbit (**B**). *(From Antinucci and Hindges (2018).)*

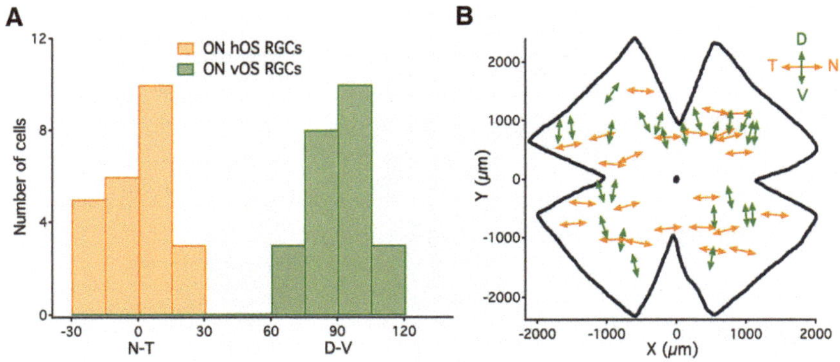

Fig. 10.2 ON OS RGCs in mouse are tuned to either horizontal or vertical orientations. (**A**) Histogram of orientation tuning of ON OS RGCs recorded in mouse retina. (**B**) The position of each cell on the retina is shown with a line indicating its preferred orientation. *(Modified from Nath and Schwartz (2016).)*

Mechanisms for OS in the retina

The most straightforward mechanism for OS in an RGC is for it to sample the mosaic of bipolar cells (BCs) using elongated dendrites. There is a close relationship between the shape of an RGC's dendritic field and its receptive field. Most RGCs have circularly symmetric dendritic fields, but several types have dendritic fields with elongated shapes that could support OS. Horizontally selective OS RGCs with prominent horizontally elongated

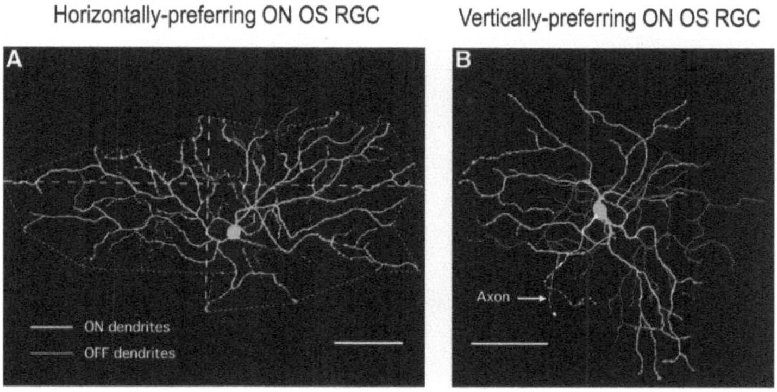

Fig. 10.3 Dendritic morphology of ON OS RGCs in mouse. Images of ON OS RGCs preferring horizontal (**A**) or vertical (**B**) orientations. *Dotted lines* in (**A**) show the outline of the dendritic tree and the major and minor axes used to compute dendritic elongation. Dendrites in the inner (ON) half of the IPL are colored *cyan*, while those in the outer (OFF) half are colored *magenta*. Scale bars = 100 μm. *(Modified from Nath and Schwartz (2016).)*

dendrites have been found in both rabbit (Venkataramani and Taylor, 2016) and mouse (Nath and Schwartz, 2016) (Fig. 10.3). Elongated dendrites alone, however, cannot account for the degree of OS observed in horizontally selective ON OS RGCs, and vertically selective ON OS RGCs in mouse have symmetric dendritic trees (Nath and Schwartz, 2016) as do both orientations of OFF OS RGCs (Nath and Schwartz, 2017), so circuit mechanisms must play a role in the computation.

The first study on the circuit mechanisms of retinal OS used the $GABA_A$ and $GABA_C$ receptor antagonist, picrotoxin, to alter inhibition throughout the retina and reported that OS spike responses in RGCs were abolished (Caldwell and Daw, 1978a). The authors suggested that amacrine cells (ACs) were part of the OS circuit, providing inhibition directly onto RGCs. Other possible circuit mechanisms include OS inhibition onto BC terminals, resulting in orientation preference in the excitatory synaptic input to RGCs and gap junctions between OS ACs and RGCs. All three of these circuit mechanisms, in addition to the intrinsic mechanism of elongated dendrites, work together in the retina to compute OS (Fig. 10.4). However, their relative contributions vary between different types of OS RGCs.

Following the work of Caldwell and Daw, Venkataramani and Taylor (2016) performed a detailed study of the OS mechanism in horizontally selective ON OS RGCs in rabbit. They also found that $GABA_A$ receptors were

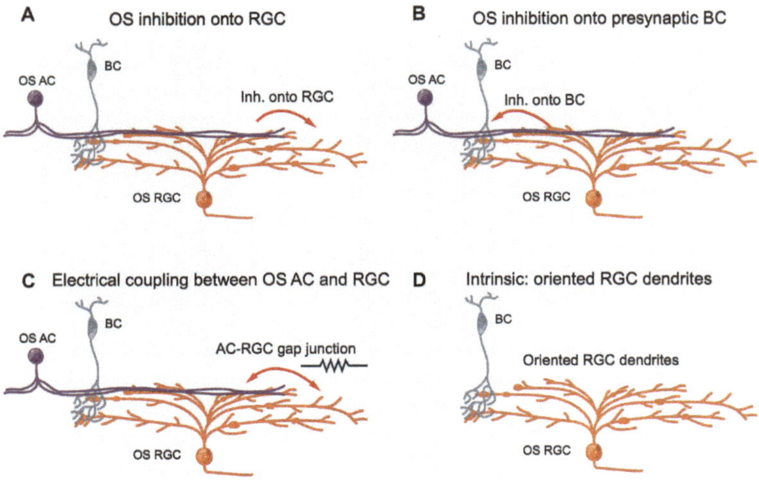

Fig. 10.4 Possible mechanisms of OS. (**A**) OS ACs provide inhibition to an RGC to suppress responses in the null orientation. (**B**) OS ACs provide inhibition to BC terminals presynaptic to the RGC. (**C**) OS ACs are electrically coupled to the OS RGC via gap junctions. (**D**) OS RGCs have oriented dendrites so OS synaptic inputs are not required. Note that these mechanisms are not mutually exclusive.

critical to OS. Voltage-clamp experiments revealed that both excitatory and inhibitory inputs to the RGCs were tuned. Excitation was largest in the preferred orientation, while inhibition was largest in the orthogonal (null) orientation. Orientation tuning in both types of synaptic input was altered by the $GABA_A$ receptor antagonist, gabazine, suggesting that OS inhibition arrives both directly onto RGCs and presynaptically on BC terminals. Both horizontal and vertical ON OS RGCs were found in mice, and as in the rabbit, they also had tuned excitation and inhibition (Fig. 10.5). However, the pharmacological results were different. GABA receptor antagonists reduced the overall amplitude of inhibition, but orientation tuning remained intact. The glycine receptor antagonist, strychnine, similarly reduced the amplitude of inhibition without disrupting orientation tuning. Even with a block of both GABA and glycine receptors, orientation tuning in the excitatory input to mouse ON OS RGCs remained (Nath and Schwartz, 2016).

OFF OS RGCs have similar morphologies in rabbit and mouse retina, and the mechanism leading to OS has been studied in both species. Pharmacological block of either GABA or glycine receptors had mixed results in rabbit OFF OS RGCs, never completely abolishing OS. The authors

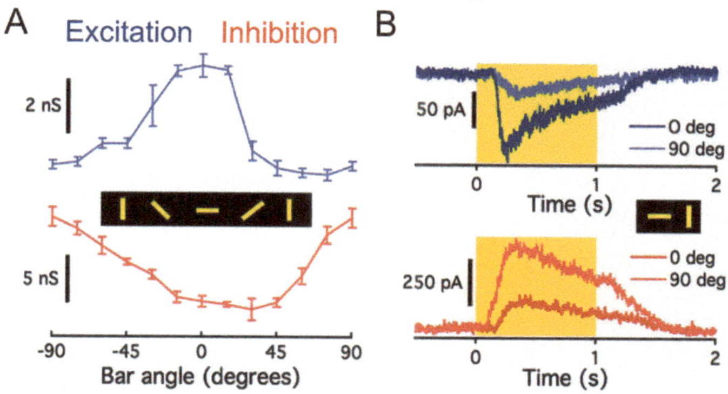

Fig. 10.5 Both excitatory and inhibitory synaptic conductances to ON OS RGCs are orientation tuned. (**A**) Peak excitatory *(blue)* and inhibitory *(red)* conductances as a function of bar angle for a horizontally selective ON OS RGC. (**B**) Example traces of excitation and inhibition for bars at the preferred (0 degrees) and null (90 degrees) orientations. *(Modified from Nath and Schwartz (2016).)*

interpreted their results as suggesting a role for presynaptic inhibition in the OS circuit upstream of these cells (Venkataramani and Taylor, 2010). In mouse retina, Nath and Schwartz (2017) showed that OFF OS RGCs actually inherit their orientation tuning from OS ACs to which they are electrically coupled via gap junctions (Fig. 10.6). Inhibition onto OFF OS RGCs, while present, was not orientation tuned, and excitation was completely absent under this study's stimulus conditions. Thus, pharmacological block of GABA and glycine receptors did not alter OS, while blocking gap junctions, including with the internal connexin-36 antagonist, quinine, dramatically reduced OS (Nath and Schwartz, 2017). Interestingly, the rabbit homolog of vertically selective OFF OS RGCs in mouse, called the G3 cell, is also coupled to highly oriented ACs (Hoshi and Mills, 2009), suggesting that the mechanism of OS might be shared between species. Vertically selective OFF OS RGCs in mouse also go by the name J-RGCs based on their selective labeling in the JAM-B transgenic line (Kim et al., 2008). They were originally identified as direction selective (Kim et al., 2008), but it was later shown that their direction selectivity is quite weak and largely the result of their receptive field shape and their rod-cone color opponency within a range of light levels (Joesch and Meister, 2016). OS in these RGCs is robust across a wide range of sizes and speeds of stimuli and at least seven orders of magnitude of background luminance (Nath and Schwartz, 2017).

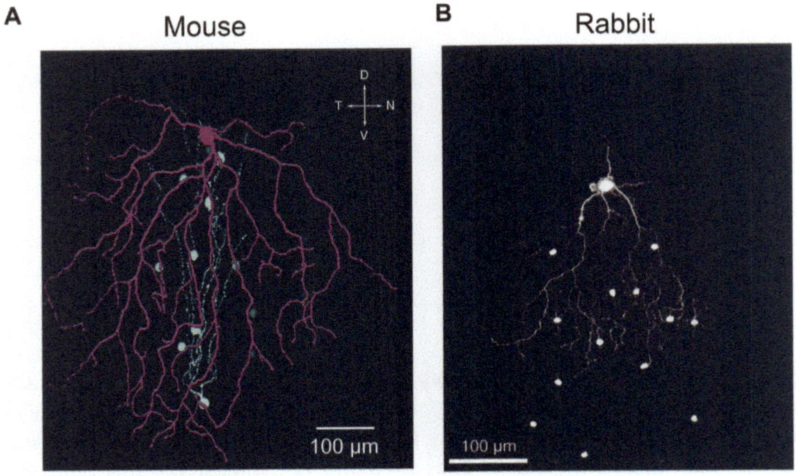

Fig. 10.6 OFF OS RGCs are electrically coupled to oriented ACs via gap junctions. Vertically selective OFF OS RGCs in mouse (**A**) and rabbit (**B**) are both coupled to ACs with oriented neurites. (**A**) The RGC *(magenta)* was filled with neurobiotin and the coupled ACs were traced and pseudocolored *(cyan)*. (**B**) The RGC and coupled ACs were labeled with the gap-junction permeable fluorescent dye Po-Pro-1. *(Modified from (A) Nath and Schwartz (2017) and (B) Hoshi and Mills (2009).)*

OS amacrine cells

All the circuit mechanisms for retinal OS described so far involve ACs. These ACs must themselves be selective for stimulus orientation, whether they convey that information via inhibition to RGCs, via inhibition to BC terminals, or via gap junctions. Several studies have identified OS ACs in the rabbit (Murphy-Baum and Rowland Taylor, 2015; Bloomfield, 1991, 1994), zebrafish (Antinucci et al., 2013, 2016), and mouse (Nath and Schwartz, 2017). OS ACs have a variety of morphologies, suggesting that they represent multiple cell types. At least three morphological types have been observed in the rabbit, including both radially symmetric cells and those with elongated neurites along the preferred orientation axis (Fig. 10.7). There is also an extremely well-conserved OS AC type found across phyla from teleost fish to rabbits to mice. We found this cell coupled to OFF OS RGCs, and we named it the "comet" AC based on its distinctive shape that looks like a comet with a long tail (Nath and Schwartz, 2017). Many different researchers have identified very similar-looking cells working in different species and with different techniques (Fig. 10.8), suggesting that some aspects of retinal OS circuitry are evolutionarily ancient. Only a few studies have reported on the mechanisms of OS in ACs, and they generally mirror the findings in RGCs

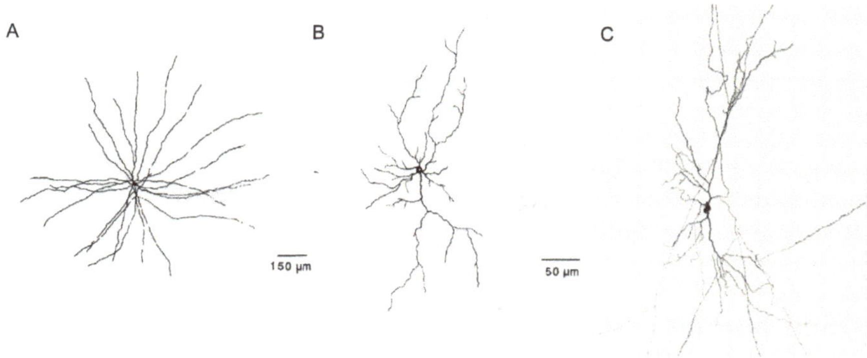

Fig. 10.7 OS ACs in rabbit. (**A–C**) Three different morphological types of OS ACs in the rabbit. *(Modified from (A, B) Bloomfield (1994) and (C) Murphy-Baum and Rowland Taylor (2015).)*

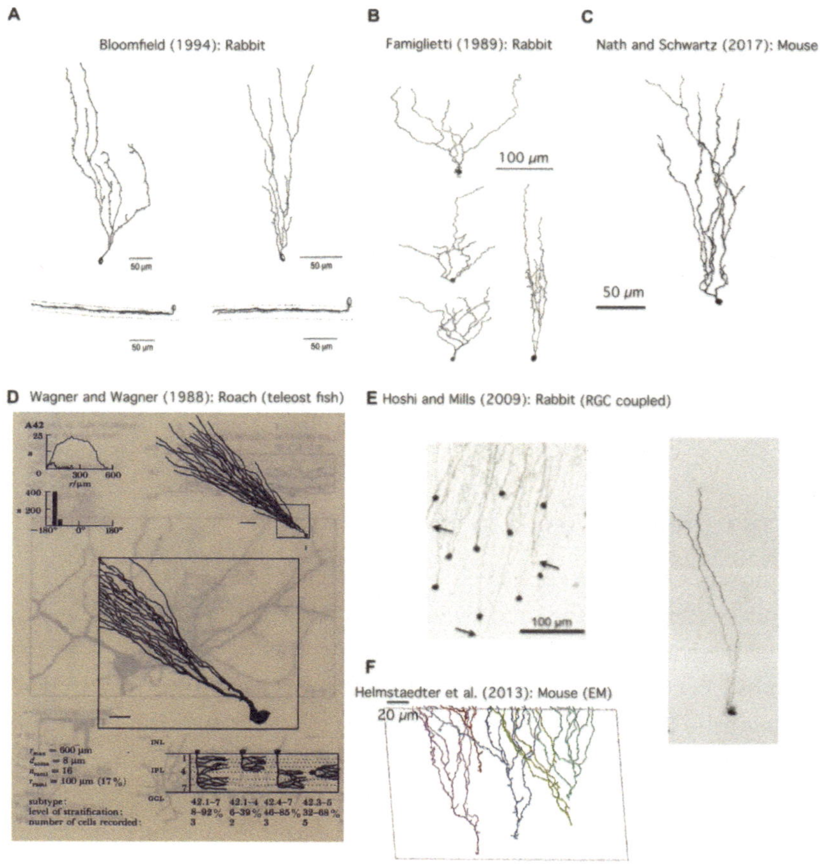

Fig. 10.8 The comet AC is conserved from fish to mammals. Images of comet ACs from different studies and different species as indicated above each panel. *(Modified from (A) Bloomfield (1994), (B) Famiglietti (1989), and (C) Nath and Schwartz (2017). (D) From Wagner and Wagner (1988). Modified from (E) Hoshi and Mills (2009) and (F) Helmstaedter et al. (2013).)*

with some differences in the details. ACs with extremely oriented neurites, like the comet ACs, can achieve strong OS from their morphology alone, while more symmetric types rely on OS synaptic input, presumably mediated by other ACs (Bloomfield, 1991, 1994; Murphy-Baum and Rowland Taylor, 2015).

OS in bipolar cell outputs

As described earlier, orientation tuning has been measured in the excitatory input to RGCs and OS ACs. This result implies that BC outputs can carry OS information. Orientation tuning in BC outputs was confirmed directly in zebrafish retina by imaging glutamate release from BC terminals (Johnston et al., 2019). In most cases where it has been measured, pharmacology has shown that OS excitation in BCs relies on inhibition from ACs (Bloomfield, 1991; Johnston et al., 2019; Venkataramani and Taylor, 2010). However, OS excitation to ON OS RGCs in mouse was unaffected by blockers of GABA and glycine receptors (Nath and Schwartz, 2016).

Several important questions remain unanswered with so few studies on the origin of OS in BC terminals. Which AC types inhibit BC terminals? Are they the same cells that directly inhibit OS RGCs? If so, how is specific wiring achieved so that the excitatory and inhibitory inputs to OS RGCs have opposite tuning? Is there a dedicated OS BC type, or do orientation tuned and untuned terminals coexist on the same BC? Finally, is there a mechanism relying on gap junctions or a non-GABA and non-glycine receptor synaptic pathway that accounts for OS excitation in the mouse retina?

How is retinal OS used in the brain?

Even though OS in the retina was discovered in the 1960s, work on OS in the visual system over the next five decades was dominated by the assumption that the computation began in the visual cortex. While there is certainly a form of OS computed in the cortex through a mechanism very similar to that originally proposed by Hubel and Wiesel (Lien and Scanziani, 2013), a set of studies over the past decade has identified OS responses in subcortical visual areas that do not rely on feedback from the cortex, suggesting a retinal origin. In the dorsal lateral geniculate nucleus (dLGN) of the mouse, three separate groups reported OS responses tuned to the cardinal orientations (Fig. 10.9) (Marshel et al., 2012; Piscopo et al., 2013; Zhao et al., 2013). OS responses were also found in the koniocellular layer of marmoset dLGN (Cheong et al., 2013).

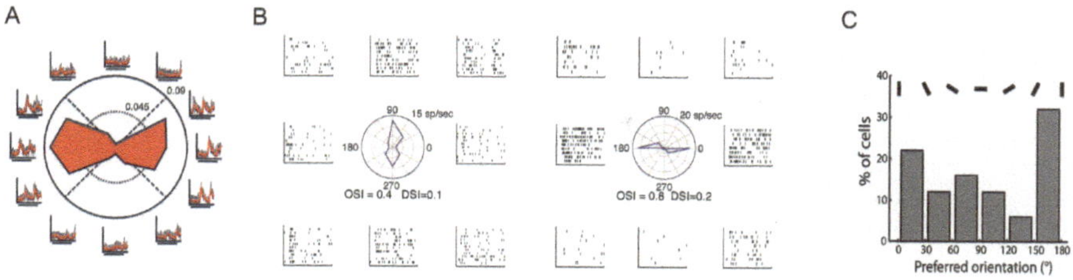

Fig. 10.9 OS is present at the level of dLGN. (**A**) Calcium responses from a neuron in mouse dLGN to gratings drifting in different directions. (**B**) Spike responses from mouse dLGN neurons to drifting gratings show vertical *(left)* or horizontal *(right)* OS. (**C**) Distribution of preferred orientation among neurons recorded in mouse dLGN. *(Modified from (A) Marshel et al. (2012), (B) Piscopo et al. (2013), and (C) Zhao et al. (2013).)*

OS responses have also been identified in the brain's other main retinal target: superior colliculus (SC). Unlike the cardinal tuning found in dLGN, OS neurons in SC represent a range of orientations (Ahmadlou and Heimel, 2015; de Malmazet et al., 2018; Feinberg and Meister, 2015) raising the possibility that local computations within SC alter OS from the retina. A relationship between position in visual space and orientation preference in SC has led to the suggestion that OS in this region is used to represent optic flow and aid in navigation (Fig. 10.10) (Ahmadlou and Heimel, 2015; de Malmazet et al., 2018).

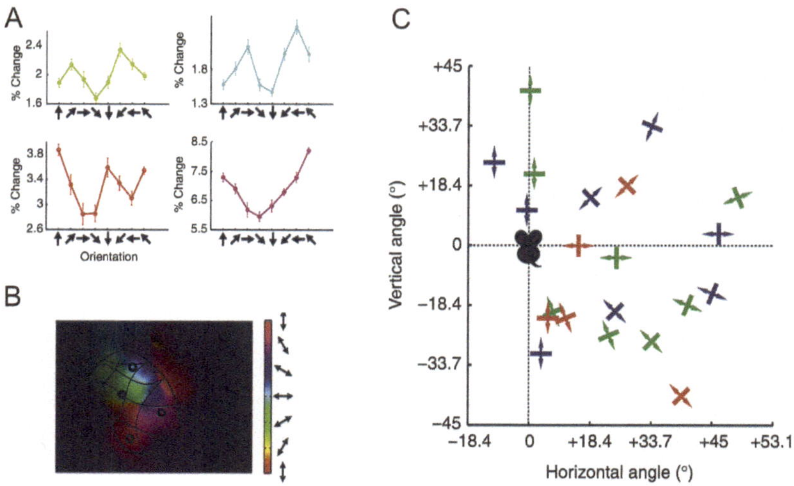

Fig. 10.10 Orientation tuning in SC depends on retinal position. (**A**) Tuning curves of four different neurons in mouse SC in response to drifting gratings. Measurements are made with a calcium sensor. (**B**) Map of orientation tuning across the surface of SC. (**C**) Each cell is plotted at the position of its RF in visual space. *Bars and arrows* depict the orientation of the bars of a grating stimulus and its axis of motion, respectively, for the preferred orientation. *(Modified from Ahmadlou and Heimel (2015).)*

Landmark paper: Bloomfield, S.A. (1991). Two types of orientation-sensitive responses of amacrine cells in the mammalian retina. Nature *350*(6316), 347–350.

Discussion questions:

1. What is the key distinction between the two types of OS ACs described in this study? List the other differences between these types, both functional and morphological, in a table. Speculate about how these differences could support different functional roles.
2. The "symmetric" OS amacrine cells shown in Fig. 1 of this paper (and Fig. 10.7A) have a distinctive morphology with very long, nonbranching radial neurites. What advantages could this morphology confer for either gathering or transmitting OS information in the retina?
3. The author recorded at the soma and made no distinction between input and output neurites in OS ACs. How could you distinguish inputs from outputs with modern techniques? What might you speculate about the locations of inputs and outputs based on the data in this chapter and the presumed function of these cells?

How is retinal OS used in the brain?

1. What are your speculations on the different roles of retinal OS and cortical OS on behavior? Can you design experiments that would test your hypotheses?
2. While I noted the four main types of OS RGCs found in mammals, horizontal and vertical types of both ON and OFF polarities, I made no distinction between the four types in discussing their projections in the brain. That is because almost nothing is currently known about the specific projection patterns of different OS RGCs. How do you think horizontal versus vertical and ON versus OFF OS information could be either integrated in common target regions or used differently in separate brain regions?

Programming exercise

To access the programming exercise, please visit https://www.elsevier.com/books-and-journals/book-companion/9780128198964.

This exercise implements a model of the receptive field of an ON OS RGC to explore how a variety of factors affect OS. The factors that you will explore include (1) elongation of RGC dendrites in the ON and OFF layers, (2) receptive field size, shape, spacing, nonlinearities of BC subunits providing excitation to the RGC and AC subunits providing inhibition, and (3) the balance of excitation and inhibition in tuning the spike output of the RGC.

The file "ON_OS_model_package.zip" contains the code for this model.

Open the script "ON_OS_model.m." The "parameters" block at the top will enable you to change the relevant parameters in the simulation.

1. Run the code block that creates the RGC dendritic filters across a range of eccentricities from 0 to 0.95. Plot the resulting filters to get a sense of how the value of eccentricity relates to the morphologies of OS cells pictured in this chapter and in the papers it references.
2. As you vary the eccentricity of the ON dendrites, what happens to the orientation selectivity index (OSI) for a linear excitatory model? Is this relationship consistent with the data from mouse ON OS RGCs (Nath and Schwartz, 2016)? Why or why not?
3. Now run the code block that computes the linear response of each BC and AC subunit. Try using different values of eccentricity for BC and AC subunits and see how they affect responses across the subunit array. Use the following command to visualize the output of the subunit array:
   For BCs: scatter(bpX, bpY, [], bipolar_resp(:,i)).
   For ACs: scatter(amX, amY, [], amacrine_resp(:,i)).

   In both cases, *i* is the index (from 1 to 12) that determines which bar angle was presented to the model. Use the "subplot" command to plot all 12 responses on the same graph, labeling the angle on each plot. Make sure you make the color scale the same in all the graphs. What observations can you make about how eccentricity, receptive field size (of the subunits), and subunit spacing contribute to the responses?

Advanced

4. Implement the subunit nonlinearities as cumulative Gaussians with the included functions. Can nonlinear subunit output improve OS? Can you use "nlinfit" to fit the nonlinearity to achieve the OSIs for excitation and inhibition observed in the mouse data (Nath and Schwartz, 2016)?
5. Try combining the excitatory and inhibitory input in a simple leaky integrate-and-fire (LIF) model of the ON OS RGC. What does this model tell you about the relative importance of excitatory and inhibitory tuning for OS in the RGC? Can spike threshold increase the OSI? Is this change in OSI from the LIF model consistent with the mouse data (Nath and Schwartz, 2016)?

References

Ahmadlou, M., and Heimel, J.A. (2015). Preference for concentric orientations in the mouse superior colliculus. Nature Communications *6*, 6773.

Amthor, F.R., Takahashi, E.S., and Oyster, C.W. (1989). Morphologies of rabbit retinal ganglion cells with complex receptive fields. The Journal of Comparative Neurology *280*(1), 97–121.

Antinucci, P., and Hindges, R. (2018). Orientation-selective retinal circuits in vertebrates. Frontiers in Neural Circuits *12*, 11.

Antinucci, P., Nikolaou, N., Meyer, M.P., and Hindges, R. (2013). Teneurin-3 specifies morphological and functional connectivity of retinal ganglion cells in the vertebrate visual system. Cell Reports *5*(3), 582–592.

Antinucci, P., Suleyman, O., Monfries, C., and Hindges, R. (2016). Neural mechanisms generating orientation selectivity in the retina. Current Biology *26*(14), 1802–1815.

Bloomfield, S.A. (1991). Two types of orientation-sensitive responses of amacrine cells in the mammalian retina. Nature *350*(6316), 347–350.

Bloomfield, S.A. (1994). Orientation-sensitive amacrine and ganglion cells in the rabbit retina. Journal of Neurophysiology *71*(5), 1672–1691.

Bonhoeffer, T., and Grinvald, A. (1991). Iso-orientation domains in cat visual cortex are arranged in pinwheel-like patterns. Nature *353*(6343), 429–431.

Caldwell, J.H., and Daw, N.W. (1978a). Effects of picrotoxin and strychnine on rabbit retinal ganglion cells: changes in centre surround receptive fields. The Journal of Physiology *276*(1), 299–310.

Caldwell, J.H., and Daw, N.W. (1978b). New properties of rabbit retinal ganglion cells. The Journal of Physiology *276*(1), 257–276.

Cheong, S.K., Tailby, C., Solomon, S.G., and Martin, P.R. (2013). Cortical-like receptive fields in the lateral geniculate nucleus of marmoset monkeys. Journal of Neuroscience *33*(16), 6874–6876.

de Malmazet, D., Kühn, N.K., and Farrow, K. (2018). Retinotopic separation of nasal and temporal motion selectivity in the mouse superior colliculus. Current Biology *28*(18), 2961–2969.e4.

Famiglietti, E.V. (1989). Structure organization and development of dorsally-directed (vertical) asymmetrical amacrine cells in rabbit retina. In: Weiler, R., Osborne, N.N. (Eds.), Neurobiology of the Inner Retina. Springer, Berlin, pp. 169–180.

Feinberg, E.H., and Meister, M. (2015). Orientation columns in the mouse superior colliculus. Nature *519*(7542), 229–232.

Helmstaedter, M., Briggman, K.L., Turaga, S.C., Jain, V., Seung, H.S., and Denk, W. (2013). Connectomic reconstruction of the inner plexiform layer in the mouse retina. Nature *500*,168–174.

Hoshi, H., and Mills, S.L. (2009). Components and properties of the G3 ganglion cell circuit in the rabbit retina. Journal of Comparative Neurology *513*(1), 69–82.

Hubel, D.H., and Wiesel, T.N. (1962). Receptive fields, binocular interaction and functional architecture in the cat's visual cortex. The Journal of Physiology *160*, 106–154.

Joesch, M., and Meister, M. (2016). A neuronal circuit for colour vision based on rod-cone opponency. Nature *532*(7598), 236–239.

Johnston, J., Seibel, S.-H., Darnet, L.S.A., Renninger, S., Orger, M., and Lagnado, L. (2019). A retinal circuit generating a dynamic predictive code for oriented features. Neuron *102*(6), 1211–1222.e3.

Kim, I.J., Zhang, Y., Yamagata, M., Meister, M., and Sanes, J.R. (2008). Molecular identification of a retinal cell type that responds to upward motion. Nature *452*(7186), 478–482.

Levick, W.R. (1967). Receptive fields and trigger features of ganglion cells in the visual streak of the rabbit's retina. The Journal of Physiology *188*(3), 285–307.

Lien, A.D., and Scanziani, M. (2013). Tuned thalamic excitation is amplified by visual cortical circuits. Nature Neuroscience *16*(9), 1315–1323.

Marshel, J.H., Kaye, A.P., Nauhaus, I., and Callaway, E.M. (2012). Anterior-posterior direction opponency in the superficial mouse lateral geniculate nucleus. Neuron *76*(4), 713–720.

Maturana, H.R., and Frenk, S. (1963). Directional movement and horizontal edge detectors in the pigeon retina. Science *142*(3594), 977–979.

Murphy-Baum, B.L., and Rowland Taylor, W. (2015). The synaptic and morphological basis of orientation selectivity in a polyaxonal amacrine cell of the rabbit retina. Journal of Neuroscience *35*(39), 13336–13350.

Nath, A., and Schwartz, G.W. (2016). Cardinal orientation selectivity is represented by two distinct ganglion cell types in mouse retina. Journal of Neuroscience: The Official Journal of the Society for Neuroscience *36*(11), 3208–3221.

Nath, A., and Schwartz, G.W. (2017). Electrical synapses convey orientation selectivity in the mouse retina. Nature Communications *8*(1), 2025.

Piscopo, D.M., El-Danaf, R.N., Huberman, A.D., and Niell, C.M. (2013). Diverse visual features encoded in mouse lateral geniculate nucleus. Journal of Neuroscience *33*(11), 4642–4656.

Priebe, N.J. (2016). Mechanisms of orientation selectivity in the primary visual cortex. Annual Review of Vision Science *2*(1), 85–107.

Venkataramani, S., and Taylor, W.R. (2010). Orientation selectivity in rabbit retinal ganglion cells is mediated by presynaptic inhibition. Journal of Neuroscience *30*(46), 15664–15676.

Venkataramani, S., and Taylor, W.R. (2016). Synaptic mechanisms generating orientation selectivity in the ON pathway of the rabbit retina. The Journal of Neuroscience: The Official Journal of the Society for Neuroscience *36*(11), 3336–3349.

Wagner, H.J., and Wagner, E. (1988). Amacrine cells in the retina of a teleost fish, the roach (*Rutilus rutilus*): a Golgi study on differentiation and layering. Philosophical Transactions of the Royal Society B: Biological Sciences *321*, 263–324.

Zhao, X., Chen, H., Liu, X., and Cang, J. (2013). Orientation-selective responses in the mouse lateral geniculate nucleus. Journal of Neuroscience *33*(31), 12751–12763.

PART 4

Motion

Visual systems throughout the animal kingdom extract information about the direction of motion, either for individual objects like single butterflies or for the global motion of an entire kaleidoscope. The computation of direction selectivity begins in the retina. In this static image, motion is implied both by thin trajectory lines and by differential blur to indicate speed.

CHAPTER 11

Direction selectivity

Benjamin L. Murphy-Baum[a], Gregory William Schwartz[b,c,d], and Gautam B. Awatramani[a]
[a]Biology Department, University of Victoria, Victoria, BC, Canada
[b]Departments of Ophthalmology and Physiology, Feinberg School of Medicine, Northwestern University, Chicago, IL, United States
[c]Department of Neurobiology, Weinberg College of Arts and Sciences, Northwestern University, Chicago, IL, United States
[d]Northwestern Interdepartmental Neuroscience Program (NUIN), Northwestern University, Chicago, IL, United States

The story of direction selectivity (DS) in the retina began nearly 60 years ago and has remained a highly active research area ever since. DS is arguably the most well-studied computation in the nervous system and therefore provides a fascinating timeline of how our scientific understanding has evolved. Throughout the history of DS research, technical advancements have routinely propelled the field to new levels of understanding. From the introduction of single-unit retinal recordings (1950s) to improved electrode construction and stimulus presentation (1960s), computerized data acquisition (1970s), the whole-mount retinal preparation and supporting media (1980s), the whole-cell patch clamp (1980s to 1990s), two-photon microscopy and Ca^{2+} imaging (1990s to 2000s), serial block-face electron microscopy (2000s to 2010s), and genetically encoded activity sensors (2010 to present), DS research has traversed most of the journey and has benefited from it all. The long trail of studies that emerged over the years and how they wandered, branched, and came together over time is an illustrative example of how the scientific process builds on itself toward understanding complex systems.

A history of direction selectivity in the vertebrate retina

Direction selectivity in the retina was first reported in the frog (Maturana et al., 1960), rabbit (Barlow and Hill, 1963), and pigeon retinas (Maturana and Frenk, 1963) in the early 1960s. These early investigators identified distinct types of retinal ganglion cells (RGCs) that fire a train of action potentials in response to motion in a certain "preferred" direction, but not in the opposite "null" direction (Fig. 11.1; see "Multiple types of DSGCs in the retina"). These studies

Retinal Computation
https://doi.org/10.1016/B978-0-12-819896-4.00012-3

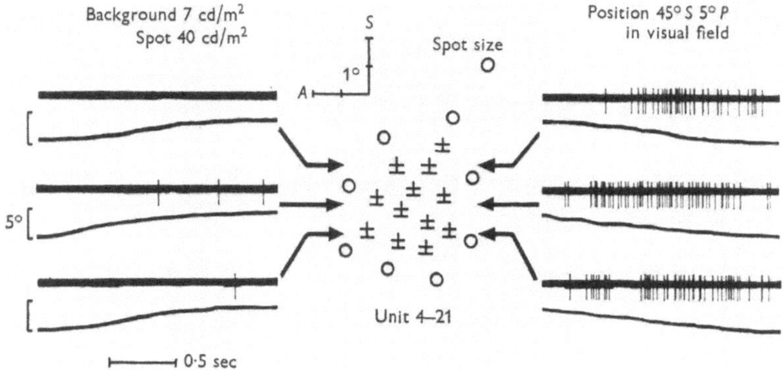

Fig. 11.1 Direction selectivity in rabbit RGCs. Direction-selective spiking from a rabbit DSGC in response to motion in different directions. *(From Barlow and Levick (1965).)*

firmly established the idea that RGCs can extract complex and specific information from the visual scene (Lettvin et al., 1959). This was a significant departure from the idea that complex feature extraction was the province of higher brain areas (Hubel and Wiesel, 1959, 1962), and that RGCs were contrast detectors with simple center-surround receptive fields (Hartline, 1940; Kuffler, 1953).

Insights into the synaptic mechanisms underlying retinal DS first emerged from the pioneering work of Horace Barlow and colleagues (Barlow et al., 1964; Barlow and Levick, 1965; Barlow and Hill, 1963). Their initial work drew comparisons between retinal DS and cortical DS mechanisms, which they argued were fundamentally different. In the cortex, DS neurons were thought to acquire their selectivity from asymmetrically distributed ON and OFF receptive fields that were activated in a particular sequence (Hubel and Wiesel, 1962). However, this mechanism could not be applied to the retina since direction-selective ganglion cells (DSGCs) were found to have coextensive ON and OFF receptive fields, and they retained their directional preference for stimuli of either contrast. Moreover, direction-selective spiking could be elicited for very small movements—about 1/4 degrees of visual angle in rabbits, or ~35 μm on the retina—at nearly any position within the receptive field (Barlow and Levick, 1965). Barlow and Levick proposed that a dozen or more of these "subunits" would cover the receptive field of a given DSGC and that direction-selective spiking could be driven by any of them.

In theory, DS within each receptive field subunit could arise via excitatory or inhibitory mechanisms (Fig. 11.2). Both mechanisms compare activity between two adjacent points in visual space by delaying one input relative to the other, allowing them to coincide only during certain activation sequences. Out of the two schemes, spiking can be either enhanced

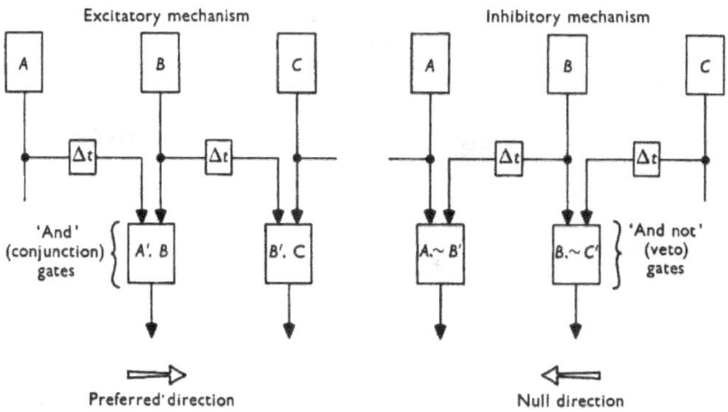

Fig. 11.2 Two potential mechanisms that could produce direction selectivity. *Left*, motion in the preferred direction evokes excitation at point A, which, after a delay, coincides with excitation at point B. These inputs would not coincide and summate for motion in the null direction. *Right*, Barlow and Levick favored an inhibitory mechanism *(right)*, in which delayed—or laterally projected—inhibition from point A coincides with excitation at point B during motion in the null direction. *(From Barlow and Levick (1965).)*

during preferred motion due to the coincidence of two excitatory signals (Reichardt, 1987) or suppressed during null motion due to the coincidence of excitation and inhibition (Fig. 11.2). Barlow and Levick's results from two-slit sequence discrimination experiments favored an inhibitory mechanism for DS, in which laterally projected inhibition coincides with and "vetoes" locally driven excitation during motion in the null, but not preferred, direction (Fig. 11.2 right). Barlow and Levick tentatively suggested that the local excitation was provided by bipolar cells, which were inhibited by laterally projecting horizontal cells during motion in a DSGC's null direction. However, the finding that bipolar cells did not appear to be direction selective (Werblin, 1970) led to the proposal that excitation and inhibition interact at the level of DSGCs. Furthermore, the DSGC receptive field's subunit structure led to the paradigm-shifting notion that synaptic integration may occur within "dendritic subunits" spread across the arbor of DSGCs, rather than at the soma. Indeed such dendritic subunits could be predicted from the cable properties and natural branching patterns of DSGC dendrites (Koch et al., 1983; Marchiafava, 1979; Torre and Poggio, 1978) (see "Dendritic integration of excitation and inhibition").

The idea that an inhibitory mechanism generated DS was bolstered by several studies demonstrating that DS was significantly reduced after exposing the retina to GABA receptor antagonists (Ariel and Daw, 1982; Caldwell et al., 1978; Wyatt and Daw, 1976). Importantly, this inhibition was spatially offset from the more symmetrical excitatory receptive field, in keeping with

Barlow and Levick's original proposal that laterally projected inhibition underlies DS (Ariel and Daw, 1982; Wyatt and Daw, 1975). These studies also revealed two unusual features of the excitatory pathway to DSGCs. First, about half of the excitatory drive was mediated by nicotinic acetylcholine (ACh) receptors (Ariel and Daw, 1982; Masland and Ames, 1976). Second, glutamatergic transmission from bipolar cells appeared to be mediated in large part by NMDA receptors rather than AMPA receptors. Although blocking ACh and NMDA receptors in combination reduced the DSGC spiking response by ~90%, it left DS relatively intact (Kittila and Massey, 1997). Together, these data supported the idea that asymmetric GABAergic inhibition was the most critical factor in generating DS.

After confirming GABA's role in shaping DS in ganglion cells, work over the next 20 years largely focused on pinpointing the source of inhibition. Multiple lines of research using a variety of techniques were required to finally identify the starburst amacrine cell (SAC) as the source of asymmetric GABAergic inhibition. Initially, SACs were not known as GABAergic neurons; autoradiographic studies first described them because of their distinct ability to synthesize ACh (Baughman and Bader, 1977; Hayden et al., 1980; Masland and Mills, 1979). The identified cholinergic neurons comprised two populations located on either side of the inner plexiform layer (IPL). Around the same time, anatomists using Golgi staining methods had identified "starburst" amacrine cells with a radiating dendritic morphology (Fig. 11.3) that also were mirror symmetric about the IPL (Famiglietti, 1983b; Perry and Walker, 1980). Serendipitously, the same population of cholinergic neurons could be selectively labeled by the DAPI stain (Masland et al., 1984b), which allowed SACs to finally be identified as the mirror-symmetric cholinergic neurons (Vaney, 1984).

Interest in SACs grew stronger when it was realized that their dendrites co-fasciculate with those of DSGCs (Famiglietti, 1987, 1983a). This

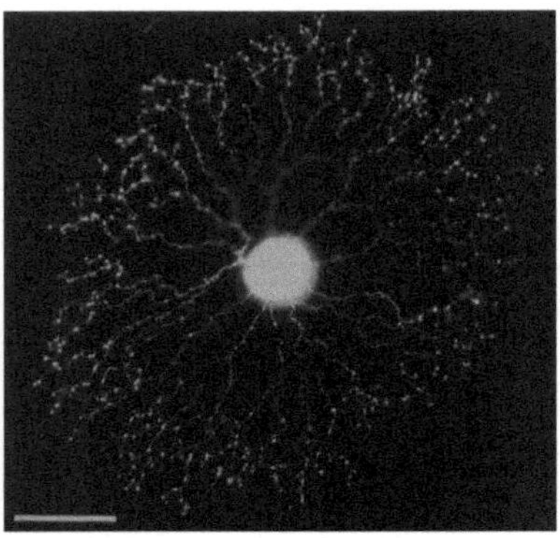

Fig. 11.3 Morphology of the starburst amacrine cell. DAPI-labeled starburst amacrine cell that was filled with a fluorescent dye in the rabbit retina, showing their characteristic radial symmetry, thin proximal dendrites, and distal varicosities. Scale bar = 50 μm. *(From Vaney (1984).)*

suggested that SACs provide synaptic input to DSGCs, which was consistent with functional results showing that DSGCs receive a strong cholinergic excitatory drive (Ariel and Daw, 1982; Masland et al., 1984a; Masland and Ames, 1976). Thus, SACs were first thought to be a major source of excitation to DSGCs rather than the source of asymmetric GABAergic inhibition. However, this line of thinking changed when it was discovered that SACs synthesize both ACh and GABA (Brecha et al., 1988; Kosaka et al., 1988; Vaney and Young, 1988).

The idea that SACs release GABA provided a different context for interpreting their anatomical features and made SACs attractive candidates for mediating DS. For instance, while the excitatory inputs to SACs are distributed throughout their dendritic arbors, their output synapses are localized to their distal varicosities, making them ideal substrates for the laterally projected inhibition required by the Barlow-Levick model of DS (Famiglietti, 1983a, 1991). Since SAC dendrites project radially, each dendrite must make specific connections with DSGCs, as became evident many years later (see "Selective anatomical connectivity between SACs and DSGCs"). Insightful mathematical models based on linear cable theory (Rall, 1964) and the morphology of SACs further suggested that each SAC dendrite would act as an independent unit (Miller and Bloomfield, 1983), and would respond best for motion that traveled from the soma toward the dendritic tips (Borg-Graham and Gryzwacz, 1992). Together, these studies seeded the hypothesis that DS in the retina originates within individual SAC dendrites (see "Mechanisms of DS in the starburst amacrine cell").

At the turn of the century, many of the previously mentioned predictions regarding SAC function were proven to be mostly correct using newly developed techniques. First, selective ablation of SACs using genetic or chemical methods identified SACs as the source of the null-direction inhibition to DSGCs (Amthor et al., 2002; Yoshida et al., 2001). Second, patch-clamp recordings from rabbit DSGCs revealed that the inhibitory currents are significantly larger during null motion, whereas the excitatory currents are slightly larger during preferred motion (Taylor and Vaney, 2002) (Fig. 11.4, see "Mechanisms of direction selectivity in ganglion cells"), consistent with the idea that SACs provide synaptic input to DSGCs that is already direction selective (Borg-Graham, 2001; Borg-Graham and Gryzwacz, 1992). Third, measurements of Ca^{2+} signals in distal SAC varicosities demonstrated that their dendritic tips were indeed direction selective according to their angle relative to the soma (Euler et al., 2002) (Fig. 11.5).

In summary, in the 50 years after the discovery of retinal DS, researchers combined anatomical, electrophysiological, pharmacological,

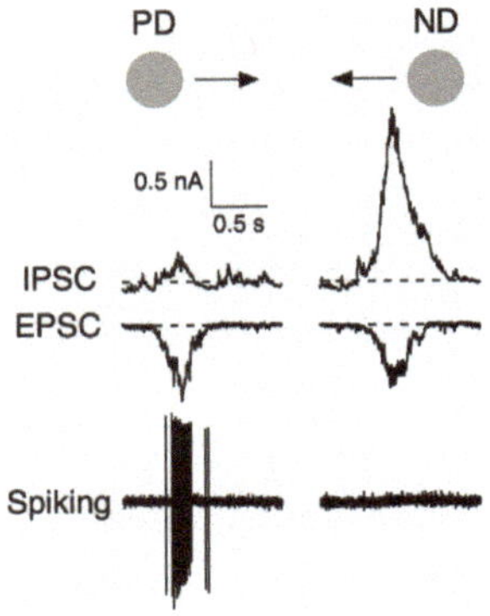

Fig. 11.4 Directional tuning of the input currents to ON-OFF DSGCs and the resulting spiking response. Motion in the null direction evokes larger inhibitory currents, whereas motion in the preferred direction often evokes slightly larger excitatory currents. *(Modified from Jain et al. (2020).)*

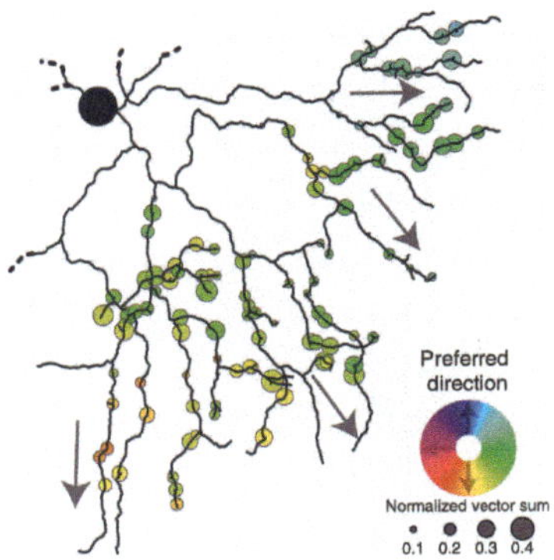

Fig. 11.5 The directional tuning of individual SAC varicosities measured using two-photon calcium imaging. The preferred angle of each varicosity is correlated with the angle that dendrite makes with the soma. *(Modified from Morrie and Feller (2018).)*

and mathematical techniques to build a circuit map that can generate the direction-selective computation. The original theories of DS proved to be remarkably accurate, although the true robustness and complexity of the system could not begin to be fully appreciated until the early 2000s when technical advances allowed experimental studies to catch up with the theoretical predictions. The major mechanistic conclusions from this period were that excitation was suppressed postsynaptically by offset, direction-selective GABAergic inhibition during null motion and that SACs were the likely source of this inhibition. Although researchers had provided a critical explanation for how DS spiking responses are generated, there were many questions about the circuit that had yet to be tested. For instance, how do SAC dendrites acquire their DS? Does their predicted asymmetric connectivity with DSGCs hold true? With direction-selective GABAergic inhibition established, what are the roles of ACh and co-transmission in general? Finally, how are excitatory and inhibitory inputs integrated in DSGC dendrites, and what does it mean for visual processing? Work over the last two decades has provided important insights into these questions and is discussed in detail in the following sections.

Multiple types of DSGCs in the retina

In some of the earliest recordings of DSGCs, it was recognized that they come in both ON and ON-OFF varieties. ON-OFF DSGCs fire action potentials transiently in response to both positive and negative contrast, whereas ON DSGCs exhibit a sustained firing pattern only in response to positive contrast (Barlow et al., 1964; but see Kanjhan and Sivyer, 2010). The dendritic arbors of ON DSGCs are generally much larger than those of ON-OFF DSGCs, and they also have weaker receptive field surrounds. Consequently, ON-OFF DSGCs tend to have better spatial selectivity but respond poorly to globally coherent motion (Ölveczky et al., 2003).

DSGCs in either category do not all have the same preferred direction. A systematic analysis showed that the ON-OFF DGSCs prefer either of four cardinal directions (anterior, posterior, superior, and inferior motion), whereas the ON DSGCs prefer either of three directions (anterior, superior, and inferior; Oyster, 1968; Oyster et al., 1972; Oyster and Barlow, 1967; Fig. 11.6). However, the distribution of ON DSGCs has recently been challenged to include four subtypes (Sabbah et al., 2017). The specific directions preferred by ON and ON-OFF DSGCs were recognized to have a close association with the axes of the extraocular muscles (Oyster, 1968; Oyster

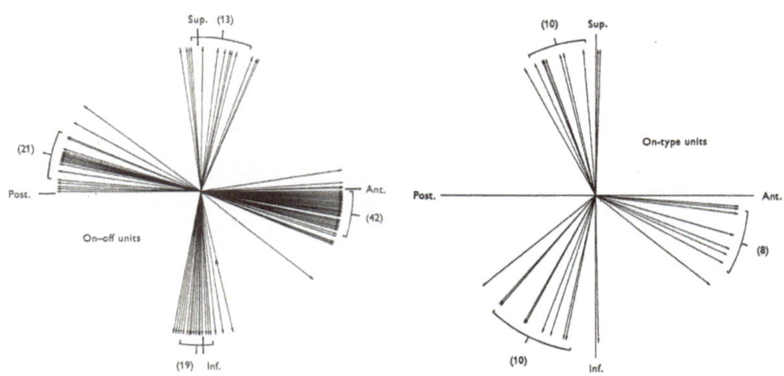

Fig. 11.6 Directional preferences of DSGCs. Distribution of directional preferences in the rabbit retina for ON-OFF *(left)* and ON *(right)* DSGCs. *(From Oyster (1968).)*

and Barlow, 1967) as well as with the axes of the vestibular semicircular canals (Dhande et al., 2013), suggesting that DSGCs may play a role in controlling reflexive eye movements. Relevant to this theory was the finding that ON and ON-OFF DSGCs are sensitive to different motion velocities, with ON DSGCs responding best to slower motion ($<1°/s$ for rabbits) and ON-OFF DSGCs responding better to faster motion ($>1°/s$; Oyster, 1968; Oyster et al., 1972; Sivyer et al., 2010). Their specific velocity tunings corresponded well with the slow and fast phases of the optokinetic reflex (OKR), in which the eye tracks a moving object (slow phase) before resetting to its original position (fast phase; Collewijn, 1969; Collewijn et al., 1972; Oyster, 1968; Oyster et al., 1972). Indeed, disrupting DS in the retina abolishes the OKR in mice, suggesting that DSGCs play an important role (Yonehara et al., 2009; Yoshida et al., 2001).

The idea that DSGCs contribute to the OKR is supported by the central projection patterns of ON DSGCs, which primarily innervate the nuclei of the accessory optic system (AOS) that are thought to generate the OKR (Dhande et al., 2013; Kay et al., 2011; Masseck and Hoffmann, 2009). ON-OFF DSGCs also make projections to nuclei of the AOS, but their primary targets appear to be the superior colliculus and dorsal lateral geniculate nucleus of the thalamus (Cruz-Martín et al., 2014; Dhande et al., 2013; Kay et al., 2011; Rivlin-Etzion et al., 2011). With such diverse projection patterns, directional information from DSGCs may be used for multiple purposes.

In addition to the conventional DSGCs, the molecularly identified JAM-B cells in the mouse are an OFF-type RGC that prefers upward motion in certain stimulus conditions due to their asymmetric receptive field along the same axis (Kim et al., 2008). The F-mini-ON and F-mini-OFF RGCs in the mouse retina also show direction-selective responses. Similar to the JAM-B cells, the dendritic arbors of F-mini RGCs tend to point in their preferred direction, suggesting that DS arises from morphological asymmetries (Rousso et al., 2016). As we will see in the following section, similar dendritic asymmetries likely contribute to DS in SACs, which provide direction-selective input to conventional DSGCs.

Mechanisms of DS in the starburst amacrine cell

SAC dendrites are the first points in the visual system where DS is observed and are, therefore, critical in understanding the direction-selective computation. Their DS is dendrite specific, in which larger dendritic Ca^{2+} signals are measured in the distal dendrites during motion from the soma toward that particular dendritic tip (centrifugal motion; Euler et al., 2002; Morrie and Feller, 2018; Poleg-Polsky et al., 2018; Fig. 11.5). There are many ways that this directional asymmetry might arise, and researchers have identified both cell-autonomous and synaptic mechanisms that may contribute to SAC behavior (reviewed by Taylor and Smith, 2012; Vaney et al., 2012; Wei, 2018).

The cell-autonomous mechanisms—in which SAC dendrites are direction selective due to their membrane properties and morphology—are generally based on the pioneering work of Rall (1964). Rall provided a mathematical framework for understanding neurons' electrical properties and showed that dendrites are sensitive to the spatiotemporal pattern of synaptic input that they receive. Applying the principles of Rall's cable theory to the direction-selective circuit resulted in two general conclusions about SACs. First, their thin, tapering dendrites make it likely that there is significant electrical filtering between the soma and the distal dendrites, where synaptic output occurs (Miller and Bloomfield, 1983). Second, voltage attenuation across this variable resistance will be significantly stronger for signals propagating centripetally than centrifugally (Poznanski, 1992). Consequently, during centrifugal motion, the proximal excitatory inputs are low-pass filtered to produce slow, sustained depolarizations at the distal output sites. These filtered depolarizations sum effectively with the distal excitatory inputs that are subsequently activated as the stimulus moves toward the dendritic tips (Borg-Graham and Gryzwacz, 1992; Tukker et al., 2004; Fig. 11.7). Recently, Vlasits et al. (2016) used glutamate uncaging techniques to sequentially activate excitatory inputs along SAC dendrites, providing direct support for a cell-autonomous mechanism. However, this mechanism is only effective because the synaptic outputs are localized to the distal dendrites, whereas the synaptic inputs are located more proximally (Ding et al., 2016; Vlasits et al., 2016).

The DS arising from passive models is modest compared with both the direction-selective Ca^{2+} signals experimentally measured in SAC dendrites and the direction-selective inhibition measured in DSGCs. Indeed, a variety of additional cell-autonomous mechanisms could act in concert to improve SAC DS (Vaney et al., 2012). Amplification by voltage-gated Ca^{2+} (Hausselt et al., 2007; Tukker et al., 2004) or TTX-insensitive Na^{+} channels (Oesch and Taylor, 2010), improved dendritic isolation through the activation of voltage-gated K^{+} channels (Ozaita et al., 2004), as well as gradients in dendritic voltage (Hausselt et al., 2007) and dendritic chloride

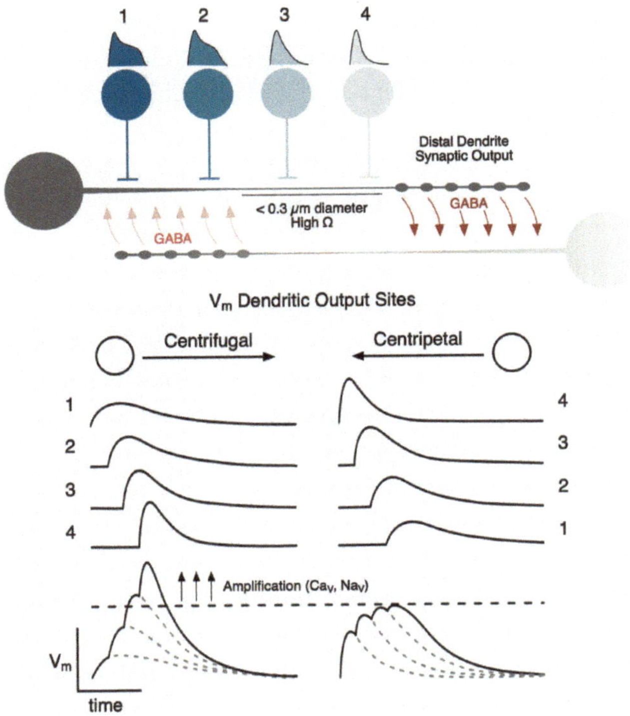

Fig. 11.7 Multiple mechanisms underlie SAC DS. *Top*, a reduced SAC network model indicating the specific wiring of kinetically distinct bipolar cells and reciprocal inhibitory inputs from overlapping SACs. *Bottom*, the voltage response at distal SAC output sites evoked by different sequences of excitatory inputs, as indicated by their numerical order. The input kinetics are shaped presynaptically by different types of bipolar cells or postsynaptically by the filtering properties of the SAC dendrites. In either case, optimal temporal summation at the distal output sites will occur during centrifugal motion. In addition, excitation will arrive at the dendritic tips before reciprocal inhibition during centrifugal motion, whereas during centripetal motion, inhibition and excitation are expected to arrive together, thus cancelling each other.

(Enciso et al., 2010; Gavrikov et al., 2003, 2006) have all been implicated. All of these mechanisms could potentially serve to amplify differences in voltage caused by dendritic filtering mechanisms.

In addition to cell-intrinsic properties, both inhibitory and excitatory network mechanisms could contribute to DS in SAC dendrites. For example, SACs are known to cross-inhibit each other using GABAergic transmission, and this reciprocal inhibition has been shown to suppress responses to centripetal motion (Lee and Zhou, 2006; Münch and Werblin, 2006). An excitatory synaptic mechanism has also gained traction in recent years. In this model, proximal SAC dendrites receive input from distinct sets of sustained bipolar cells, while more distal dendrites receive input from transient bipolar cells (Ding et al., 2016; Fransen and Borghuis, 2017; Greene et al., 2016; Kim et al., 2014; but see Stincic et al., 2016). This arrangement promotes centrifugal responses in a manner similar to the original dendritic filtering mechanism that we discussed earlier (Fig. 11.7).

Owing to the multitude of parallel mechanisms contributing to SAC DS, determining the "core" mechanism appears to be a futile endeavor. Indeed, many of the previously mentioned mechanisms are *sufficient* for generating DS in SAC dendrites, but no single mechanism appears to be strictly *necessary*. For example, reciprocal SAC-SAC inhibition can generate DS (Lee and Zhou, 2006; Münch and Werblin, 2006), but blocking GABA receptors does not abolish DS in SACs (Chen et al., 2016; Euler et al., 2002; Hausselt et al., 2007; Oesch and Taylor, 2010; Poleg-Polsky et al., 2018). Similarly, bipolar cell mechanisms alone can generate SAC DS (Kim et al., 2014), but SAC output remains DS in the absence of bipolar cell input when SACs are directly stimulated using optogenetics (Sethuramanujam et al., 2016). Interestingly, under conditions in which both inhibitory and excitatory network mechanisms are blocked, SAC output is not direction selective (Hanson et al., 2019). At first pass, these findings would imply that the remaining dendritic filtering mechanisms are not important. However, it is likely that the dendritic filtering mechanisms were inadvertently compromised during the optogenetic stimulation used in this study. Thus, it appears that multiple mechanisms contribute to DS in parallel, making the computation more robust over a large variety of stimulus conditions (Chen and Wei, 2018; Vaney et al., 2012; Wei, 2018).

Selective anatomical connectivity between SACs and DSGCs

For a SAC to generate asymmetric inhibition in downstream DSGCs, each of its dendritic branches must selectively connect to particular DSGCs encoding the opposite direction (Borg-Graham and Gryzwacz, 1992; Vaney, 1990; Vaney et al., 1989). Dual patch-clamp recordings revealed that stimulating SACs located on the null, but not preferred, side of a DSGC's receptive field evoked strong inhibitory currents in the DSGC, providing clear evidence for a functional asymmetry in GABAergic connectivity (Fried et al., 2002; Lee et al., 2010). In a landmark study, Briggman et al. (2011) reconstructed a collection of ON-OFF DSGCs and many of their

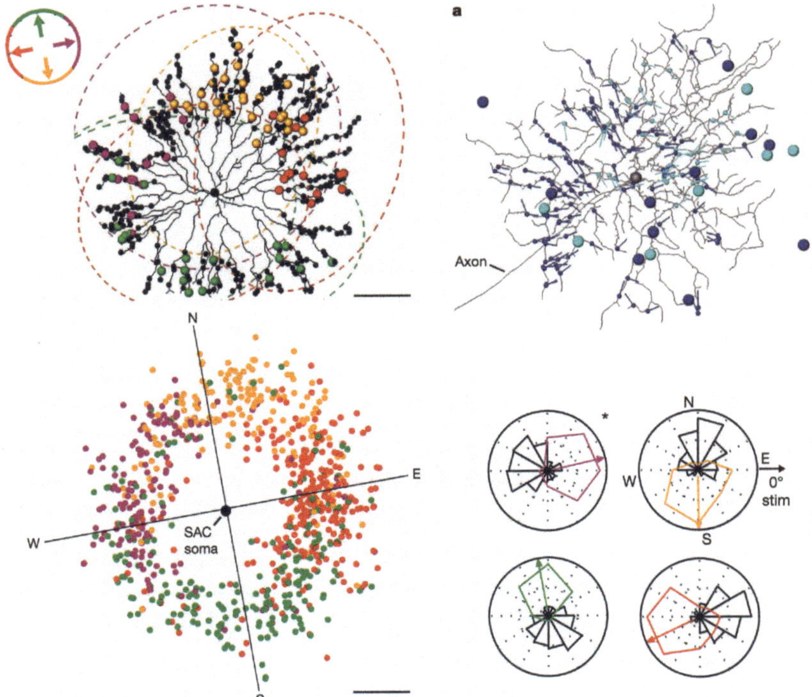

Fig. 11.8 Asymmetric connectivity between SACs and DSGCs. *Top left*, an electron microscopy reconstruction of a SAC with many of its output varicosities labeled according to the directional preference of its postsynaptic DSGCs. *Top right*, reconstructed DSGC with its presynaptic SAC synapses labeled. The angle of the vector at each synapse indicates the angle of the synapse relative to the SAC soma. *Bottom left*, population data from multiple SACs colored as in the *top-left* example. *Bottom right*, directional tuning plots from four different DSGCs. The colors indicate the DSGC's directional preference as measured using Ca^{2+} imaging, and the black traces indicate the distribution of SAC input synapse angles. The angle of SAC inputs predicts the directional preference of the DSGC. Scale bars = 50 μm. *(Modified from Briggman et al. (2011).)*

SAC inputs using serial block-face electron microscopy (SBEM; Fig. 11.8) and provided concrete evidence for a selective "anti-parallel" wiring between SAC dendrites and each subtype of DSGC. For example, they demonstrated that SAC dendrites that point to the left make large "wrap-around" synapses primarily with DSGCs that prefer rightward motion, ensuring that GABA is released onto the DSGC most strongly during motion in its null direction. Thus, a single SAC provides input to all subtypes of DSGC.

The asymmetrically distributed wraparound synapses explain the asymmetric GABAergic inhibition measured electrophysiologically. However, electrophysiological evidence suggests that unlike GABA, the cholinergic input to DSGCs is mostly symmetrical, or at best is weakly biased toward preferred motion (Ariel and Daw, 1982; Brombas et al., 2017; Chen et al., 2016; Lee et al., 2010; but see Pei et al., 2015). This apparent discrepancy between functional and anatomical connectivity led researchers to hypothesize that ACh is transmitted over a broader spatial scale (Briggman et al., 2011; Brombas et al., 2017; Sethuramanujam et al., 2016). Further SBEM analysis revealed that SAC varicosities also make "peripheral" contacts nonselectively with DSGC dendrites, in addition to their specific asymmetric "wraparound" synapses, together constituting a "tripartite" complex (Sethuramanujam et al., 2020). These peripheral contacts do not bear the typical pre- and postsynaptic ultrastructural features, explaining why they were not counted in the previous EM analysis. Nevertheless, these tripartite complexes appear to meditate rapid, synchronized, quantal cholinergic excitatory currents (~1 ms rise time) in pairs of DSGCs encoding opposite directions. Thus, SACs can differentially transmit ACh and GABA to DSGCs through distinct synaptic mechanisms and produce symmetrical excitation patterns but asymmetrical inhibition.

Mechanisms of direction selectivity in ganglion cells

While early work clearly established a central role for inhibition in shaping ganglion cells' direction-selective responses, precisely how inhibition is implemented has been a matter of changing opinion. Early models of DS relied purely on the relative timing of otherwise equivalent excitatory and inhibitory conductances (Barlow and Levick, 1965; Taylor et al., 2000). However, voltage-clamp results emphasized the amplitude differences in inhibition and excitation. Timing differences between excitation and inhibition were often noted, but they were not consistent across the ON and OFF phases of the DSGC's response or across different studies (Fried et al., 2002; Taylor and Vaney, 2002). Importantly, robust DS spiking responses were observed even when timing differences between excitation and inhibition could not be detected (Taylor and Vaney, 2002). Additionally, excitatory currents were linear as a function of voltage, suggesting that they were largely mediated by AMPA-type glutamate receptors (Taylor and Vaney, 2002). Together, these results helped establish a model in which DS GABA release from SAC dendrites plays a critical role in sculpting nondirectional excitatory responses mediated by bipolar cells to generate DS in the dendrites of DSGCs, with ACh playing a more modulatory role (Borst and Euler, 2011; Chen et al., 2016; Demb, 2007; Li et al., 2015; Morrie and Feller, 2018; Park et al., 2014; Yonehara et al., 2013).

Based on the previously mentioned model, it might have been expected that disrupting DS in SAC dendrites would lead to a loss of DS in ganglion cells. On the contrary, under conditions in which GABA release from SACs is rendered nondirectional, DS can still be generated purely based on timing offsets between ACh and GABA (Hanson et al., 2019). Under these conditions, inhibition coincides in time with cholinergic excitation during null-direction motion, presumably because SACs on the null-side of a DSGC's receptive field co-transmit ACh and GABA. However, during preferred direction motion, inhibition arrives later than excitation, providing a window in which robust spiking can occur (Fig. 11.9). These results helped consolidate the idea that both the relative timing and amplitude of inhibition and

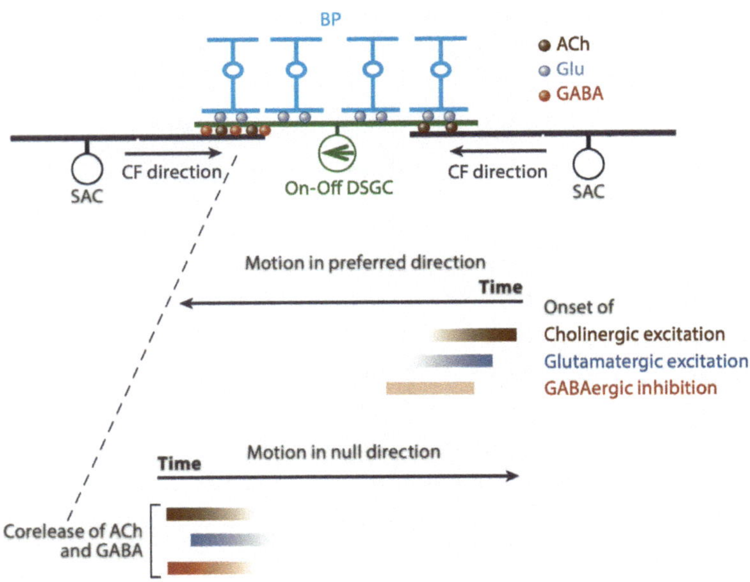

Fig. 11.9 Timing differences arise from receptive field offsets in the DS circuit. *Top*, schematic of the offset SAC connectivity with DSGCs. Note, SACs make contact with DSGCs throughout their dendritic arbor; they are shown on the outskirts of the DSGC's receptive field here for visual clarity. *Bottom*, timing diagram of the different neurotransmitter inputs during preferred and null motion. Note, during preferred motion, both excitatory sources arrive earlier than inhibition, whereas during null motion, inhibition is earlier and directly counters excitation. *(Modified from Chen and Wei (2018).)*

excitation contribute to DS in ganglion cells (Borg-Graham, 2001; Fried et al., 2002; Taylor and Vaney, 2002).

While direction-selective inhibition has been consistently observed, the idea that excitation is direction selective is more controversial. Although voltage-clamp recordings often show excitatory currents that are larger during preferred motion, a computational analysis suggested that this asymmetry was due to voltage-clamp error induced by the large inhibitory conductance evoked by null motion (Poleg-Polsky and Diamond, 2011). However, excitatory currents remained directional after inhibition was strongly reduced in DSGCs (Pei et al., 2015), and no correlation between the amplitude of inhibition and the directionality of excitation was found (Percival et al., 2019). Direction-selective excitatory currents remain after blocking ACh receptors, suggesting that the asymmetry may arise from glutamatergic input from bipolar cells (Percival et al., 2019). Studies using glutamate sensors initially supported the idea that individual glutamatergic inputs to DSGCs are not direction selective in their amplitude (Park et al., 2014; Yonehara et al., 2013), but recent evidence may suggest otherwise (Matsumoto et al., 2020).

Overall, in the current working model of DS, the SAC network produces both amplitude and timing differences between excitation and inhibition that depend on the direction of motion, resulting in direction-selective spiking in postsynaptic DSGCs. The amplitude differences rely on both the DS of SAC dendrites and their asymmetric connectivity with DSGCs. The timing differences are hard-wired into the circuit by the asymmetric GABAergic and cholinergic connections between SACs and DSGCs, as described in the previous section. Finally, it is interesting to note that DS in other sensory areas, including the auditory and somatosensory systems, may also rely on temporal and amplitude asymmetries between excitation and inhibition (Li et al., 2015; Ye et al., 2010; Zhang et al., 2003), but whether these asymmetries are shaped by similar mechanisms remains to be tested.

Dendritic integration of excitation and inhibition

Early mathematical models suggested that DSGCs might integrate synaptic information locally within their dendrites, resulting in multiple computational compartments that operate independently (Koch et al., 1982, 1983; Torre and Poggio, 1978). These studies primarily modeled the ability of inhibition to suppress excitation in a passive dendritic tree. They found that inhibition was most effective at suppressing excitation if the conductances arrived within ~10 ms of each other and if inhibition arrived either at the same dendritic location as excitation or between the excitatory site and the soma (Koch et al., 1983). Importantly, these studies proposed that inhibition acts nonlinearly on excitation as a shunt since the resting membrane potential of DSGCs is expected to be close to the chloride reversal potential, where inhibition has no driving force. Here, any depolarization driven by excitation immediately provides a driving force for open GABA receptors, which then pass chloride, thereby resisting the change in membrane potential. Shunting inhibition is inherently local and nonlinear compared with the more linear effect that hyperpolarizations have on depolarizations when they summate, since open conductances are visible at only half the length constant of voltage (Koch et al., 1990; Rall, 1967; Williams, 2004).

These modeling studies force us to think about when and where synaptic inputs arrive on a DSGC's dendritic arbor. Excitatory inputs on one side of the arbor are unlikely to be suppressed by an inhibitory input on the other side. Likewise, inhibition will fail to suppress excitation if it does not arrive at the appropriate time. To illustrate the problem, consider that activated bipolar cells will drive excitatory inputs onto DSGC dendritic branches directly underneath their receptive fields. By contrast, activated SACs will drive excitatory and inhibitory inputs that are laterally offset from their receptive fields. How is it that inhibition and excitation from different sources arrive close together enough on the dendritic tree that they can interact?

These considerations are important to recognize in the context of more recent work showing that many types of DSGCs initiate action potentials in their dendrites, as opposed to only at the soma (Fig. 11.10). Dendritic spikes are revealed in electrophysiological recordings when somatic action potentials are blocked by locally applying the voltage-gated Na^+ channel blocker, TTX (Oesch et al., 2005; Trenholm et al., 2014). Alternatively, paired somatodendritic patch-clamp recordings were made from rabbit ON DSGCs to record dendritic action potential generation directly (Brombas et al., 2017; Sivyer and Williams, 2013). These studies demonstrated two important aspects of action potential generation in DSGCs. First, dendritic action potentials are already direction selective (Oesch et al., 2005; Sivyer and Williams, 2013). This is a critical feature since it suggests that during null motion, inhibition and excitation are interacting locally in the dendrites, as opposed to at the soma. Second, once a dendritic spike is initiated, it triggers a somatic action potential with high efficiency (Brombas et al., 2017; Sivyer and Williams, 2013), even after unreasonably large inhibitory inputs are activated on its propagation path to the soma (Schachter et al., 2010). Thus, during null motion, dendritic action potentials must be suppressed before they

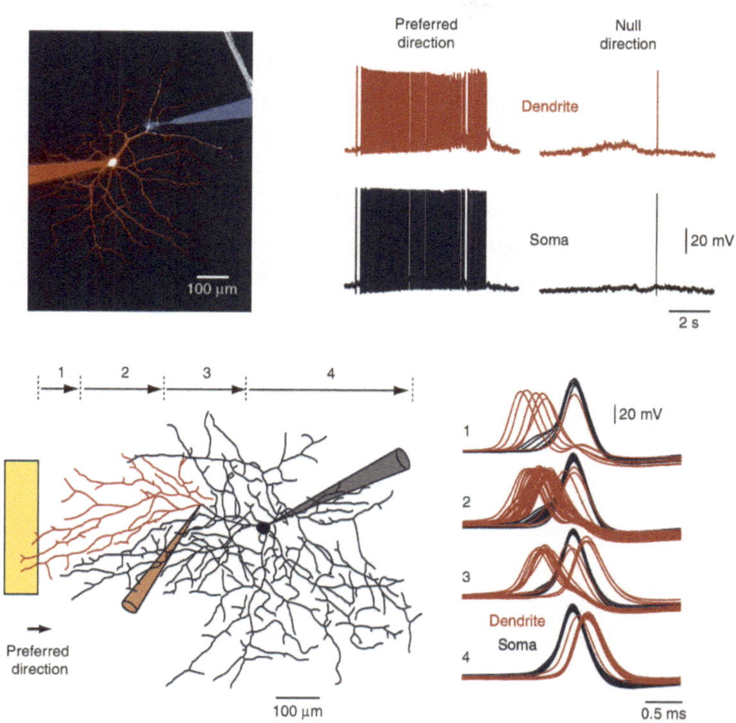

Fig. 11.10 Direction-selective dendritic action potentials in ON DSGCs. Paired somatodendritic recordings in a rabbit ON DSGC show direction-selective dendritic and somatic spikes. *Bottom*, dendritic spikes precede somatic spikes when the stimulus motion is close to the dendritic recording site (positions 1–3). As the stimulus traverses to the other side of the dendritic arbor (position 4), the dendritic electrode is mostly recording back-propagating somatic action potentials. *(Modified from Sivyer and Williams (2013).)*

are initiated, further emphasizing the tight spatiotemporal requirements for excitation and inhibition.

Additional multiplicative nonlinearities in DSGC dendrites are driven by NMDA receptor activity (Poleg-Polsky and Diamond, 2016; Sethuramanujam et al., 2017) which mediates a significant portion of the glutamatergic input to ON-OFF DSGCs (Kittila and Massey, 1997; Tjepkes and Amthor, 2000; but see Percival et al., 2019; Taylor and Vaney, 2002). Interestingly, recent evidence suggests that the cholinergic input from SACs gates NMDA receptor activity by providing the receptors with the necessary depolarization to relieve their Mg^{2+} block (Sethuramanujam et al., 2016), thereby gating local dendritic action potential generation (Brombas et al., 2017). Such a mechanism seems to suggest that during null motion—when there is little depolarization—GABAergic inhibition may not have to directly suppress glutamatergic excitation since those NMDA receptors will remain under Mg^{2+} block as long as GABA successfully suppresses ACh-driven depolarizations (Tjepkes and Amthor, 2000). To this end, GABA and ACh are co-transmitted from the same SAC varicosities (Sethuramanujam et al., 2020) ensuring that their spatiotemporal coupling on the postsynaptic DSGC is tight enough for an effective shunt. Indeed, selectively ablating SACs only disrupts DS locally on DSGC dendrites, suggesting that inhibition and excitation interact over small dendritic domains (Jain et al., 2020).

In summary, two major nonlinearities dominate synaptic integration in DSGC dendrites. First, shunting inhibition nonlinearly *divides* the excitatory input, whereas amplification by voltage-gated conductances nonlinearly *multiplies* the excitatory input (Brombas et al., 2017; Koch et al., 1982, 1983; Oesch et al., 2005; Poleg-Polsky and Diamond, 2016; Sethuramanujam et al., 2016; Sivyer and Williams, 2013; Torre and Poggio, 1978; Vaney et al., 2012). There is evidence that both of these nonlinearities operate locally within the dendritic tree, implying that DSGCs may act as parallel processors: multiple comparisons between excitation and inhibition can occur simultaneously and independently in the dendrites. If this is the case, DSGCs may be capable of resampling the direction of motion multiple times as a stimulus traverses its receptive field, thereby providing the brain with a real-time update on local directional information.

Conclusions and future directions

Over the past 60 years, one of the overarching themes of DS research has been that multiple parallel mechanisms contribute to DS (Chen and Wei, 2018). In the SAC, DS may be generated by dendritic filtering, input-output segregation, bipolar cell distributions, voltage-gated channel expression, and reciprocal inhibition, among other possible mechanisms. In the DSGC, both the relative timing and amplitude of excitation and inhibition play a role in forming direction-selective spiking. However, these parallel mechanisms are not necessarily all active simultaneously. Reciprocal inhibition between antiparallel SAC dendrites, for instance, is more prominent during noisy or more globally active visual conditions than it is during stimulation with small moving objects (Chen et al., 2016). One future goal is to understand how these mechanisms contribute to DS under different visual conditions, particularly in the context of stimuli that an animal is likely to encounter in its natural environment.

As research into the mechanisms of retinal DS progresses, we have begun to appreciate the finer details of the circuit architecture and how it impacts its function. Broad ideas about how excitation and inhibition might generate DS were eventually refined into more specific ideas about where the excitatory and inhibitory synaptic inputs come from, the neurotransmitters involved, the visual conditions under which they are released, the dendritic locations that they are released to, and the particular circuit wiring that allows it to happen. In short, the direction-selective circuit has forced us to think about the "what," "when," "where," and "how" of synaptic transmission more than any other circuit in the retina. The question of "why," however, has only been partially answered, and there is still plenty of work to be done. As was discussed in an earlier section, DSGCs likely contribute to visual behaviors like the OKR. The same level of evidence has yet to be provided for a potential DSGC contribution to the visual component of the vestibulo-ocular reflex (VOR). Interestingly, the OKR and VOR produce ocular movements that are directionally opposed: the OKR follows visual motion, whereas the VOR opposes it. If DSGCs were to contribute to the VOR, it would require differential wiring from that of the OKR, or perhaps an extra inhibitory element. In both instances, however, the goal seems to be to minimize motion on the retina using the ocular muscles, which is compatible with the type of signals that DSGCs deliver. This viewpoint has been around for some time, and DSGCs have been thus treated as velocity-correcting servos, delivering metadata about the visual world rather than forming an image of it (Oyster and Barlow, 1967). Of course, significantly more work is required in circuits downstream of DSGCs to elucidate the full bandwidth of functionality that directional information has.

Finally, our knowledge of the direction-selective circuit is at a point where it makes sense to use DSGCs as a tool for asking more general questions about neuronal function and information processing. It is rare to have a neuron that performs a complex computation on information that researchers can so precisely understand and control in a preparation that is so intact and accessible to both electrophysiological and optical recordings. This allows retinal scientists to ask much more detailed and precise questions than they otherwise could. For instance, if we consider DSGCs as processing directional information, open questions about how DSGCs process opposing information (preferred versus null-direction stimuli) or changing information (changes in direction) are relatively simple to address. Moreover, we can view these computations and how they are disrupted or changed on a dendritic level at predictable locations within the dendritic arbor. The advent of quality two-photon compatible voltage indicators could allow us to dive even deeper into questions of how information is relayed and transformed as it flows through neural circuits.

Landmark paper: Vlasits, A.L., Morrie, R.D., Tran-Van-Minh, A., Bleckert, A., Gainer, C.F., DiGregorio, D.A., and Feller, M.B. (2016). A role for synaptic input distribution in a dendritic computation of motion direction in the retina. Neuron *89*(6), 1317–1330.

Discussion questions:

1. What is the central message of Fig. 1? How does this result in SACs contrast with similar measurements in other retinal neurons like bipolar and ganglion cells?
2. If currents are transmitted very poorly through SAC dendrites, could this be an alternative explanation for the data in Fig. 1? How do the data in Fig. 2 argue against this alternative explanation?
3. What details of the data in Figs. 3 and 4 suggest that the authors measured individual synaptic sites? Where are there disagreements between the two methods of measuring synapses, and what do you think accounts for the differences?
4. What do the simulations in Fig. 5 reveal about how the excitatory synapse distribution impacts DS in the SAC dendrite?
5. Do you think the ~2–3 mV depolarization difference for inward vs. outward motion is sufficient to explain the degree of DS in GABA release? Why or why not? Do the data in Fig. 6 support your conclusion? Why or why not?
6. What accounts for the remaining DS in Fig. 5 with a regular distribution of excitatory inputs?
7. What role do the authors attribute to reciprocal GABAergic inhibition between SACs in the generation of DS in the SAC dendrites? What additional experiments could help test this further? Note: a subsequent study used a genetic strategy to remove SAC-SAC inhibitory synapses (Chen et al., 2016).

Landmark paper: Briggman, K.L., Helmstaedter, M., and Denk, W. (2011). Wiring specificity in the direction-selectivity circuit of the retina. Nature *471*(7337), 183–190.

Discussion questions:

1. Explain in your own words what makes this paper so significant to the field. How did it transform a component of the DS circuit from speculation into fact?

2. Why was it essential to make functional measurements from the exact same cells used in the EM reconstruction? Would separate functional and EM datasets (even much larger ones) have been able to come to the same conclusions?
3. The authors are very careful in their use of the words "contact" and "synapse." What is the difference? How good is the correspondence between contacts and synapses? What figure tests this? Why was the tissue used in the main dataset not "conventionally" stained?
4. What additional detail about wiring specificity is explored in Fig. 6? What do the authors conclude?
5. How do you think this circuit is wired up in development? Is subcellular localization of molecules required?

Landmark paper: Hanson, L., Sethuramanujam, S., de Rosenroll, G., Jain, V., and Awatramani, G.B. (2019). Retinal direction selectivity in the absence of asymmetric starburst amacrine cell responses. eLife *8*, e42392.

Discussion questions:

1. Explain how Fig. 2 shows that both excitatory and inhibitory network mechanisms are responsible for DS in SAC dendrites. Are these results consistent with those in Vlasits et al. (2016)? Why or why not?
2. Why is the Gabra2 KO/ChR2 manipulation (Fig. 2D) such a useful tool for dissecting the DS circuit? How is it different from all previous manipulations of the circuit?
3. What elements of the DS circuit remain intact in the experiments shown in Fig. 3? Given your answer, explain why the finding that DSGCs retain their selectivity in these conditions is so important.
4. What would you predict would happen to the DSGC responses if the authors applied hexamethonium in the Gabra2 KO/ChR2 condition? Why? Would this experiment strengthen the conclusions of the paper?
5. What do the authors conclude about how the two "core" DS mechanisms work together to complement each other? What data support these conclusions? Do you have other ideas about stimulus conditions that might favor one mechanism or the other?

Programming exercise

To access the programming exercise, please visit https://www.elsevier.com/books-and-journals/book-companion/9780128198964.

This exercise examines how selective connectivity and DS in starburst amacrine cell (SAC) dendrites both contribute to the DS computation.

The file "DS_script.m" contains the code. It calls the function "runDSModel," which implements the model, and it loads a temporal filter found in "temporalFilter.mat." The first four sections of DS_script contain the parameters that you will modify to explore their effects on DS.

- SAC_pref_null_ratio defines DS in SAC dendrites. A value of 1.0 makes the dendrites non-DS. Larger values specify the ratio of response amplitude in the centrifugal direction (soma to dendrites) to those in the opposite direction.
- Null_Pref_GABA_ratio is the selective connectivity parameter. It specifies the ratio of SAC GABAergic connectivity on the null side of the DSGC dendritic arbor to that on the preferred side. A value of 1.0 makes SAC connectivity nonselective. ACh input is always nonselective.
- Ach_glut_ratio is the amplitude ratio of ACh (from SACs) versus glutamate (from BCs) excitatory inputs to the DSGC.
- GABA_Ach_ratio is the amplitude ratio of GABA to ACh from SACs onto DSGCs. Larger values create stronger inhibition.
- RF sizes for BCs, SACs, and the DSGC are specified as 1 sigma values in microns.
- Stimulus bar length and speed are specified in microns and microns per second, respectively.

Problems

1. Run the script with its default parameters and look at the resulting plots. What is the evidence of DS spiking? How is it quantified? How does the degree of selectivity of the spiking response compare to that of the glutamate, ACh, and GABA inputs? Explain how this is achieved referencing the plot of interneuron weights.
2. Now remove the DS in SAC dendrites by setting SAC_pref_null_ratio to 1. What happens to inhibition? Is the spike response still DS? Why or why not?
3. Set SAC_pref_null_ratio back to 1.5 and now remove selective wiring by setting Null_Pref_GABA_ratio to 1. What happens to inhibition? Is the spike response DS now? Why or why not? What if you make inhibition stronger by increasing GABA_Ach_ratio while keeping it nonselective? Does DS return?
4. Demonstrate how the ACh to glutamate ratio in the circuit affects DS by collecting DSI values across a range of this parameter and showing representative example plots.
5. How does the role of each part of the DS circuit depend on bar speed? Show plots to back up your conclusions.
6. What happens to DS if the SAC RF is substantially larger or substantially smaller? Explain why.

7. Explore another parameter or set of parameters not explicitly varied in the preceding problems and show how it (they) influences DS. Explain your intuition for these results.

References

Amthor, F.R., Keyser, K.T., and Dmitrieva, N.A. (2002). Effects of the destruction of starburst-cholinergic amacrine cells by the toxin AF64A on rabbit retinal directional selectivity. Visual Neuroscience *19*(4), 495–509.

Ariel, M., and Daw, N.W. (1982). Pharmacological analysis of directionally sensitive rabbit retinal ganglion cells. The Journal of Physiology *324*(1), 161–185.

Barlow, H.B., and Hill, R.M. (1963). Selective sensitivity to direction of movement in ganglion cells of the rabbit retina. Science *139*(3553), 412–414.

Barlow, H.B., and Levick, W.R. (1965). The mechanism of directionally selective units in rabbit's retina. The Journal of Physiology *178*(3), 477–504.

Barlow, H.B., Hill, R.M., and Levick, W.R. (1964). Retinal ganglion cells responding selectively to direction and speed of image motion in the rabbit. The Journal of Physiology *173*(3), 377–407.

Baughman, R.W., and Bader, C.R. (1977). Biochemical characterization and cellular localization of the cholinergic system in the chicken retina. Brain Research *138*(3), 469–485.

Borg-Graham, L.J. (2001). The computation of directional selectivity in the retina occurs presynaptic to the ganglion cell. Nature Neuroscience *4*(2), 176–183.

Borg-Graham, L.J., and Gryzwacz, N.M. (1992). A Model of the Directional Selectivity Circuit in Retina: Transformations by Neurons Singly and in Concert. In Single Neuron Computation, T. McKenna, J. Davis, and S.F. Zornetzer, eds. (Boston: Academic Press), pp. 347–375.

Borst, A., and Euler, T. (2011). Seeing things in motion: models, circuits, and mechanisms. Neuron *71*(6), 974–994.

Brecha, N., Johnson, D., Peichl, L., and Wassle, H. (1988). Cholinergic amacrine cells of the rabbit retina contain glutamate decarboxylase and γ-aminobutyrate immunoreactivity. Proceedings of the National Academy of Sciences of the United States of America *85*(16), 6187–6191.

Briggman, K.L., Helmstaedter, M., and Denk, W. (2011). Wiring specificity in the direction-selectivity circuit of the retina. Nature *471*(7337), 183–190.

Brombas, A., Kalita-De Croft, S., Cooper-Williams, E.J., and Williams, S.R. (2017). Dendro-dendritic cholinergic excitation controls dendritic spike initiation in retinal ganglion cells. Nature Communications *8*, 15683.

Caldwell, J.H., Daw, N.W., and Wyatt, H.J. (1978). Effects of picrotoxin and strychnine on rabbit retinal ganglion cells: lateral interactions for cells with more complex receptive fields. The Journal of Physiology *276*(1), 277–298.

Chen, Q., and Wei, W. (2018). Stimulus-dependent engagement of neural mechanisms for reliable motion detection in the mouse retina. Journal of Neurophysiology *120*(3), 1153–1161.

Chen, Q., Pei, Z., Koren, D., and Wei, W. (2016). Stimulus-dependent recruitment of lateral inhibition underlies retinal direction selectivity. eLife *5*, e21053.

Collewijn, H. (1969). Optokinetic eye movements in the rabbit: input-output relations. Vision Research *9*(1), 117–132.

Collewijn, H., Oyster, C.W., and Takahashi, E. (1972). Rabbit optokinetic reactions and retinal direction-selective cells. A preliminary model. Bibliotheca ophthalmologica: supplementa ad ophthalmologica *82*, 280–287.

Cruz-Martín, A., El-Danaf, R.N., Osakada, F., Sriram, B., Dhande, O.S., Nguyen, P.L., Callaway, E.M., Ghosh, A., and Huberman, A.D. (2014). A dedicated circuit links direction-selective retinal ganglion cells to the primary visual cortex. Nature *507*(7492), 358–361.

Demb, J.B. (2007). Cellular mechanisms for direction selectivity in the retina. Neuron *55*(2), 179–186.

Dhande, O.S., Estevez, M.E., Quattrochi, L.E., El-Danaf, R.N., Nguyen, P.L., Berson, D. M., and Huberman, A.D. (2013). Genetic dissection of retinal inputs to brainstem nuclei controlling image stabilization. Journal of Neuroscience *33*(45), 17797–17813.

Ding, H., Smith, R.G., Poleg-Polsky, A., Diamond, J.S., and Briggman, K.L. (2016). Species-specific wiring for direction selectivity in the mammalian retina. Nature *535*(7610), 105–110.

Enciso, G.A., Rempe, M., Dmitriev, A.V., Gavrikov, K.E., Terman, D., and Mangel, S.C. (2010). A model of direction selectivity in the starburst amacrine cell network. Journal of Computational Neuroscience *28*(3), 567–578.

Euler, T., Detwiler, P.B., and Denk, W. (2002). Directionally selective calcium signals in dendrites of starburst amacrine cells. Nature *418*(6900), 845–852.

Famiglietti, E.V. (1983a). On and off pathways through amacrine cells in mammalian retina: the synaptic connections of "starburst" amacrine cells. Vision Research *23*(11), 1265–1279.

Famiglietti, E.V. (1983b). "Starburst" amacrine cells and cholinergic neurons: mirror-symmetric ON and OFF amacrine cells of rabbit retina. Brain Research *261*(1), 138–144.

Famiglietti, E.V. (1987). Starburst amacrine cells in cat retina are associated with bistratified, presumed directionally selective, ganglion cells. Brain Research *413*(2), 404–408.

Famiglietti, E.V. (1991). Synaptic organization of starburst amacrine cells in rabbit retina: analysis of serial thin sections by electron microscopy and graphic reconstruction. Journal of Comparative Neurology *309*(1), 40–70.

Fransen, J.W., and Borghuis, B.G. (2017). Temporally diverse excitation generates direction-selective responses in ON- and OFF-type retinal starburst amacrine cells. Cell Reports *18*(6), 1356–1365.

Fried, S.I., Münch, T.A., and Werblin, F.S. (2002). Mechanisms and circuitry underlying directional selectivity in the retina. Nature *420*(6914), 411–414.

Gavrikov, K.E., Dmitriev, A.V., Keyser, K.T., and Mangel, S.C. (2003). Cation-chloride cotransporters mediate neural computation in the retina. Proceedings of the National Academy of Sciences of the United States of America *100*(26), 16047–16052.

Gavrikov, K.E., Nilson, J.E., Dmitriev, A.V., Zucker, C.L., and Mangel, S.C. (2006). Dendritic compartmentalization of chloride contransporters underlies directional responses of starburst amacrine cells in retina. Proceedings of the National Academy of Sciences of the United States of America *103*(49), 18793–18798.

Greene, M.J., Kim, J.S., and Seung, H.S. (2016). Analogous convergence of sustained and transient inputs in parallel on and off pathways for retinal motion computation. Cell Reports *14*(8), 1892–1900.

Hanson, L., Sethuramanujam, S., de Rosenroll, G., Jain, V., and Awatramani, G.B. (2019). Retinal direction selectivity in the absence of asymmetric starburst amacrine cell responses. eLife *8*, e42392.

Hartline, H.K. (1940). The receptive fields of optic nerve fibers. American Journal of Physiology-Legacy Content *130*(4), 690–699.

Hausselt, S.E., Euler, T., Detwiler, P.B., and Denk, W. (2007). A dendrite-autonomous mechanism for direction selectivity in retinal starburst amacrine cells. PLoS Biology *5*(7), 1474–1493.

Hayden, S.A., Mills, J.W., and Masland, R.M. (1980). Acetylcholine synthesis by displaced amacrine cells. Science *210*(4468), 435–437.

Hubel, D.H., and Wiesel, T.N. (1959). Receptive fields of single neurones in the cat's striate cortex. The Journal of Physiology *148*(3), 574–591.

Hubel, D.H., and Wiesel, T.N. (1962). Receptive fields, binocular interaction and functional architecture in the cat's visual cortex. The Journal of Physiology *160*(1), 106–154.

Jain, V., Murphy-Baum, B.L., Derosenroll, G., Sethuramanujam, S., Delsey, M., Delaney, K., and Awatramani, G.B. (2020). The functional organization of excitation and inhibition in the dendrites of mouse direction-selective ganglion cells. eLife *9*, e52949.

Kanjhan, R., and Sivyer, B. (2010). Two types of ON direction-selective ganglion cells in rabbit retina. Neuroscience Letters *483*(2), 105–109.

Kay, J.N., de la Huerta, I., Kim, I.J., Zhang, Y., Yamagata, M., Chu, M.W., Meister, M., and Sanes, J.R. (2011). Retinal ganglion cells with distinct directional preferences differ in molecular identity, structure, and central projections. Journal of Neuroscience *31*(21), 7753–7762.

Kim, I.J., Zhang, Y., Yamagata, M., Meister, M., and Sanes, J.R. (2008). Molecular identification of a retinal cell type that responds to upward motion. Nature *452*(7186), 478–482.

Kim, J.S., Greene, M.J., Zlateski, A., Lee, K., Richardson, M., Turaga, S.C., Purcaro, M., Balkam, M., Robinson, A., Behabadi, B.F., et al. (2014). Space-time wiring specificity supports direction selectivity in the retina. Nature *509*(7500), 331–336.

Kittila, C.A., and Massey, S.C. (1997). Pharmacology of directionally selective ganglion cells in the rabbit retina. Journal of Neurophysiology 77(2), 675–689.

Koch, C., Poggio, T., and Torre, V. (1982). Retinal ganglion cells: a functional interpretation of dendritic morphology. Philosophical Transactions of the Royal Society of London. Series B, Biological Sciences *298*(1090), 227–263.

Koch, C., Poggio, T., and Torre, V. (1983). Nonlinear interactions in a dendritic tree: localization, timing, and role in information processing. Proceedings of the National Academy of Sciences of the United States of America *80*(9 I), 2799–2802.

Koch, C., Douglas, R., and Wehmeier, U. (1990). Visibility of synaptically induced conductance changes: theory and simulations of anatomically characterized cortical pyramidal cells. Journal of Neuroscience *10*(6), 1728–1744.

Kosaka, T., Tauchi, M., and Dahl, J.L. (1988). Cholinergic neurons containing GABA-like and/or glutamic acid decarboxylase-like immunoreactivities in various brain regions of the rat. Experimental Brain Research *70*(3), 605–617.

Kuffler, S.W. (1953). Discharge patterns and functional organization of mammalian retina. Journal of Neurophysiology *16*(1), 37–68.

Lee, S., and Zhou, Z.J. (2006). The synaptic mechanism of direction selectivity in distal processes of starburst amacrine cells. Neuron *51*(6), 787–799.

Lee, S., Kim, K., and Zhou, Z.J. (2010). Role of ACh-GABA cotransmission in detecting image motion and motion direction. Neuron *68*(6), 1159–1172.

Lettvin, J., Maturana, H., McCulloch, W., and Pitts, W. (1959). What the frog's eye tells the frog's brain. Proceedings of the IRE *47*(11), 1940–1951.

Li, Y.T., Liu, B.H., Chou, X.L., Zhang, L.I., and Tao, H.W. (2015). Strengthening of direction selectivity by broadly tuned and spatiotemporally slightly offset inhibition in mouse visual cortex. Cerebral Cortex *25*(9), 2466–2477.

Marchiafava, P.L. (1979). The responses of retinal ganglion cells to stationary and moving visual stimuli. Vision Research *19*(11), 1203–1211.

Masland, R.H., and Ames, A. (1976). Responses to acetylcholine of ganglion cells in an isolated mammalian retina. Journal of Neurophysiology *39*(6), 1220–1235.

Masland, R.H., and Mills, J.W. (1979). Autoradiographic identification of acetylcholine in the rabbit retina. Journal of Cell Biology *83*(1), 159–178.

Masland, R.H., Mills, J.W., and Cassidy, C. (1984a). The functions of acetylcholine in the rabbit retina. Proceedings of the Royal Society of London B: Biological Sciences *223*(1230), 121–139.

Masland, R.H., Mills, J.W., and Hayden, S.A. (1984b). Acetylcholine-synthesizing amacrine cells: identification and selective staining by using radioautography and fluorescent markers. Proceedings of the Royal Society of London B: Biological Sciences *223*(1230), 79–100.

Masseck, O.A., and Hoffmann, K.P. (2009). Comparative neurobiology of the optokinetic reflex. Annals of the New York Academy of Sciences *1164*, 430–439.

Matsumoto, A., Agbariah, W., Solveig Nolte, S., Andrawos, R., Levi, H., Sabbah, S., and Yonehara, K. (2020). Synapse-specific direction selectivity in retinal bipolar cell axon terminals. BioRxiv. 10.1101/2020.10.12.335810.

Maturana, H.R., and Frenk, S. (1963). Directional movement and horizontal edge detectors in the pigeon retina. Science *142*(3594), 977–979.

Maturana, H.R., Lettvin, J.Y., McCulloch, W.S., and Pitts, W.H. (1960). Anatomy and physiology of vision in the frog (*Rana pipiens*). Journal of General Physiology *43*(6), 129–175.

Miller, R.F., and Bloomfield, S.A. (1983). Electroanatomy of a unique amacrine cell in the rabbit retina. Proceedings of the National Academy of Sciences of the United States of America *80*(10 I), 3069–3073.

Morrie, R.D., and Feller, M.B. (2018). A dense starburst plexus is critical for generating direction selectivity. Current Biology *28*(8), 1204–1212.e5.

Münch, T.A., and Werblin, F.S. (2006). Symmetric interactions within a homogeneous starburst cell network can lead to robust asymmetries in dendrites of starburst amacrine cells. Journal of Neurophysiology *96*(1), 471–477.

Oesch, N.W., and Taylor, W.R. (2010). Tetrodotoxin-resistant sodium channels contribute to directional responses in starburst amacrine cells. PLoS ONE *5*(8), e12447.

Oesch, N., Euler, T., and Taylor, W.R. (2005). Direction-selective dendritic action potentials in rabbit retina. Neuron *47*(5), 739–750.

Ölveczky, B.P., Baccus, S.A., and Meister, M. (2003). Segregation of object and background motion in the retina. Nature *423*(6938), 401–408.

Oyster, C.W. (1968). The analysis of image motion by the rabbit retina. The Journal of Physiology *199*(3), 613–635.

Oyster, C.W., and Barlow, H.B. (1967). Direction-selective units in rabbit retina: distribution of preferred directions. Science *155*(3764), 841–842.

Oyster, C.W., Takahashi, E., and Collewijn, H. (1972). Direction-selective retinal ganglion cells and control of optokinetic nystagmus in the rabbit. Vision Research *12*(2), 183–193.

Ozaita, A., Petit-Jacques, J., Völgyi, B., Ho, C.S., Joho, R.H., Bloomfield, S.A., and Rudy, B. (2004). A unique role for Kv3 voltage-gated potassium channels in starburst amacrine cell signaling in mouse retina. Journal of Neuroscience *24*(33), 7335–7343.

Park, S.J.H., Kim, I.J., Looger, L.L., Demb, J.B., and Borghuis, B.G. (2014). Excitatory synaptic inputs to mouse on-off direction-selective retinal ganglion cells lack direction tuning. Journal of Neuroscience *34*(11), 3976–3981.

Pei, Z., Chen, Q., Koren, D., Giammarinaro, B., Ledesma, H.A., and Wei, W. (2015). Conditional knock-out of vesicular GABA transporter gene from starburst amacrine cells reveals the contributions of multiple synaptic mechanisms underlying direction selectivity in the retina. Journal of Neuroscience *35*(38), 13219–13232.

Percival, K.A., Venkataramani, S., Smith, R.G., and Taylor, W.R. (2019). Directional excitatory input to direction-selective ganglion cells in the rabbit retina. Journal of Comparative Neurology *527*(1), 270–281.

Perry, V.H., and Walker, M. (1980). Amacrine cells, displaced amacrine cells and interplexiform cells in the retina of the rat. Proceedings of the Royal Society of London, Series B: Biological Sciences *208*(1173), 415–431.

Poleg-Polsky, A., and Diamond, J.S. (2011). Imperfect space clamp permits electrotonic interactions between inhibitory and excitatory synaptic conductances, distorting voltage clamp recordings. PLoS ONE *6*(4), e19463.

Poleg-Polsky, A., and Diamond, J.S. (2016). NMDA receptors multiplicatively scale visual signals and enhance directional motion discrimination in retinal ganglion cells. Neuron *89*(6), 1277–1290.

Poleg-Polsky, A., Ding, H., and Diamond, J.S. (2018). Functional compartmentalization within starburst amacrine cell dendrites in the retina. Cell Reports *22*(11), 2898–2908.

Poznanski, R.R. (1992). Modelling the electrotonic structure of starburst amacrine cells in the rabbit retina: a functional interpretation of dendritic morphology. Bulletin of Mathematical Biology *54*(6), 905–928.

Rall, W. (1964). Theoretical Significance of Dendritic Trees for Neuronal Input-Output Relations. In Neural Theory and Modeling, R.F. Reiss, ed. (Stanford University Press).

Rall, W. (1967). Distinguishing theoretical synaptic potentials computed for different soma-dendritic distributions of synaptic input. Journal of Neurophysiology *30*(5), 1138–1168.

Reichardt, W. (1987). Computation of optical motion by movement detectors. Biophysical Chemistry *26*(2–3), 263–278.

Rivlin-Etzion, M., Zhou, K., Wei, W., Elstrott, J., Nguyen, P.L., Barres, B.A., Huberman, A.D., and Feller, M.B. (2011). Transgenic mice reveal unexpected diversity of on-off direction-selective retinal ganglion cell subtypes and brain structures involved in motion processing. Journal of Neuroscience *31*(24), 8760–8769.

Rousso, D.L., Qiao, M., Kagan, R.D., Yamagata, M., Palmiter, R.D., and Sanes, J.R. (2016). Two pairs of ON and OFF retinal ganglion cells are defined by intersectional patterns of transcription factor expression. Cell Reports *15*(9), 1930–1944.

Sabbah, S., Gemmer, J.A., Bhatia-Lin, A., Manoff, G., Castro, G., Siegel, J.K., Jeffery, N., and Berson, D.M. (2017). A retinal code for motion along the gravitational and body axes. Nature *546*(7659), 492–497.

Schachter, M.J., Oesch, N., Smith, R.G., and Rowland Taylor, W. (2010). Dendritic spikes amplify the synaptic signal to enhance detection of motion in a simulation of the direction-selective ganglion cell. PLoS Computational Biology *6*(8), e1000899.

Sethuramanujam, S., McLaughlin, A.J., DeRosenroll, G., Hoggarth, A., Schwab, D.J., and Awatramani, G.B. (2016). A central role for mixed acetylcholine/GABA transmission in direction coding in the retina. Neuron *90*(6), 1243–1256.

Sethuramanujam, S., Yao, X., DeRosenroll, G., Briggman, K.L., Field, G.D., and Awatramani, G.B. (2017). "Silent" NMDA synapses enhance motion sensitivity in a mature retinal circuit. Neuron *96*(5), 1099–1111.e3.

Sethuramanujam, S., Matsumoto, A., McIntosh, J.M., Jing, M., Li, Y., Berson, D., Yonehara, K., and Awatramani, G.B. (2020). Rapid 'multi-directed' cholinergic transmission at central synapses. BioRxiv. 10.1101/2020.04.18.048330.

Sivyer, B., and Williams, S.R. (2013). Direction selectivity is computed by active dendritic integration in retinal ganglion cells. Nature Neuroscience *16*(12), 1848–1856.

Sivyer, B., van Wyk, M., Vaney, D.I., and Taylor, W.R. (2010). Synaptic inputs and timing underlying the velocity tuning of direction-selective ganglion cells in rabbit retina. Journal of Physiology *588*(17), 3243–3253.

Stincic, T., Smith, R.G., and Taylor, W.R. (2016). Time course of EPSCs in ON-type starburst amacrine cells is independent of dendritic location. Journal of Physiology *594*(19), 5685–5694.

Taylor, W.R., and Smith, R.G. (2012). The role of starburst amacrine cells in visual signal processing. Visual Neuroscience *29*(1), 73–81.
Taylor, W.R., and Vaney, D.I. (2002). Diverse synaptic mechanisms generate direction selectivity in the rabbit retina. Journal of Neuroscience *22*(17), 7712–7720.
Taylor, W.R., He, S., Levick, W.R., and Vaney, D.I. (2000). Dendritic computation of direction selectivity by retinal ganglion cells. Science *289*(5488), 2347–2350.
Tjepkes, D.S., and Amthor, F.R. (2000). The role of NMDA channels in rabbit retinal directional selectivity. Visual Neuroscience *17*(2), 291–302.
Torre, V., and Poggio, T. (1978). A synaptic mechanism possibly underlying directional selectivity to motion. Proceedings of the Royal Society of London, Series B: Biological Sciences *202*(1148), 409–416.
Trenholm, S., McLaughlin, A.J., Schwab, D.J., Turner, M.H., Smith, R.G., Rieke, F., and Awatramani, G.B. (2014). Nonlinear dendritic integration of electrical and chemical synaptic inputs drives fine-scale correlations. Nature Neuroscience *17*(12), 1759–1766.
Tukker, J.J., Taylor, W.R., and Smith, R.G. (2004). Direction selectivity in a model of the starburst amacrine cell. Visual Neuroscience *21*(4), 611–625.
Vaney, D.I. (1984). "Coronate" amacrine cells in the rabbit retina have the "starburst" dendritic morphology. Proceedings of the Royal Society of London, Series B: Biological Sciences *220*(1221), 501–508.
Vaney, D.I. (1990). Chapter 2 The mosaic of amacrine cells in the mammalian retina. Progress in Retinal Research *9*, 49–100.
Vaney, D.I., and Young, H.M. (1988). GABA-like immunoreactivity in cholinergic amacrine cells of the rabbit retina. Brain Research *438*(1–2), 369–373.
Vaney, D.I., Collin, S.P., and Young, H.M. (1989). Dendritic Relationships Between Cholinergic Amacrine Cells and Direction-Selective Retinal Ganglion Cells. In Neurobiology of the Inner Retina, R. Weiler and N.N. Osborne, eds. (Springer), pp. 157–168.
Vaney, D.I., Sivyer, B., and Taylor, W.R. (2012). Direction selectivity in the retina: symmetry and asymmetry in structure and function. Nature Reviews Neuroscience *13*(3), 194–208.
Vlasits, A.L., Morrie, R.D., Tran-Van-Minh, A., Bleckert, A., Gainer, C.F., DiGregorio, D. A., and Feller, M.B. (2016). A role for synaptic input distribution in a dendritic computation of motion direction in the retina. Neuron *89*(6), 1317–1330.
Wei, W. (2018). Neural mechanisms of motion processing in the mammalian retina. Annual Review of Vision Science *4*, 165–192.
Werblin, F.S. (1970). Response of retinal cells to moving spots: intracellular recording in *Necturus maculosus*. Journal of Neurophysiology *33*(3), 342–350.
Williams, S.R. (2004). Spatial compartmentalization and functional impact of conductance in pyramidal neurons. Nature Neuroscience 7(9), 961–967.
Wyatt, H.J., and Daw, N.W. (1975). Directionally sensitive ganglion cells in the rabbit retina: specificity for stimulus direction, size and speed. Journal of Neurophysiology *38*(3), 613–626.
Wyatt, H.J., and Daw, N.W. (1976). Specific effects of neurotransmitter antagonists on ganglion cells in rabbit retina. Science *191*(4223), 204–205.
Ye, C.Q., Poo, M.M., Dan, Y., and Zhang, X.H. (2010). Synaptic mechanisms of direction selectivity in primary auditory cortex. Journal of Neuroscience *30*(5), 1861–1868.
Yonehara, K., Ishikane, H., Sakuta, H., Shintani, T., Nakamura-Yonehara, K., Kamiji, N.L., Usui, S., and Noda, M. (2009). Identification of retinal ganglion cells and their projections involved in central transmission of information about upward and downward image motion. PLoS ONE *4*(1), e4320.
Yonehara, K., Farrow, K., Ghanem, A., Hillier, D., Balint, K., Teixeira, M., Jüttner, J., Noda, M., Neve, R.L., Conzelmann, K.K., et al. (2013). The first stage of cardinal

direction selectivity is localized to the dendrites of retinal ganglion cells. Neuron *79*(6), 1078–1085.

Yoshida, K., Watanabe, D., Ishikane, H., Tachibana, M., Pastan, I., and Nakanishi, S. (2001). A key role of starburst amacrine cells in originating retinal directional selectivity and optokinetic eye movement. Neuron *30*(3), 771–780.

Zhang, L.I., Tan, A.Y.Y., Schreiner, C.E., and Merzenich, M.M. (2003). Topography and synaptic shaping of direction selectivity in primary auditory cortex. Nature *424*(6945), 201–205.

Features of the visual scene rarely move in isolation. As you gaze at a migrating flock of birds in the sky, the clouds in the background move as well—as does the entire retinal image as you move your eyes and head. Object motion sensitivity is a retinal computation that helps to differentiate foreground from background and motion of objects in the world from self-motion.

CHAPTER 12

Object motion sensitivity

Gregory William Schwartz[a,b,c] and David Swygart[c]
[a]Departments of Ophthalmology and Physiology, Feinberg School of Medicine, Northwestern University, Chicago, IL, United States
[b]Department of Neurobiology, Weinberg College of Arts and Sciences, Northwestern University, Chicago, IL, United States
[c]Northwestern Interdepartmental Neuroscience Program (NUIN), Northwestern University, Chicago, IL, United States

Motion is ubiquitous in vision, but there is a fundamental dichotomy in its source with important behavioral implications. Some motion on the retina is caused by self-motion of the eyes, head, or body, and other motion on the retina represents the actual movement of objects in the visual scene. Distinguishing between self and object motion is required for visually guided navigation through the world, recognizing the movements of predators and prey, and segmentation of objects.

The brain has its own set of circuits for distinguishing self-motion from object motion across all sensory modalities. These circuits route copies of self-generated efferent commands from the motor system to sensory systems to be incorporated into ongoing predictions of sensory input. The visual system certainly contains efferent copies of motor commands, most studied in the context of saccades (Wurtz, 2018), but these postretinal computations are beyond the scope of this book. Instead, this chapter focuses on a retinal computation that may play a role in disambiguating self-motion from object motion.

Retinal circuits do not have direct access to the motor commands that drive eye, head, or body movements. The fixational drift eye movements that jitter the retinal image between saccades are independent in the two eyes and are unlikely to be represented in the efferent system. Therefore, the retina must rely on a simpler (albeit imperfect) visual proxy for self-motion vs. object motion: global vs. local motion. Self-generated movements create global motion across the retina, whereas objects move differently from their surroundings. Importantly, the background need not be static in the case of object motion. If, for example, your eyes are panning horizontally, causing a global leftward motion on the retina, and an animal suddenly runs in the opposite direction, causing a rightward motion on the retina, this

Retinal Computation
https://doi.org/10.1016/B978-0-12-819896-4.00013-5

"differential" motion should be recognized as that of an object. Recognizing global versus object motion may also be important in contexts beyond self-motion. For example, one can more easily spot a fish swimming opposite the direction of a stream than a fish swimming with the current. Such motion pop-out effects are well documented in human psychophysics (Franconeri and Simons, 2003).

Discovery of object motion sensitivity in the retina

Ölveczky et al. (2003) first described object motion sensitivity (OMS) as a retinal computation. They studied how retinal ganglion cells (RGCs) in salamander and rabbit responded to the motion of grating stimuli in the context of simulated fixational eye movements (Fig. 12.1). In their "eye only" (global motion) condition, both a central spot and the surrounding regions moved in the same random walk, simulating global motion due to fixational eye movements. In their "eye + object" (differential motion) condition, the center and surround again moved in a random walk but differed in their random seed. Hence, the motion followed a different pattern in the two regions. This stimulus was meant to simulate the movement of a central object and the ongoing motion of the surround due to fixational eye movements. Subsets of RGCs in both species responded robustly in the eye + object condition and were almost silent in the eye-only condition. Also, cells under the object responded synchronously, a property likely to facilitate the segmentation of object from background. The authors labeled these RGCs as OMS cells.

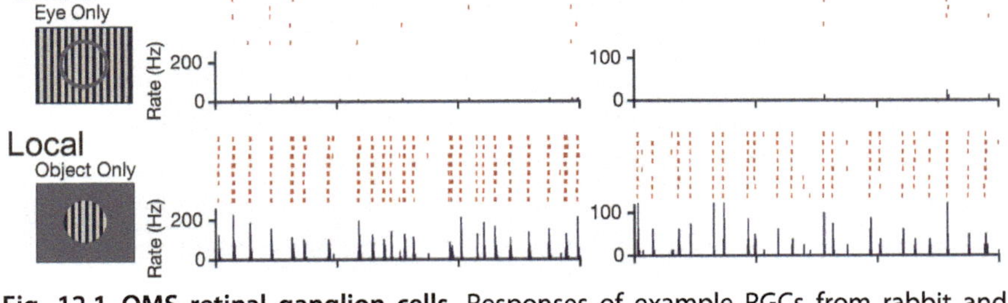

Fig. 12.1 OMS retinal ganglion cells. Responses of example RGCs from rabbit and salamander to differential, global, and local motion. *(Modified from Ölveczky et al. (2003).)*

Circuit mechanism

Ölveczky et al. (2003) proposed a model for the OMS circuit that was elaborated and tested in a subsequent paper by the same group (Baccus et al., 2008). The model has two core components: nonlinear subunits in the receptive field (RF) center and rapid inhibition from a population of nonlinear amacrine cells (ACs) in the ganglion cell RF surround (Fig. 12.2).

The receptive field (RF) center of OMS RGCs is provided by bipolar cells (BCs) with strong rectification at their output. As discussed in Chapter 7, this results in nonlinear spatial subunits. In the context of motion, the nonlinear subunits make the OMS RGC invariant to the phase or precise pattern of motion. For example, any movement of the many dark bars within the RF center of a fast OFF OMS RGC in salamander will activate some of the BC subunits with their much smaller and overlapping RFs. Consequently, shifting the starting phase of the grating or the direction or other statistics of the motion has little effect on the summed response at the level of the RF center of the RGC. For OMS RGCs to respond to rapid motion, as is the case at least in salamander, the BC inputs in their RF center must also respond at high temporal frequencies (Baccus et al., 2008).

Nonlinear subunits (also presumably BCs) provide input to wide-field amacrine cells (ACs) whose influence covers a vast region of visual space. To affect such wide-field inhibition, ACs must either have a large RF or, in the case of polyaxonal amacrine cells, a small RF coincident with their dendritic arbor and a large output region, or projective field, covered by long axons. For ACs with a large RF, the RF is not necessarily defined by the dendritic extent of individual ACs because strong electrical coupling can spread activation laterally within the population, similar to the coupling

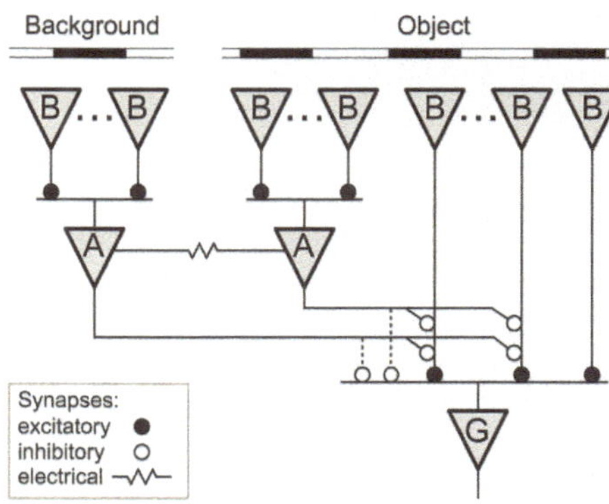

Fig. 12.2 Circuit model for OMS. Schematic of a circuit motif that can account for OMS. All BC synapses are rectified, leading to nonlinear spatial integration in the RF center and surround. Wide-field ACs provide inhibition presynaptically onto BCs, postsynaptically onto RGCs, or both. *(From Baccus et al. (2008).)*

among horizontal cells in the outer retina (Fig. 12.2, see Chapter 8). Wide-field ACs provide fast and temporally precise inhibition to the RF center circuit, either onto the BC terminals, the OMS RGC dendrites, or both.

Baccus et al. (2008) tested this model by recording responses of various retinal interneurons to global and differential motion (Fig. 12.3). Several aspects of their data were consistent with the model. First, while somatic voltage responses of BCs were well described by a linear spatiotemporal filter, the postsynaptic responses in OMS RGCs showed nonlinear spatial integration, suggesting rectification at the BC output synapse. Second, they identified a class of polyaxonal ACs with long axons and the appropriate response properties: invariance to grating phase and rapid and temporally precise depolarization that were coincident with those of OMS RGCs. Finally, they performed technically challenging experiments in which they injected current into the polyaxonal ACs while simultaneously recording OMS RGCs on a multi-electrode array. These experiments demonstrated functional connectivity between the ACs and the RGCs, though the location of inhibitory synapses (onto BC terminals versus RGC dendrites) could not be definitively identified.

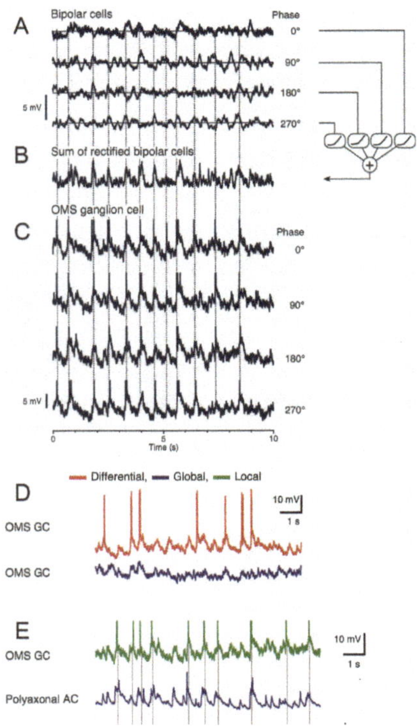

Fig. 12.3 Responses of retinal neurons to different types of motion stimuli. (**A**) Intracellular recordings of BCs show dependence on the spatial phase of the motion stimulus. (**B**) Sum of the traces in (**A**) after rectification to retain only depolarizations. (**C**) Intracellular responses of an OMS RGC to the same four stimuli at different phases (spikes truncated). The RGC's response is independent of stimulus phase. (**D**) Intracellular recording of an OMS RGC to the same RF center motion pattern in differential *(red)* or global *(blue)* mode. (**E**) Intracellular recording of an OMS RGC to a local motion stimulus (*green*) and a polyaxonal AC to the same stimulus pattern in global mode (*blue*). *(Modified from Baccus et al. (2008).)*

Identifying specific components of OMS circuits in mouse retina

The W3 mouse line labels an OMS retinal ganglion cell

More recent work on OMS circuits has largely been performed in the mouse, where knowledge of cell types and genetic tools have made detailed circuit-dissection experiments possible. Zhang et al. (2012) described an OMS RGC type that was fluorescently labeled in the W3 mouse line. Although the W3 line labeled multiple retinal ganglion cell types, the authors recorded from the brightest cells and described them as a single cell type called W3b ("bright"). Similar to the OMS mechanism described in salamander, the W3b RGC pooled ON and OFF rectified excitation from the center of its RF while receiving strong pre- and postsynaptic nonlinear surround suppression. This strong nonlinear surround suppression caused the cell to stay silent for wide-field visual stimuli but respond robustly to local contrast-reversing gratings and motion of small objects (Fig. 12.4). Notably, the authors compared local motion with global motion in this study, but they did not explicitly test the differential motion stimuli used in previous reports of OMS RGCs. Subsequent work on the OMS mechanism mouse has focused on the cells identified as "W3" or "W3b" and local vs. global motion. We will return to consider whether these RGCs represent a single type and the generality of the results for differential motion.

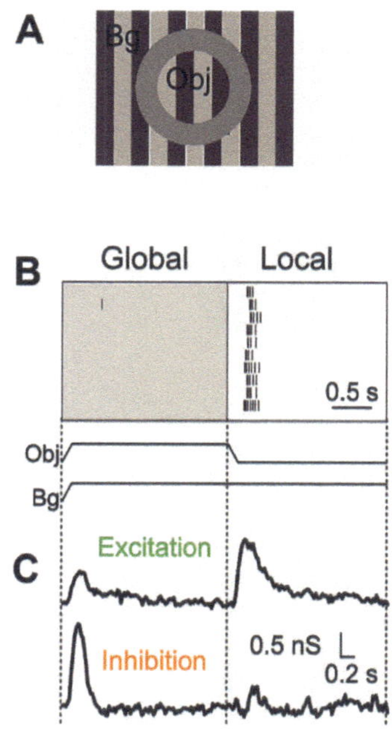

Fig. 12.4 Global and local motion responses of W3b RGCs in mouse retina. (**A**) Spatial configuration dividing the stimulus into center (Obj) and surround (Bg) regions. (**B**) Spike raster of an RGC across the global and local motion phases of the stimulus. (**C**) Excitatory and inhibitory conductances for the cell in (**B**). *(Modified from Zhang et al. (2012).)*

TH-2 ACs provide inhibition to some OMS RGCs

Pharmacology experiments by Zhang et al. (2012) showed that the ACs providing both presynaptic and postsynaptic inhibition in the W3b circuit were spiking since their effects were largely eliminated by the voltage-gated sodium channel antagonist, tetrodotoxin (TTX). Kim and Kerschensteiner (2017) used a different mouse line (TH-Cre) in which a subset of ACs was labeled to study OMS circuits. This line inserted Cre into the promoter for tyrosine hydroxylase (TH), a necessary enzyme for dopamine production. While dopaminergic ACs were labeled in this line (Vuong et al., 2015), Kim and Kerschensteiner (2017) reported that another AC, called TH-2, which is actually TH-negative, provided direct inhibition to OMS RGCs. Importantly, TH-2 ACs are nonspiking, and the authors reported no significant changes in their light responses with TTX. They also identified RGCs as OMS (called "W3" in their paper) by morphology and light responses, but they did not use the W3 line. Thus, it is possible that largely nonoverlapping populations of OMS RGCs were investigated in these two studies with potentially different upstream circuit elements.

Nonetheless, Kim and Kerschensteiner (2017) demonstrated direct connectivity between TH-2 ACs and OMS RGCs using optogenetics. They also showed that knocking out the vesicular GABA transporter, VGAT, specifically in Cre-expressing cells in this line, largely eliminated OMS in the recorded RGCs (Fig. 12.5). The efficacy of TH-2 ACs in this circuit for

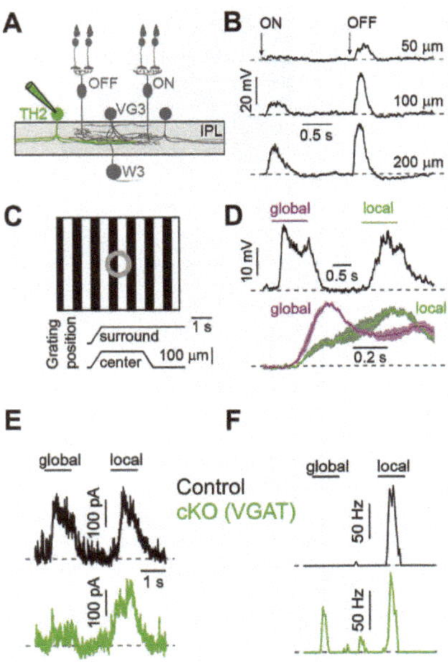

Fig. 12.5 Inhibition from TH-2 ACs is part of the OMS circuit. (**A**) Schematic of recording configuration for (**B**) and (**D**). (**B**) Current-clamp responses of a TH-2 AC for spots of different sizes. (**C**) Stimulus configuration for global and local motion similar to that in Fig. 12.4. (**D**) Responses of TH-2 ACs to global and local motion separately *(top)* and superimposed at a magnified scale *(bottom)*. (**E**) Inhibitory currents in an OMS RGC in control *(black)* and in a mouse line in which GABA release from TH ACs is disrupted *(green)*. (**F**) Spike responses in an OMS RGC for the same conditions as in (**E**). *(Modified from Kim and Kerschensteiner (2017).)*

suppressing responses to global motion depended more on response timing than it did on amplitude. Although the amplitude of the response for the two forms of motion was similar in the TH-2 AC, global motion elicited much faster responses than local motion. This timing difference was essential to the computation because, while the fast inhibition from global motion was well aligned to cancel excitation, the slow inhibition from local motion was not. Thus, rapid GABA release from TH-2 ACs onto at least a subpopulation of OMS RGCs represents an important component of some OMS circuits in the mouse retina.

Delayed and OMS excitation from a glutamatergic AC

The most surprising element discovered in OMS circuits has been the VGluT3 AC. We encountered this cell in Chapter 6, for its role in providing glycinergic inhibition to suppressed-by-contrast RGCs. Here, its role is completely different. These ACs are unique in their expression of the type 3 vesicular glutamate transporter (VGluT3), and in this circuit, they actually release glutamate to excite OMS RGCs. Studies from three different groups demonstrated this noncanonical excitatory pathway from VGluT3 ACs in which they release glutamate onto the dendrites of OMS/W3 RGCs at both light ON and OFF (Fig. 12.6; Kim et al., 2015; Krishnaswamy et al., 2015; Lee et al., 2014). Two of these studies, as well as one previous paper (Grimes et al., 2011), reported the RF properties of the VGluT3 ACs themselves. VGluT3 ACs receive excitation from ON and OFF BCs and strong surround suppression both presynaptically and postsynaptically from other ACs (Grimes et al., 2011; Kim et al., 2015; Lee et al., 2014). This is the same circuit motif described earlier for OMS, and indeed VGluT3 ACs are themselves OMS, with inhibition provided by spiking ACs (Kim et al., 2015).

But why does the W3b retinal ganglion cell require this unusual circuit element, since simply having a nonlinear center and surround can be enough to generate OMS? One possible explanation is that the delay of the excitatory signal to the W3b by the addition of the VGluT3 AC allows for a more complete silencing by the distant lateral inhibition during global motion, thus allowing for even better OMS. Indeed, Zhang et al. (2012) showed that W3b RGCs tend to have slower excitatory input filters than other mouse RGCs. Additionally, if VGluT3 ACs are already OMS, then this could provide a substrate on which the computation is further enhanced at the level of RGCs.

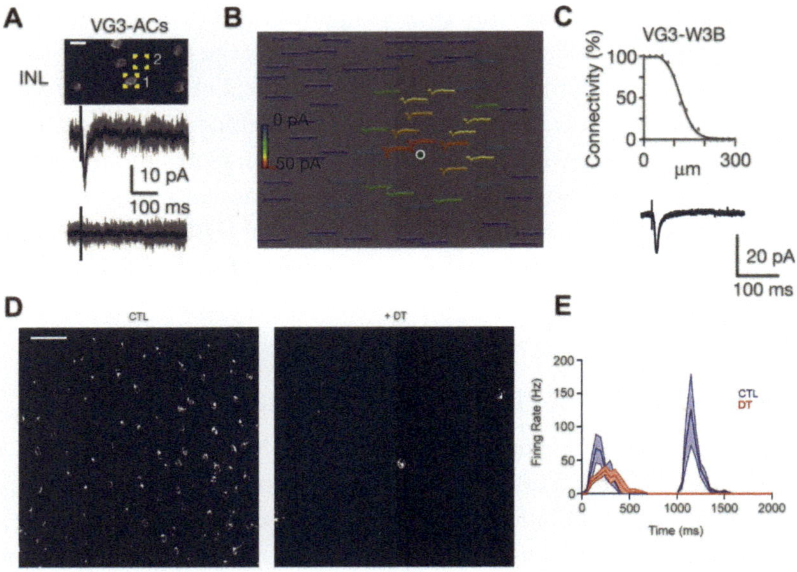

Fig. 12.6 OMS RGCs receive excitatory input from VGluT3 ACs. (**A**) Two-photon stimulation of channelrhodopsin in VGluT3 ACs while recording a postsynaptic OMS RGC. Laser stimulation location 1 on a VGluT3 AC elicits an excitatory current in the RGC *(top trace)*, but nearby location 2 does not *(bottom trace)*. Scale bar = is 10 μm. (**B**) Map of postsynaptic responses elicited by stimulation of VGluT3 ACs at different locations relative to the RGC soma *(black spot)*. (**C**) Fraction of VGluT3 ACs eliciting a response in the RGC as a function of their distance from the RGC soma. Example response shown below. (**D**) Image of a region of retina in a VGluT3 AC reporter line expressing the diphtheria toxin receptor in these ACs *(left)* and the same line following diphtheria toxin administration. Scale bar = 40 μm. (**E**) Average spike responses from OMS RGCs responding to a light step (ON at 0 s, OFF at 1 s) in control conditions and with VGluT3 ACs ablated by diphtheria toxin. *(Modified from Krishnaswamy et al. (2015).)*

How many RGC types are OMS?

Although we have nearly complete knowledge of RGC types in the mouse retina, no comprehensive survey has been completed to determine which of them are OMS. At first glance, the ability to detect object motion seems like a fairly complicated computation and thus might be limited to one or two dedicated cell types. However, we have seen that OMS can arise simply by canceling the nonlinear receptive field center with nonlinear surround suppression. Nonlinear surround suppression is widespread among RGCs (see Chapter 8) and perhaps also among ACs (Kim et al., 2015), raising the possibility that OMS could be a common functional feature of retinal circuits.

Indeed, Jacoby and Schwartz (2017) described four different RGC types in the mouse with small receptive fields and strong surround suppression (Fig. 12.7). The morphological similarities between these four types add to

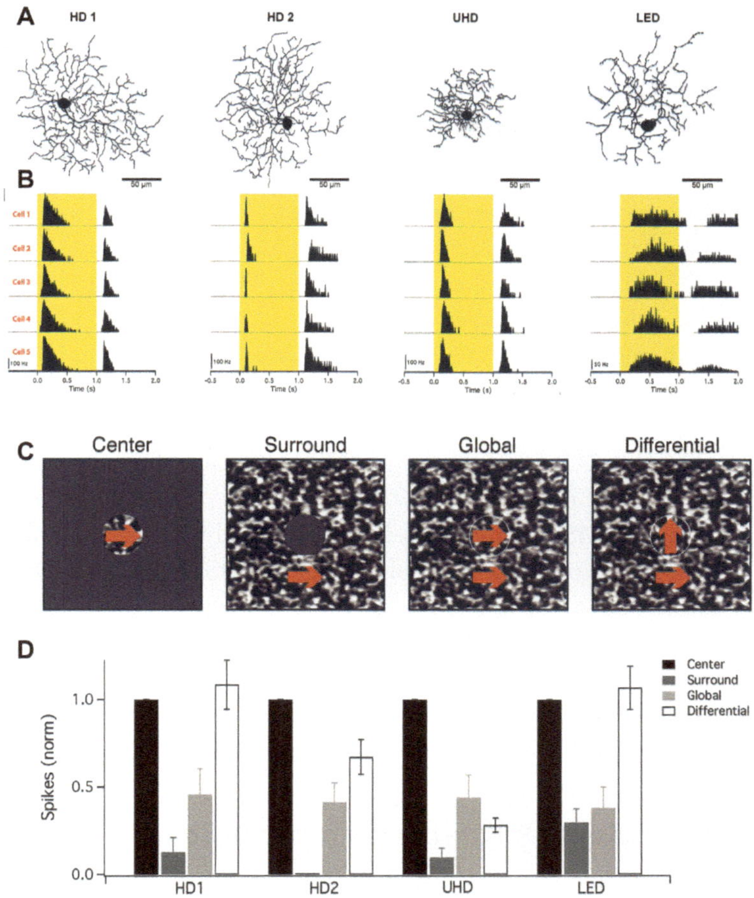

Fig. 12.7 Four different RGC types are all suppressed by global motion relative to local motion. (**A**) Morphologies of four small-RF, surround-suppressed RGCs in the mouse retina. (**B**) Five examples of spike responses from each RGC type to the presentation of a small spot centered on the RF. (**C**) Motion stimulus paradigms in which textures are drifted in the center, surround, globally, or differentially by shifting the center motion direction 90 degrees relative to that of the surround. (**D**) Mean responses of the four RGC types to each type of motion stimulus. *(Modified from Jacoby and Schwartz (2017).)*

the possibility that they could have been erroneously grouped in previous studies. Each cell type had nonlinear centers and surrounds. When tested with a stimulus in which textures either moved synchronously across both center and surround (global motion) or with the center and surround offset 90 degrees in their movement direction (differential motion), the authors found a spread of OMS. Two of the cell types were highly OMS (LED and HD1), one cell type was only slightly OMS (HD2), and the last type was not at all OMS (UHD).

However, all four of these cells would have been considered OMS under the definition imposed by other mouse studies since they all responded more strongly to local motion (RF center alone) than to global motion.

Is direction selectivity related to OMS?

One might expect OMS to be a useful feature in RGCs that are also direction-selective (DS) since both computations deal with motion. Ölveczky et al. (2003) showed that ON-OFF DS cells in the rabbit were OMS cells, indicating that they would respond to differential motion in a preferred direction under naturalistic conditions. Kuhn and Gollisch (2016) reported a group of OMS-DS RGCs in salamander that encoded a specific direction of motion (Fig. 12.8). The OMS-DS RGC type in this study had smaller RFs than the non-OMS DS RGCs, further implicating surround suppression in OMS.

The lack of OMS in most DS cells might actually be important for their behavioral role. In mouse and rabbit, both ON DS RGCs and ON-OFF DS RGCs have been implicated in sensing the direction of optic flow induced by locomotion (Cafaro et al., 2020; Sabbah et al., 2017). An OMS RGC would be suppressed by the global motion caused by optic flow. Perhaps, DS cells specifically avoid OMS by receiving linear or weak surround suppression or by timing excitation and inhibition to avoid suppression during global motion.

Why have multiple RGC types that are OMS?

Movement is varied and ubiquitous within natural visual scenes. Having multiple OMS RGC types could allow cell types to occupy different niches in stimulus space, such as object size, acceleration, or velocity. The RGC types described by Jacoby and Schwartz (2017) each had distinct size and speed tuning. Whether these cells are silenced by particular background motion depends on matching the temporal selectivity of excitation in the center and inhibition from the background. Therefore, different OMS RGCs types might then be better suited for supporting different behaviors, such as canceling out activity caused by random jitter from fixational eye movements, tracking an object during locomotion-induced smooth optic flow, canceling out activity caused by fixational eye movements, or promoting the perceptual suppression that briefly occurs following saccades. More comprehensive studies of the OMS computation, aided by our improved knowledge of mouse RGC types, will be important in parsing out the subtleties in what appears to be a fairly general circuit motif in the retina.

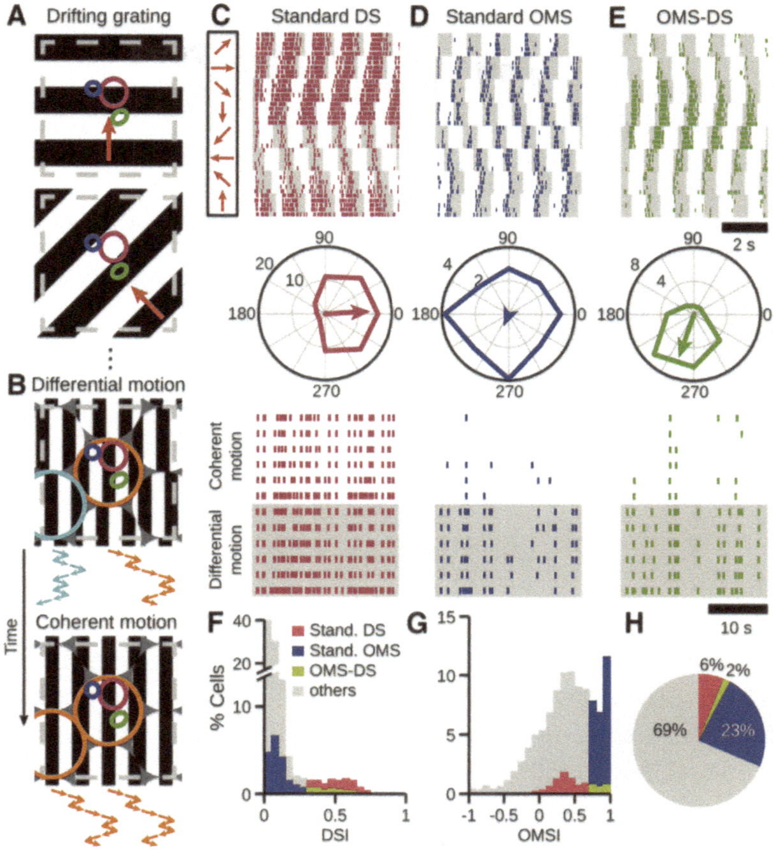

Fig. 12.8 A population of RGCs in salamander is both OMS and direction selective. (**A**) Global drifting grating stimulus presented at different directions. (**B**) Differential motion stimulus in which separate, circular patches of the stimulus move with either different trajectories *(top)* or coherently *(bottom)*. (**C**) Example spike raster *(top)* and polar plot *(bottom)* of a "Standard DS" RGC stimulated with drifting gratings. *Bottom*, spike rasters for the coherent and differential motion stimuli shown in (**B**). (**D**) Same as (**C**) but for a "Standard OMS" RGC. (**E**) Same as (**C**) but for an "OMS-DS" RGC. (**F**) Distribution of direction selectivity index (DSI) for all cells including the three types above (corresponding colors) and other RGCs *(gray)*. (**G**) Same as (**F**) but for object motion sensitivity index (OMSI). (**H**) Pie chart showing the percentage of each type in the recorded population. *(From Kuhn and Gollisch (2016).)*

Landmark paper: Baccus, S.A., Ölveczky, B.P., Manu, M., and Meister, M. (2008). A retinal circuit that computes object motion. Journal of Neuroscience *28*(27), 6807–6817.

Discussion questions:

1. What type of natural motion was their random walk stimuli supposed to simulate (Fig. 1)? In what ways were their stimuli altered to make analysis simpler? Give an example of a natural scene that a wild salamander might

encounter that would stimulate its RGCs similarly to the stimuli in Fig. 1B. Give a natural example for the stimuli used in Fig. 1C.

2. How does the OMS ganglion cell response change when the grating phase is shifted by 180° (Fig. 1D)? How does the bipolar cell response change to the same phase shift (Fig. 3B)? Why do these two cell types behave differently?
3. How do the authors determine the shape of the nonlinear function between bipolar cells and OMS ganglion cells (Fig. 4D)?
4. What does Fig. 5D tell us about the bipolar cell types that are likely to be involved in the OMS circuit? How many bipolar cell types are used in their OMS model? How does this compare with the number of ganglion cell types that are likely involved in the OMS circuit?
5. How do the authors manage to inject current into a wide-field amacrine cell while simultaneously recording from retinal ganglion cells (Fig. 8)? Why didn't the authors use two patch electrodes?

Landmark paper: Kim, T., Soto, F., and Kerschensteiner, D. (2015). An excitatory amacrine cell detects object motion and provides feature-selective input to ganglion cells in the mouse retina. eLife *4*, e08025.

Discussion questions:

1. What part of Fig. 1 shows us that the VGluT3 AC cell has surround suppression? Is the surround suppression mediated by presynaptic or postsynaptic mechanisms?
2. Why do the responses in Fig. 1K indicate that the VGluT3 AC's surround is nonlinear? How would the responses look if the surround was perfectly linear? What makes the authors conclude that the nonlinearities likely result from bipolar cells?
3. In Fig. 2, the authors test the cell's response to three different modes of motion, global, $\mathrm{diff}_{\mathrm{Ce}}$, and $\mathrm{diff}_{\mathrm{Su}}$. What is actually moving in their differential motion modes ($\mathrm{diff}_{\mathrm{Ce}}$ and $\mathrm{diff}_{\mathrm{Su}}$)? Contrast this to differential motion as defined in the previous landmark paper.
4. How are W3 retinal ganglion cells targeted and identified in this paper (Fig. 4)? Why did the authors not simply use the W3 mouse line used in the original W3 discovery paper?

Programming exercise

To access the programming exercise, please visit https://www.elsevier.com/books-and-journals/book-companion/9780128198964.

This exercise implements the core OMS circuit model in Baccus et al. (2008) so that the reader can explore how it responds to different types of motion stimuli and different values of model parameters. The script "OMS_model.m" uses a simplified spatial arrangement where individual pixels correspond to BC RFs. The script builds two kinds of spatial stimuli, gratings and random textures, and both stimuli are tested in global and differential motion conditions.

1. Run the script up to line 181. Explain how the plots demonstrate why there are more spikes for differential than global motion. What information is provided by the correlation coefficient? Which spatial pattern shows a larger difference between global and differential motion? Why do you think this is the case?
2. Now run the rest of the script. This part tests how time shifts in the motion of center versus surround affect the responses. What do the results tell you? Which end of the time shift curve would best represent global motion? Which end of the curve would best represent differential motion? What part of the model is responsible for the shapes of these curves?
3. Measure the total amount of excitation and inhibition in parts 1 and 2. Does either quantity depend on whether the motion is global or differential? Why or why not? How does this result relate to the difference in spike rate for the two conditions?
4. The last parameter in "makeMotionStim." is a 2-element vector specifying whether motion is horizontal (0) or vertical (1) for center and surround motion, respectively. Why does manipulating this parameter make sense for textures but not for gratings? Try setting this vector to [0,1] for textures so that center and surround motion is 90 degree offset but otherwise keep the parameters as in the "global" motion condition (lines 72–73). What do you expect to happen? Should this be differential motion for the model? What happened? How does this compare to the result in Jacoby and Schwartz (2017)? If there is a discrepancy, explain your theory for why.

Advanced

5. There are three nonlinearities in the model. Each BC goes through a rectification step specified as a cumulative Gaussian relationship between contrast and activation (NL_X vs. NL_exc_Y). The full RF surround is further rectified by its own nonlinearity (NL_X vs. NL_inh_Y). These nonlinearities are each specified by two parameters: an offset and a slope. Finally, spikes are counted by a

simple threshold crossing (spikeThres). Explore how each of these nonlinearities affects OMS. Within the same basic circuit motif, it is likely that different retinal circuits have different values for these nonlinearities. Could this help the OMS computation become specialized for different stimulus types in different circuits? Perform an experiment with the model (including making different stimuli if needed) and show the results to prove your point.

References

Baccus, S.A., Ölveczky, B.P., Manu, M., and Meister, M. (2008). A retinal circuit that computes object motion. Journal of Neuroscience *28*(27), 6807–6817.

Cafaro, J., Zylberberg, J., and Field, G. (2020). Global motion processing by populations of direction-selective retinal ganglion cells. The Journal of Neuroscience *40*(30), 5807–5819.

Franconeri, S.L., and Simons, D.J. (2003). Moving and looming stimuli capture attention. Perception & Psychophysics *65*(7), 999–1010.

Grimes, W.N., Seal, R.P., Oesch, N., Edwards, R.H., and Diamond, J.S. (2011). Genetic targeting and physiological features of VGLUT3+ amacrine cells. Visual Neuroscience *28*(5), 381–392.

Jacoby, J., and Schwartz, G.W. (2017). Three small-receptive-field ganglion cells in the mouse retina are distinctly tuned to size, speed, and object motion. The Journal of Neuroscience *37*(3), 610–625.

Kim, T., and Kerschensteiner, D. (2017). Inhibitory control of feature selectivity in an object motion sensitive circuit of the retina. Cell Reports *19*(7), 1343–1350.

Kim, T., Soto, F., and Kerschensteiner, D. (2015). An excitatory amacrine cell detects object motion and provides feature-selective input to ganglion cells in the mouse retina. eLife *4*, e08025.

Krishnaswamy, A., Yamagata, M., Duan, X., Hong, Y.K., and Sanes, J.R. (2015). Sidekick 2 directs formation of a retinal circuit that detects differential motion. Nature *524*(7566), 466–470.

Kuhn, N.K., and Gollisch, T. (2016). Joint encoding of object motion and motion direction in the salamander retina. Journal of Neuroscience *36*(48), 12203–12216.

Lee, S., Chen, L., Chen, M., Ye, M., Seal, R.P., and Zhou, J. (2014). An unconventional glutamatergic circuit in the retina formed by vGluT3 amacrine cells. Neuron *84*(4), 708–715.

Ölveczky, B.P., Baccus, S.A., and Meister, M. (2003). Segregation of object and background motion in the retina. Nature *423*(6938), 401–408.

Sabbah, S., Gemmer, J.A., Bhatia-Lin, A., Manoff, G., Castro, G., Siegel, J.K., Jeffery, N., and Berson, D.M. (2017). A retinal code for motion along the gravitational and body axes. Nature *546*(7659), 492–497.

Vuong, H.E., Pérez de Sevilla Müller, L., Hardi, C.N., McMahon, D.G., and Brecha, N.C. (2015). Heterogeneous transgene expression in the retinas of the TH-RFP, TH-Cre, TH-BAC-Cre and DAT-Cre mouse lines. Neuroscience *307*, 319–337.

Wurtz, R.H. (2018). Corollary discharge contributions to perceptual continuity across saccades. Annual Review of Vision Science *4*, 215–237.

Zhang, Y., Kim, I.J., Sanes, J.R., and Meister, M. (2012). The most numerous ganglion cell type of the mouse retina is a selective feature detector. Proceedings of the National Academy of Sciences of the United States of America *109*(36), E2391–E2398.

In all sensory systems, there are unavoidable delays between events in the environment and neural signals. Thus, the image currently represented in your brain captures the state of the world about 60 milliseconds in the past. For quickly moving objects, this delay is significant. If it were not accounted for, you could never hit a tennis serve and this kingfisher would never catch its prey. The anticipation of an object's future position begins in the retina and leads to many powerful visual illusions. Rather than 60 milliseconds in the past, it is more accurate to say that your current visual perception represents an extrapolation to the present.

CHAPTER 13

Motion anticipation

Gregory William Schwartz[a,b,c]
[a]Departments of Ophthalmology and Physiology, Feinberg School of Medicine, Northwestern University, Chicago, IL, United States
[b]Department of Neurobiology, Weinberg College of Arts and Sciences, Northwestern University, Chicago, IL, United States
[c]Northwestern Interdepartmental Neuroscience Program (NUIN), Northwestern University, Chicago, IL, United States

Vision seems like an extremely fast sense. Trained athletes can hit a baseball thrown at 100 mph from only 60 ft away. In daily life, we perceive ourselves as seeing the world in the present, not with a time delay. The reality is that transduction and neural processing in the human visual system is considerably slower than the auditory, tactile, or vestibular senses. How is it possible that we can hit a tennis ball, and the ball's sound hitting the racket and the feeling of the force transmitted to your wrist are both in sync with where your eyes tell you the ball is located?

The answer is a computation called "motion anticipation." Motion anticipation is a form of prediction, estimating the future position of an object from its past trajectory. While some forms of motion anticipation are purposeful, like shooting ahead of a moving target, others are completely subconscious. The form of motion anticipation that begins in the retina is one of these automatic forms of prediction, and it has been exploited in a host of visual illusions.

Perhaps the most famous of these illusions is called the "flash-lag effect" (Fig. 13.1) (Marcus Baldo and Klein, 1995; Nijhawan, 1994). One version of this illusion can be imagined on a clock face. Imagine that a spot is moving rapidly around the edge of the clock, completing a revolution every second. Each time the spot reaches 12:00, the minute hand flashes for just an instant and then disappears again. Instead of observing the minute hand and the spot at the same position on the clock, when the minute hand flashes, you will see the spot positioned substantially ahead, perhaps near 1:00. If the spot instead moves rapidly counterclockwise, you will perceive it positioned near 11:00 when the minute hand flashes at 12:00. A demonstration of the flash-lag effect can be found at https://michaelbach.de/ot/mot-flashLag/. Even if you are aware of this trick, you cannot alter the perception.

Retinal Computation
https://doi.org/10.1016/B978-0-12-819896-4.00014-7

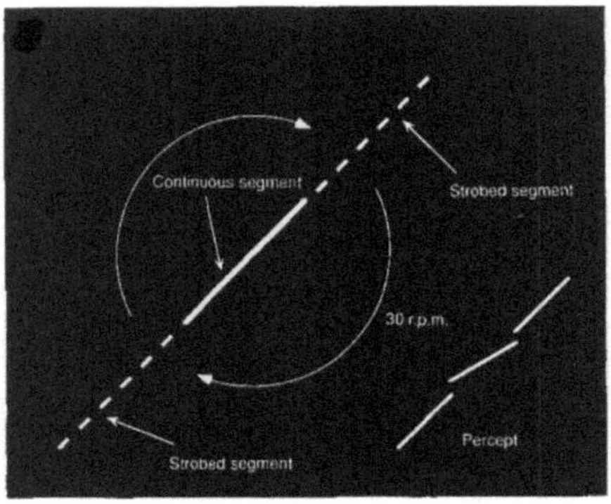

Fig. 13.1 The flash-lag effect. A line segment is rotated and two additional line segments are flashed so that they form a continuous line. Instead of seeing a continuous line, the observer perceives the central line segment as tiled in its direction of rotation. *(Modified from Nijhawan (1994).)*

How could one test whether motion anticipation occurs in the retina? One approach is to use the same stimuli used in perceptual experiments to measure the flash-lag effect, a flashed bar and a moving bar, and compare the response of the retina's output, the spikes of the RGCs. This was the method used by Berry et al. (1999) in a seminal paper on motion anticipation in the retina. Of course, you cannot ask the retina where it "perceives" the bar to be. Instead, Berry and colleagues devised a method to quantify the population spike response's spatial location, termed the "neural image."

While the term "neural image" is colorful, evoking some complex algorithm, the concept is actually quite simple, and it goes by different names in other parts of the neuroscience literature. Each RGC has its receptive field (RF) centered at a particular location in space. In the Berry et al. experiment, spikes from hundreds of RGCs were measured simultaneously on a multi-electrode array (MEA). The RF for each RGC was measured using the standard method of reverse-correlating the spikes to a random checkerboard pattern. To compute the neural image during the presentation of the bar stimuli, each RGC was first assigned a spatial location according to its RF position along the axis of bar movement. In each time bin, each RGC was given a weight proportional to the number of spikes elicited in that time bin. These weights across space are the neural image. They form a bump around the position in space where the most RGCs with nearby RFs are firing the most spikes.

When Berry and colleagues flashed a bar at a single location on the retina, the population of RGCs responded as expected. The neural image was centered in the middle of the bar, and there was a delay of about 60 ms before the RGCs reached their peak firing rate. The result was different, however, for the moving bar. Instead of lagging behind the bar, centered where it had been 60 ms in the past, the neural image tracked the bar's leading edge (Fig. 13.2).

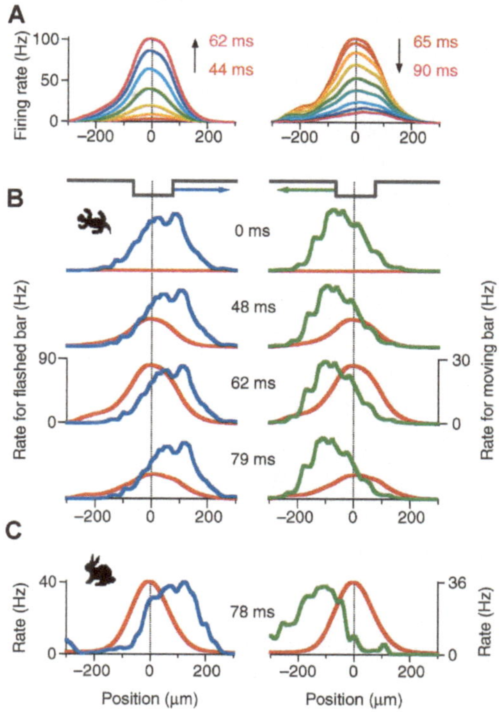

Fig. 13.2 Motion anticipation in populations of RGCs. (**A**) Neural image for a population of salamander RGCs as it evolves over time *(shown in different colors)* when the retina was presented with a flashed dark bar. (**B**) Neural image for the same population of RGCs when the retina was presented with a bar moving to the right *(blue)* or to the left *(green)*. The neural image is shown for the video frame in which the bar reached the same position as the flashed bar (0 ms) and at three subsequent time points. *Red lines* are the neural image from the flashed bar presentation, replotted from (**A**). (**C**) Similar results from a population of rabbit RGCs. *(From Berry et al. (1999).)*

The model that Berry and colleagues proposed to account for this phenomenon was remarkably simple, and it relied on a mechanism already known to exist in retinal circuits: gain control. When the bar enters the edge of an RGC's RF, it activates upstream photoreceptors and bipolar cells, culminating in action potentials in the RGC. However, this activation also engages gain control mechanisms at various locations (within cells and at synapses between them) to essentially turn down the volume on the light response. Some of these gain controls are very rapid so that by the time the bar has arrived at the center of the RGC's RF, the gain is already much lower than it had been when the bar first entered the RF. Thus, the peak response of the RGC is shifted in position to the edge of its RF from which the bar approached rather than its center. In the population, this shifts the neural image in the direction of movement, thus acting to cancel the 60 ms response delay.

The phenomenological model in the Berry et al. paper was agnostic to the precise circuit location of the gain control mechanism, but subsequent work has identified circuit motifs throughout the retina that contribute to gain control and motion anticipation. Specifically, feed-forward inhibition onto RGCs (Fig. 13.3) is common in the retina, and modeling work has shown that

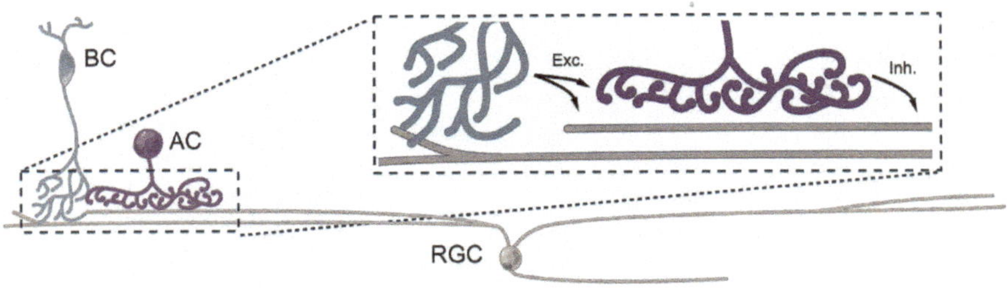

Fig. 13.3 Circuit motif for feed-forward inhibition in the retina. A BC provides excitation both an RGC and to an intervening AC that then inhibits the RGC.

it is sufficient to account for the motion anticipation measured in RGCs (Johnston and Lagnado, 2015). While the work described earlier was performed in salamander, zebrafish, rabbit, and mouse, motion anticipation also occurs in the primate retina. Populations of parasol RGCs in the macaque, which have direct homologs in human, were shown to be effective in encoding motion at various speeds and contrasts (Chichilnisky and Kalmar, 2003).

Lag normalization

While elegant in its simplicity, the gain control mechanism of motion anticipation has a serious flaw: it is inflexible to motion at different speeds. As speed increases, the object's position is further from the RF center when a response must be initiated to cancel the constant (~60 ms) delay from photon arrival to RGC spikes (Berry et al., 1999). At sufficiently high speed, the object has not even reached the edge of the RF when the response would need to be initiated. Part of the solution to this problem involves processing beyond the retina. Gain control is prevalent in the feed-forward pathways of the visual system. By virtue of their larger RFs, neurons in the primary visual cortex and beyond can anticipate motion over larger distances, and thus at higher speeds. However, a fascinating pair of papers revealed an additional, specialized mechanism within the retina that extends the speed range over which motion anticipation is effective.

Trenholm et al. (2013a,b) recorded the responses of ON-OFF direction-selective (DS) RGCs to bars moving along their preferred direction. The three ON-OFF DS RGC types encoding temporal, nasal, and dorsal movement on the retina displayed the normal motion anticipation that could not adjust to high speeds. Ventral-preferring ON-OFF DS RGCs, however, showed a phenomenon that the authors called "lag normalization." Remarkably, these RGCs responded when the bar reached a fixed location across a range of speeds (from 300 to 1800 μm/s) (Fig. 13.4A and B).

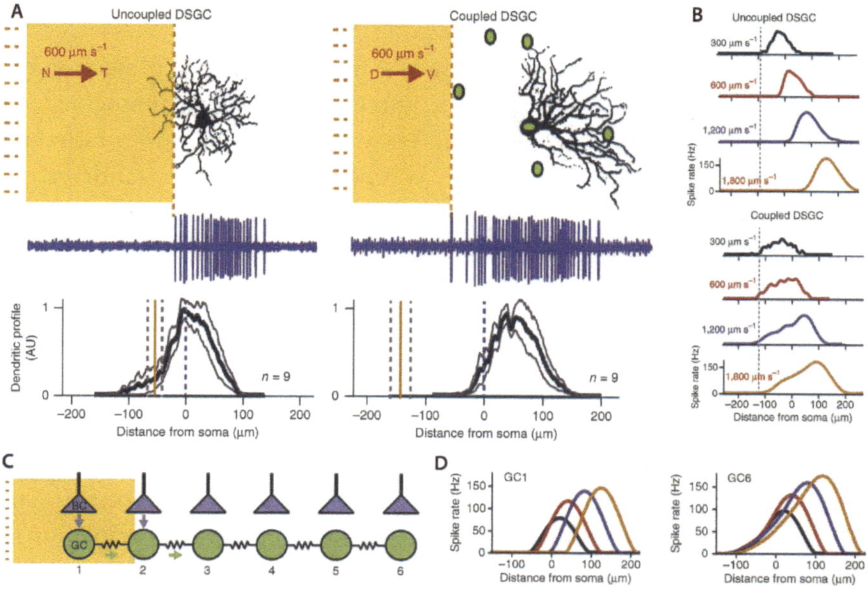

Fig. 13.4 Lag normalization in ventral-preferring ON-OFF DS RGCs relies on gap junctions. (**A**) Moving bar stimuli approaching the dendritic fields of uncoupled *(left)* and coupled *(right)* DS RGCs. Bar is shown at its position when the RGC fired its first spike. Spike traces *(blue)* are for the whole time period when the bar swept across the RF. *Bottom*, average dendritic profile across cells. *Blue dashed line* indicates soma position. *Orange vertical line* indicates the average position of the bar (± SEM in *dashed gray*) when each RGC fired its first spike. (**B**) Neural image of the DS RGC population for different motion speeds as indicated for uncoupled DS RGCs *(top)* and coupled DS RGCs *(bottom)*. (**C**) Schematic of RGC electrical coupling model. (**D**) Neural images as in (**B**) collapsed for the model cell at position 1 *(left)* and at position 6 *(right)*. *(Modified from Trenholm et al. (2013b).)*

They proved that the electrical coupling of the ventral-preferring ON-OFF DS RGCs through gap junctions was required for lag normalization (Fig. 13.4C and D). As the bar moved across the retina, not only did it activate circuits at its present location, but like a traveling wave, it spread depolarization ahead through the gap junctions. Thus, RGCs in the moving bar's path were primed to respond even before the bar reached their RFs. Lag normalization requires this traveling wave of depolarization to be stronger for higher speeds. While the model in Trenholm et al. (2013a,b) predicted this effect, it was later shown that a similar traveling wave of depolarization in the electrically coupled bipolar cell network indeed creates larger responses with speed in this range and serves as a general mechanism of motion sensitivity both in the rodent (Kuo et al., 2016) and in the primate retina (Manookin et al., 2018).

Beyond smooth motion

So far, I have only considered motion at a constant velocity. Objects in the real world, of course, move with complex trajectories, including sudden starts, stops, and changes in direction. How does the visual system account for complex trajectories? Are there components of motion anticipation beyond extrapolation of a constant velocity that also exist in the retina? Schwartz et al. (2007) investigated these questions by measuring how a population of RGCs responds to objects that suddenly reverse direction. Naturally, right after the reversal, the neural image became displaced from the object. However, the RGC population recovered in a surprising way. Cells with a wide variety of RF locations near the reversal location fired synchronously at a fixed lag after the reversal (Fig. 13.5). Modeling revealed that this synchronized firing event could be accurately distinguished from responses to smooth motion by a downstream decoder. Following this synchronized reversal response, the RGCs resumed motion anticipation to track the object moving in its new direction. A similar synchronized firing event signals the onset of movement (Chen et al., 2013; Schwartz et al., 2007), and modeling has shown that a combination of nonlinear RF subunits and gain control at the level of BC-to-RGC synapses can account for both the onset and reversal responses (Chen et al., 2013, 2014).

One way to think about the onset and reversal responses is to consider the possibility that RGCs have a multiplexed code for object motion. In cases of smooth motion, cells with RFs near the object location code for the position. In cases of sharp acceleration (or perhaps also deceleration), cells with more distant RFs fire synchronously to encode information about these velocity (and direction) changes. This is the conceptual framework that underlies a recent paper about rat RGC responses to motion trajectories that included both smooth motion and speed changes (Deny et al., 2017). The researchers showed that a single RGC type—fast OFF cells likely homologous to OFF transient alpha cells in mice—used a linear code for position when the object was near its RF and a nonlinear code for acceleration when the object was further away.

Another conceptual framework for thinking about how RGCs represent motion comes from information theory (Salisbury and Palmer, 2016). Objects in the world, particularly prey objects of interest to hunting animals, have both predictable and unpredictable components to their motion. Naturally, positions at shorter times in the future are more predictable than those further in the future. Thus, one can frame the problem of motion anticipation as a

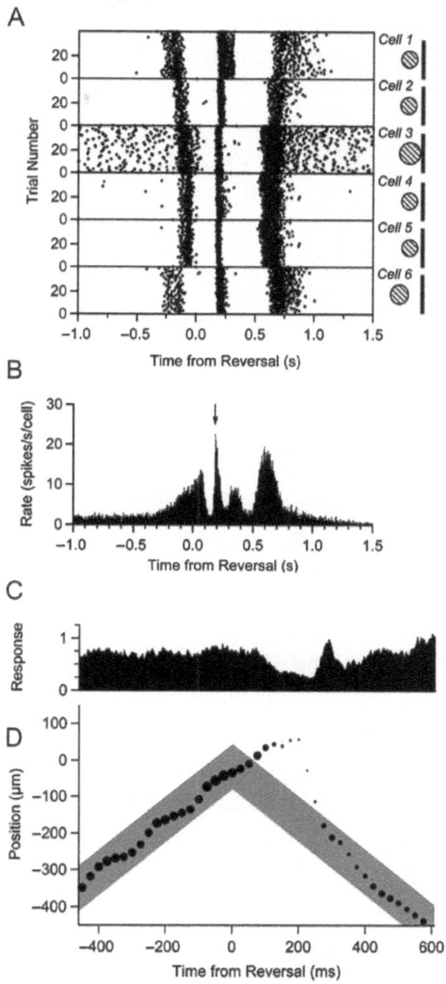

Fig. 13.5 Synchronized firing in the retina signals motion reversal. (**A**) Spike rasters for six RGCs in mouse aligned to the time that a bar reversed near their RFs. The relative locations of the RF and the position of the bar at the reversal time are depicted to the right. (**B**) Mean of the spike traces in (**A**). *Arrow* indicates the synchronous reversal response. (**C**) Population firing rate as a function of time near the reversal for all RGCs recorded. (**D**) Bar position over time around the reversal *(shaded region)* and center of the neural image *(black circles)*. Circle diameter is proportional to the total spike response across the RGC population (**C**). *(Modified from Schwartz et al. (2007).)*

problem of predictive coding. Do the spikes from a population of RGCs at time t encode information about the position of an object at time t + Δt? Stephanie Palmer and colleagues used this approach to measure the information about future object position encoded in populations of salamander RGCs. They designed the stimulus to include both predictable and random components so that that they could compute precisely how much information about the future position was possible at each time point. Remarkably, populations of RGCs approached this limit, representing nearly as much information about future position as was possible (Palmer et al., 2013; Sederberg et al., 2018). Predictive coding also offers a new perspective with which to examine standard models of RGCs. Palmer and colleagues used linear-nonlinear (LN) receptive field models of each recorded RGC (Chichilnisky, 2001) and compared their predictive coding performance to that of the real data. The LN model neurons performed far worse than the real neurons, indicating that this model does not capture key aspects of the motion anticipation computation. Intriguingly, artificially shifting the LN model back in time by ~115 ms allowed it to saturate the information bound, suggesting that this is about the timescale over which real RGCs can anticipate the future trajectory of the stimulus using mechanisms beyond this standard model (Palmer et al., 2013).

The next challenge: Causal links to behavior

From the neuroethological perspective, we really want to understand how motion anticipation serves animals' behavioral needs and whether its mechanisms in the retina can be linked directly to performance on naturalistic, visually guided tasks. Causal experiments have not linked specific manipulations of motion anticipation circuits in the retina to behavior, though such experiments are on the horizon. However, what has been accomplished is a set of correlational studies in which object tracking behavior and responses from RGC populations have been measured with the same stimuli in the same species. Measurements of snapping responses in frogs and spiking responses of their RGCs to the same stimuli suggested a role for motion anticipation in the retina even before its mechanisms had been identified (Aho et al., 1993). In the salamander, RGC responses and behavior have also been compared in an elegant set of papers. First, Leonardo and Meister (2013) stimulated the salamander retina with naturalistic motion trajectories and quantified errors in object tracking. Then Borghuis and Leonardo (2015) used high-speed videography to measure prey capture behavior in the same species of salamander, demonstrating that linear extrapolation can account for their ability to catch a fly with a tongue projection.

Due to the ever-expanding set of genetic tools, the mouse has become the model system of choice for investigating retinal circuits, but there has historically been little work on visually guided behavior in mice. Recent work has demonstrated robust prey-capture behavior in laboratory mice that relies on vision (Hoy et al., 2016; Johnson et al., 2021), creating an exciting opportunity to connect detailed circuit manipulations with natural visual behavior.

Landmark paper: Berry, M.J., Brivanlou, I.H., Jordan, T.A., and Meister, M. (1999). Anticipation of moving stimuli by the retina. Nature *398*(6725), 334–338.

Discussion questions:

1. Why does motion anticipation in the retina depend on contrast, as shown in Fig. 3? What is the evidence that the same effect is found in human psychophysics?
2. Do you think the readout of object position from RGC firing would be straightforward in the brain? What factors could add ambiguity to this representation? How realistic is it that something like the neural image is computed downstream? Based on the data in Fig. 5b, could cell type variability be an issue?
3. Is robustness to speed an important requirement for motion anticipation? Does the model in this paper achieve robustness? What additional factors or mechanisms could improve this property of the computation either in the retina or downstream?

Landmark paper: Trenholm, S., McLaughlin, A.J., Schwab, D.J., and Awatramani, G.B. (2013a). Dynamic tuning of electrical and chemical synaptic transmission in a network of motion coding retinal neurons. Journal of Neuroscience *33*(37), 14927–14938; Trenholm, S., Schwab, D.J., Balasubramanian, V., and Awatramani, G.B. (2013b). Lag normalization in an electrically coupled neural network. Nature Neuroscience *16*(2), 154–156.

Discussion questions:

1. There is a fundamental difference in how position is assumed to be read out in this paper as opposed to the Berry et al. paper. This is highlighted in Fig. 1c. Can you explain the difference? What would the neural image of a population of Hb9 RGCs look like during the movement of the edge stimulus?
2. Might it be important that this paper used a light edge rather than a narrow bar? Why or why not?
3. Where in the model do the authors make an assumption about the speed dependence of bipolar cell input to the RGCs? What is this assumption and what data supports it? Why is this important for the model? What would happen if the bipolar cells responded similarly across different stimulus speeds?

Programming exercise

To access the programming exercise, please visit https://www.elsevier.com/books-and-journals/book-companion/9780128198964.

Load the file "motion_anticipation_data.mat." It contains the following variables:

- RFlocations: vector of RF location for each of the RGCs in the simulation [100 × 1]
- timeVec: time steps of the simulation, units of seconds [1 × 201]
- stimPos: position of the leading edge of the bar (microns) at each time step
- stimMatrix: space (microns) × time (10 ms time bins) matrix of the stimulus [1000 × 201]
- RFs: 1D receptive field for each RGC, matrix of space (microns) × cell number [1000 × 100]
- RGC_linear_act: linear activation of each RGC across time during the simulation, matrix of cell number × time bin [100 × 201]
- RGC_firing_rates: firing rate of each RGC across time in a model with gain control, matrix of cell number × time bin [100 × 201]

Problems

1. Order the RGCs by their RF location and make a heatmap of their RFs.
2. Using the same RGC ordering, use "subplot" to plot in the same figure (A) the stimulus matrix, (B) the linear activation of each RGC, and (C) the firing rate of each RGC after gain control. Use the same time and space scales for each plot. What do you notice about the relationship between the linear activation of the RGCs and the stimulus? What is the response delay for these cells? How is this relationship different for the firing rates after gain control?
3. Write a function that takes as input (A) the RF locations and (B) the activity matrix of the RGCs (either linear or post gain control) and computes the neural image. Fit a Gaussian to the neural image, and make your function return its center point.
4. Use "subplot" to make a figure showing the neural images for the linear RGC activation matrix and the RGC firing rate matrix at the 1 s time point. Include your Gaussian fits. Explain what is different about the neural images and why.
5. On the same axes plot (A) the position of the stimulus over time, (B) the center of the neural image for the linear RGC activations, and (C) the center of the neural image for the RGC firing rates after gain control. How does the plot demonstrate the concept of motion anticipation? Do the RGC firing rates after gain control achieve motion anticipation immediately as soon as the bar starts moving? If not, what could account for the delay?

Advanced

6. Open NeuralImage.m. This is the script that was used to make the simulation. The variable "noiseVar" is the variance of the Gaussian noise added to RGC responses in the model. Vary this parameter to investigate the noise tolerance of motion anticipation.
7. Vary "barSpeed" from 200 μm per second to 2000 μm per second and regenerate the plots from question 5 at a variety of speeds. How robust is motion anticipation in this model to speed changes?
8. The parameter "B_gainControl" sets the strength of gain control in the model. How does changing this parameter alter the speed dependence of motion anticipation?

References

Aho, A.C., Donner, K., Helenius, S., Larsen, L.O., and Reuter, T. (1993). Visual performance of the toad (*Bufo bufo*) at low light levels: retinal ganglion cell responses and prey-catching accuracy. Journal of Comparative Physiology A *172*(6), 671–682.

Berry, M.J., Brivanlou, I.H., Jordan, T.A., and Meister, M. (1999). Anticipation of moving stimuli by the retina. Nature *398*(6725), 334–338.

Borghuis, B.G., and Leonardo, A. (2015). The role of motion extrapolation in amphibian prey capture. Journal of Neuroscience *35*(46), 15430–15441.

Chen, E.Y., Marre, O., Fisher, C., Schwartz, G., Levy, J., Silviera, R.A.D., and Berry, M.J. (2013). Alert response to motion onset in the retina. The Journal of Neuroscience *33*(1), 120–132.

Chen, E.Y., Chou, J., Park, J., Schwartz, G., and Berry, M.J. (2014). The neural circuit mechanisms underlying the retinal response to motion reversal. Journal of Neuroscience *34*(47), 15557–15575.

Chichilnisky, E.J. (2001). A simple white noise analysis of neuronal light responses. Network: Computation in Neural Systems *12*(2), 199–213.

Chichilnisky, E.J., and Kalmar, R.S. (2003). Temporal resolution of ensemble visual motion signals in primate retina. The Journal of Neuroscience *23*(17), 6681–6689.

Deny, S., Ferrari, U., Mace, E., Yger, P., Caplette, R., Picaud, S., Tkacik, G., and Marre, O. (2017). Multiplexed computations in retinal ganglion cells of a single type. Nature Communications *8*(1), 1964.

Hoy, J.L., Yavorska, I., Wehr, M., and Niell, C.M. (2016). Vision drives accurate approach behavior during prey capture in laboratory mice. Current Biology *26*(22), 3046–3052.

Johnson, K.P., Fitzpatrick, M.J., Zhao, L., Wang, B., McCracken, S., Williams, P.R., and Kerschensteiner, D. (2021). Cell-type-specific binocular vision guides predation in mice. Neuron *109*(9), 1527–1539.e4.

Johnston, J., and Lagnado, L. (2015). General features of the retinal connectome determine the computation of motion anticipation. eLife. *4*, e06250.

Kuo, S.P., Schwartz, G.W., and Rieke, F. (2016). Nonlinear spatiotemporal integration by electrical and chemical synapses in the retina. Neuron *90*(2), 320–332.

Leonardo, A., and Meister, M. (2013). Nonlinear dynamics support a linear population code in a retinal target-tracking circuit. Journal of Neuroscience *33*(43), 16971–16982.

Manookin, M.B., Patterson, S.S., and Linehan, C.M. (2018). Neural mechanisms mediating motion sensitivity in parasol ganglion cells of the primate retina. Neuron *97*(6), 1327–1340.e4.

Marcus Baldo, V.C., and Klein, S.A. (1995). Extrapolation or attention shift? Nature *378*(6557), 565–566.

Nijhawan, R. (1994). Motion extrapolation in catching. Nature *370*(6487), 256–257.

Palmer, S.E., Marre, O., Berry, M.J., and Bialek, W. (2013). Predictive information in a sensory population. Proceedings of the National Academy of Sciences of the United States of America *112*(22), 6908–6913.

Salisbury, J.M., and Palmer, S.E. (2016). Optimal prediction in the retina and natural motion statistics. Journal of Statistical Physics *162*(5), 1309–1323.

Schwartz, G., Taylor, S., Fisher, C., Harris, R., and Berry, M.J. (2007). Synchronized firing among retinal ganglion cells signals motion reversal. Neuron *55*(6), 958–969.

Sederberg, A.J., MacLean, J.N., and Palmer, S.E. (2018). Learning to make external sensory stimulus predictions using internal correlations in populations of neurons. Proceedings of the National Academy of Sciences of the United States of America *115*(5), 1105–1110.

Trenholm, S., McLaughlin, A.J., Schwab, D.J., and Awatramani, G.B. (2013a). Dynamic tuning of electrical and chemical synaptic transmission in a network of motion coding retinal neurons. Journal of Neuroscience *33*(37), 14927–14938.

Trenholm, S., Schwab, D.J., Balasubramanian, V., and Awatramani, G.B. (2013b). Lag normalization in an electrically coupled neural network. Nature Neuroscience *16*(2), 154–156.

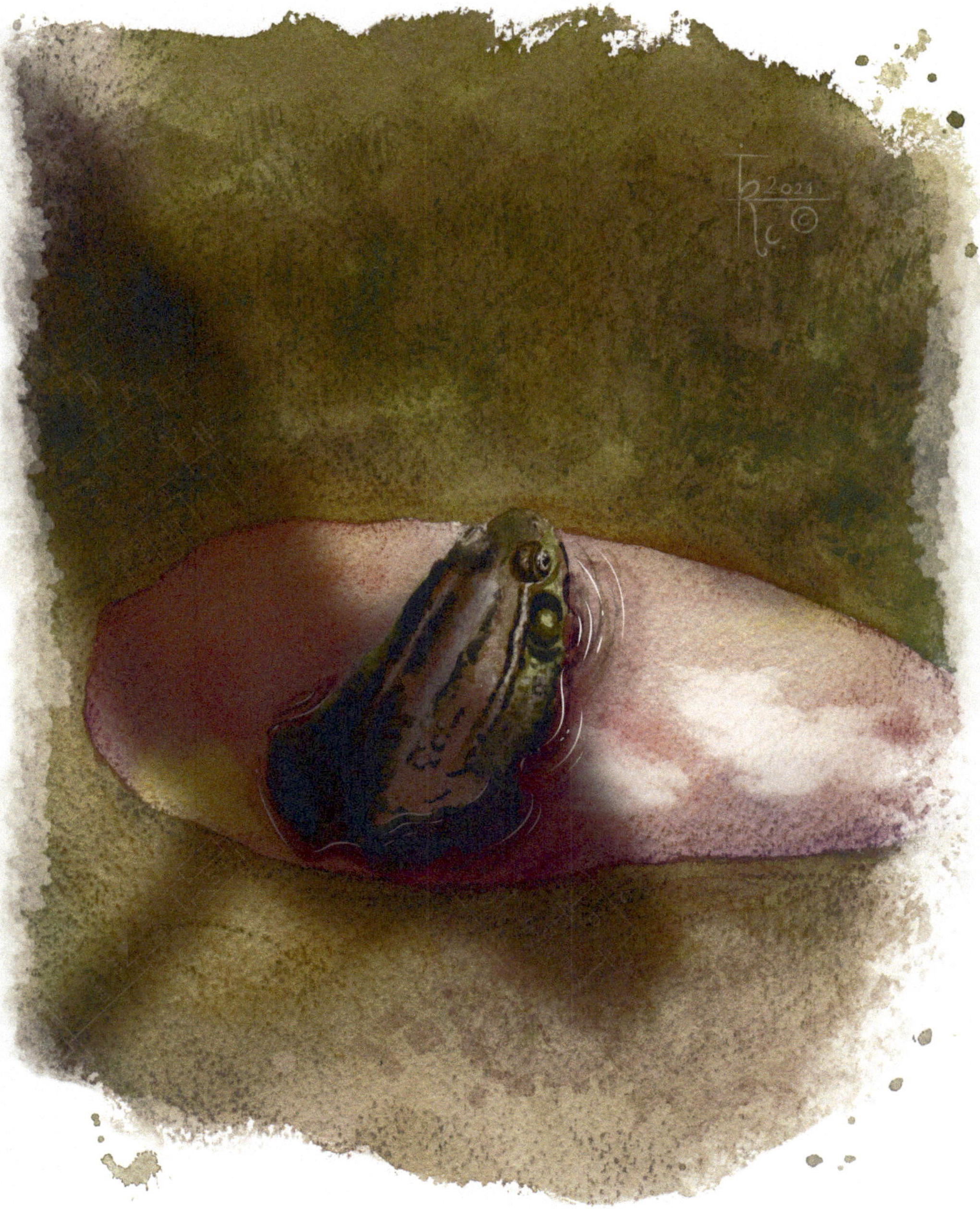

Detection of threats is one of the most significant purposes of vision from the perspective of evolution. The approach of a "looming" shadow overhead is an evolutionarily conserved danger signal that causes fear and escape responses in animals from flies, to toads, to humans. Speed is critical in threat detection, so it makes sense that much of the computation is performed within the retina and conveyed through subcortical pathways to initiate a rapid, instinctual motor program.

CHAPTER 14

Threat detection

Gregory William Schwartz[a,b,c]
[a]Departments of Ophthalmology and Physiology, Feinberg School of Medicine, Northwestern University, Chicago, IL, United States
[b]Department of Neurobiology, Weinberg College of Arts and Sciences, Northwestern University, Chicago, IL, United States
[c]Northwestern Interdepartmental Neuroscience Program (NUIN), Northwestern University, Chicago, IL, United States

What visual features specify a threat?

Classifying visual features as threatening or innocuous involves a tradeoff between selectivity and reactivity. It is instructive to consider the extremes. While a mouse should ideally discriminate an attacking cat from an uninterested dog running nearby, it is highly unlikely that retinal circuits can compute that difference. On the other hand, a mouse that mounts a full-blown escape response every time it encounters a shadow or a moving object would spend so much time in fear that it would be unable to feed itself.

Our current understanding of the visual features that engage threat-detection circuits is mostly limited to a single stimulus paradigm called "looming" described next. A fascinating area of future research is the study of mechanisms by which more complex visual features can elicit fear and escape responses. Many humans have involuntary and immediate fear responses to potentially dangerous animals like spiders or snakes, and images of these animals draw attention and elicit especially fast reaction times in psychophysics experiments (Soares et al., 2014). In fact, there is an entire theory of primate evolution built around the unique selective pressures of predation by venomous snakes (Isbell, 2006, 2019). The snake-detector threat response appears to be conserved in other mammals as well, as evidenced by the many hilarious videos you can find on the internet of cats running away from long green objects (I promise you will not be sorry if you google "cats vs. cucumbers"). For now, however, visual neuroscience has focused almost exclusively on looming as a threat stimulus.

Looming stimuli

Looming stimuli are expanding dark objects generally presented in the upper visual field (Fig. 14.1). The ethological relevance of looming is often framed

Retinal Computation
https://doi.org/10.1016/B978-0-12-819896-4.00015-9

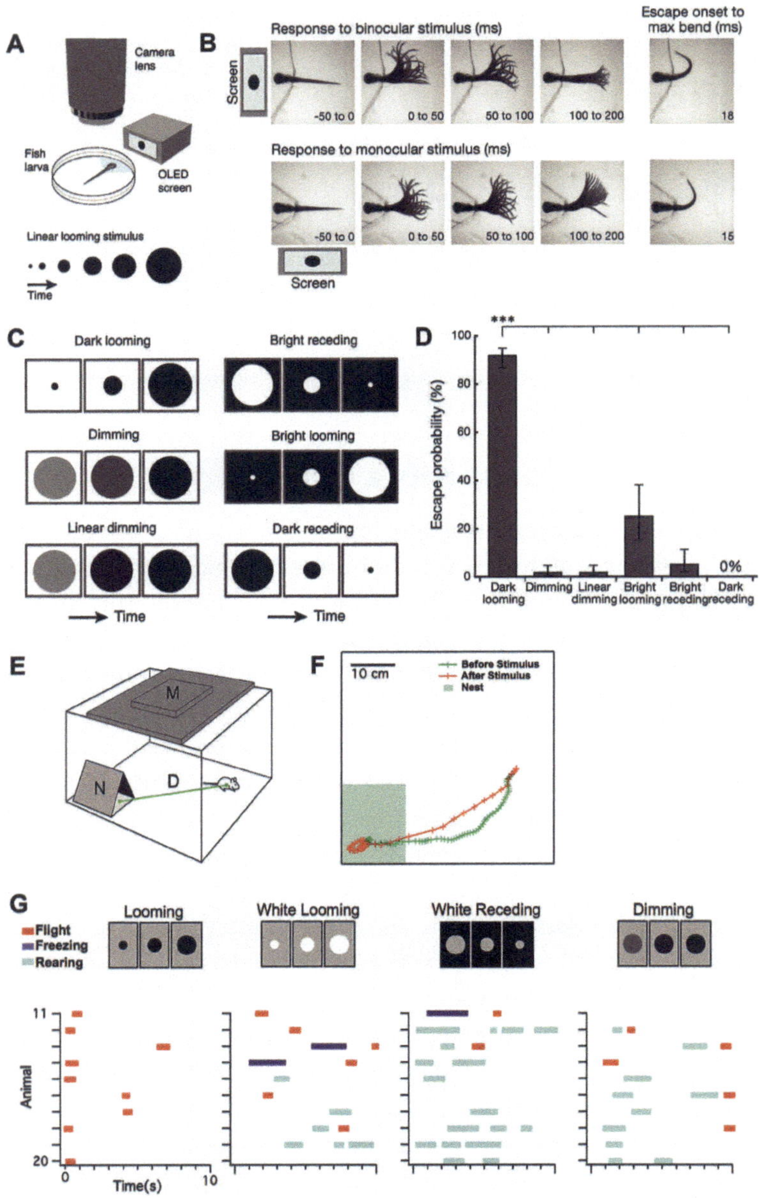

Fig. 14.1 Looming stimuli elicit defensive behaviors. (**A**) Experimental design for measuring escape behaviors in larval zebrafish with their heads embedded in agar. (**B**) Movie frames of the animal's response to a looming stimulus presented in front, in the binocular zone *(top)*, or on the side *(bottom)*. (**C**) Six different stimulus types shown to the fish. (**D**) Probability of an escape response to each of the stimuli in (**C**). (**E**) Design for measuring escape behaviors in mice. When the looming stimulus appears on an overhead monitor "M" the animal runs into its nest "N." (**F**) Example trajectory before *(green)* and after stimulus presentation *(red)*. (**G**) Time intervals color-coded by response type in 10 mice viewing four different stimulus types as indicated. *(Modified from (A–D) Temizer et al. (2015) and (E–G) Yilmaz and Meister (2013).)*

in terms of an approaching overhead predator. However, any large object on a collision course with one's head represents an imminent and universal threat. Innate fear and escape responses to looming stimuli have served as a powerful tool for investigating computation in the early visual system since at least the early 1960s when they were identified in rhesus monkeys (Schiff et al., 1962). Looming-evoked escape behaviors have been described in a host of species, including locust (Card, 2012; Fotowat and Gabbiani, 2007; Gabbiani et al., 1999, 2005), fruit fly (Card, 2012; De Vries and Clandinin, 2012; Holmqvist, 1994), zebrafish (Dunn et al., 2016; Temizer et al., 2015), goldfish (Preuss et al., 2006), frog (Ishikane et al., 2005; King et al., 1999; Nakagawa and Hongjian, 2010; Waldeck and Gruberg, 1995; Yamamoto et al., 2003), pigeon (Sun and Frost, 1998), mouse (De Franceschi et al., 2016; Yang et al., 2020; Yilmaz and Meister, 2013), non-human primate (King and Cowey, 1992; Schiff et al., 1962), human infants (Ball and Tronick, 1971), and adults (King et al., 1992). These studies have included quantification of the stimulus parameters eliciting the behavior (De Franceschi et al., 2016; Sun and Frost, 1998; Temizer et al., 2015; Yamamoto et al., 2003; Yang et al., 2020; Yilmaz and Meister, 2013), measurements of retinal ganglion cell (RGC) axon terminals in response to looming stimuli (Dunn et al., 2016; Temizer et al., 2015), and recordings of postsynaptic loom-sensitive neurons in the brain (Dunn et al., 2016; Gabbiani et al., 1999; Lee et al., 2020; Nakagawa and Hongjian, 2010; Shang et al., 2015; Sun and Frost, 1998). Less is known about circuits within the retina that support the detection of looming stimuli, but I will highlight two papers that propose retinal mechanisms involved in this computation.

Retinal circuits for looming detection

The size at which a looming stimulus reliably causes a defensive response is typically in the range of 20–70 degrees of visual angle (Ishikane et al., 2005; Temizer et al., 2015; Yang et al., 2020). This is well beyond the receptive field (RF) center size of RGCs in most species, but it makes sense in terms of the sensitivity versus reactivity tradeoff mentioned earlier. If escapes were triggered by small objects nearby, or equivalently, by large objects in the far distance, an animal would constantly live in fear in virtually any visual environment. Rather, the threshold for initiating a defensive behavior must be high enough to reduce the false-positive rate to a reasonable level. Thus, threat detection for looming stimuli is likely encoded at the level of populations of RGCs instead of in the spike trains of individual RGCs. First, I will go through a paper that explicitly recognizes this limitation but nonetheless works out the details of the circuit mechanism in "approach-sensitive" RGCs in mice that could be the building blocks of a downstream looming detector in the superior colliculus (Lee et al., 2020). Then, I will discuss a paper that makes a causal link between synchronous oscillations in a population of frog RGCs and the animal's escape behavior.

"Approach-sensitive" RGCs

Münch et al. (2009) used a transgenic mouse line that labels ~7 RGC types to target specific types for electrophysiological recordings. Among these was a cell type they called PV-5, which was later shown to be synonymous with the previously identified OFF transient alpha RGC (Baden et al., 2016; Murphy and Rieke, 2011; Pang et al., 2003; Tikidji-Hamburyan et al., 2015). Since PV-5 RGCs are OFF cells, the authors probed them with dark bars moving in three different configurations. After being flashed for 2 s at the center of the RGC's dendritic field, the bar's left and right edges were moved independently on each trial in both directions and at several speeds. Thus, the stimulus set included smooth motion (both edges moving in the same direction at the same speed), receding motion (edges moving toward each other), and approaching motion (edges moving away from each other). PV-5 cells fired at much higher rates for approaching motion than for smooth or receding motion (Fig. 14.2). Importantly, this was not simply the result of a decrease in luminance averaged over the entire RF. While PV-5 cells responded to overall dimming (as expected for any OFF RGC), they also responded to approaching motion in which the bar

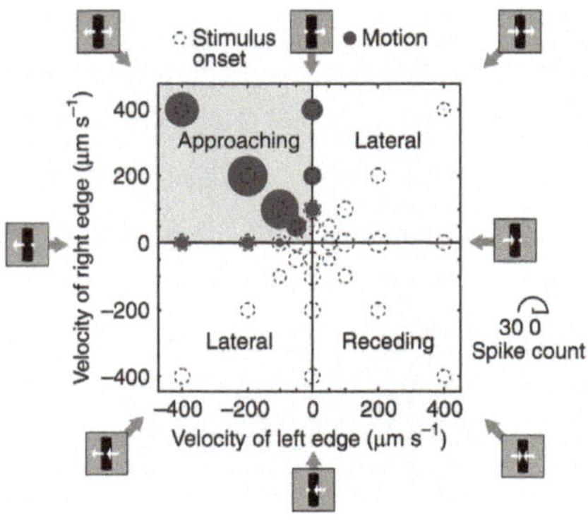

Fig. 14.2 Approach sensitivity in PV-5 RGCs. Different stimulus conditions are captured in the 2D plot of the velocity of the left edge (x-axis) versus the velocity of the right edge (y-axis). *Dashed circles* indicate stimuli presented to PV-5 RGCs. *Shaded regions* show the cell's firing rate with a larger diameter corresponding to a greater spike count. Quadrants of the graph are labeled according to motion type: lateral, receding, or approaching. *(Modified from Münch et al. (2009).)*

gradually brightened to keep the total luminance in the RF center constant during motion.

Through a series of voltage-clamp experiments with pharmacology to isolate synaptic inputs from specific circuit components, the authors were able to demonstrate that glycinergic inhibition from AII amacrine cells (ACs) directly onto the dendrites of PV-5 RGCs is critical for approach sensitivity (Fig. 14.3) (Münch et al., 2009). The inhibitory input occurs at light ON, and excitation from bipolar cells (BCs) occurs at light OFF. This is the familiar push-pull motif that we saw in Chapter 4 where it increased the slope of the contrast response function in OFF sustained alpha RGCs (Homann and Freed, 2017) and in Chapter 7, where it linearized the responses of ON parasol RGCs in response to natural scenes (Turner and Rieke, 2016). Here the same motif serves yet another computational role. Smoothly moving objects cause OFF excitation at the light-to-dark edge and ON inhibition at the dark-to-light edge. If both edges are within the RF at the same time, the excitation and inhibition largely cancel. However, approaching dark objects contain only light-to-dark edges, so the RGC receives only excitatory input and fires vigorously.

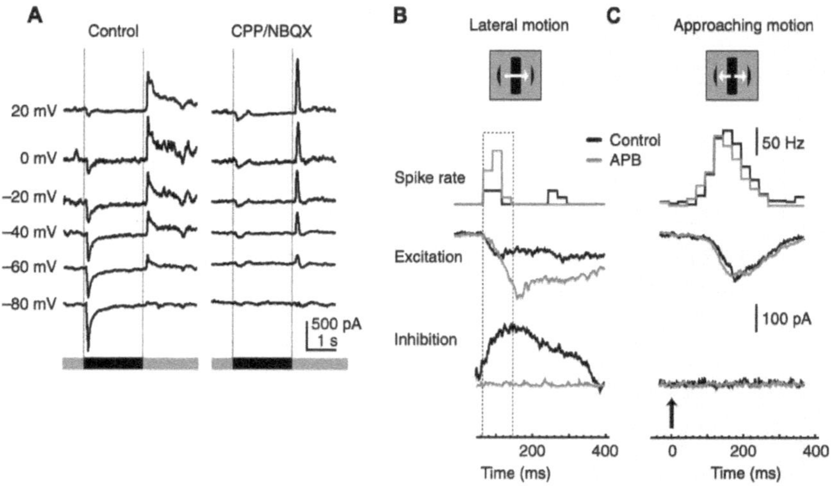

Fig. 14.3 Synaptic inputs to PV-5 RGCs. (**A**) Voltage-clamp recordings of light responses from a PV-5 RGC at different holding potentials as indicated in control solution *(left)* or in the presence of ionotropic glutamate receptor antagonists *(right)*. (**B**) Spikes, excitatory, and inhibitory currents in response to the lateral motion in control solution *(black)* or in the presence of the metabotropic glutamate receptor agonist APB to block ON pathways. (**C**) Same as (**B**) but for approaching motion. *(Modified from Münch et al. (2009).)*

Both excitation and inhibition in PV-5 RGCs come from RF subunits (see Chapter 7). Nonlinear spatial summation is essential to this mechanism because, as demonstrated by the experiment with approaching motion combined with slow brightening, the RGC responds when subunits are activated by negative contrast even if contrast averaged over the entire RF is zero. The location of inhibition onto the dendrites of the RGC rather than onto the axon terminals of presynaptic BCs is also a required circuit element to suppress responses to smooth motion. During the smooth motion of a bar, the light-to-dark edge and the dark-to-light edge appear in different parts of the RF. So inhibition locally onto the axons of nearby BCs would be ineffective at canceling excitation from BCs on the opposite side of the RF. Instead, cancelation must occur within the RGC itself, thus requiring inhibition to act on the RGC dendrites (Münch et al., 2009).

While the retinal circuit mechanisms of approach-sensitive RGCs were worked out in great detail by Münch et al., including a computational model, the link to threat detection was speculative. In this rapidly evolving field, two relevant papers were published during the final editing stages of this book. Kim et al. (2020) used a genetic strategy to ablate VGluT3 ACs in mice, resulting in dramatically reduced escape responses to looming stimuli. In addition to their roles in other circuits (see Chapters 6 and 12), VGluT3 ACs provide glycinergic inhibition to OFF transient alpha (PV-5) RGCs (Lee et al., 2016), so these authors suggested that the action of these ACs on threat detection might be relayed through the OFF transient alpha RGCs. The connection between OFF transient alpha RGCs and threat detection was tested even more directly by Wang et al. (2021) who used a new transgenic line to activate, silence, or ablate these RGCs selectively with profound effects on looming-evoked escape behaviors.

Synchronous oscillations among "dimming detectors"

A landmark study on retinal mechanisms of threat-detection in frogs included both the demonstration of a population code and a connection to behavior (Ishikane et al., 2005). Since the famous work of Lettvin et al. (1959), frog RGCs have been named by their feature selectivity. In previous work, the Tachibana lab had shown that one class of RGCs, called "dimming detectors," had synchronous oscillations in their spiking activity. The synchronized oscillations were in the gamma range ($\sim$40 Hz), and they covered vast distances of up to 2 mm on the retina (Ishikane et al., 1999).

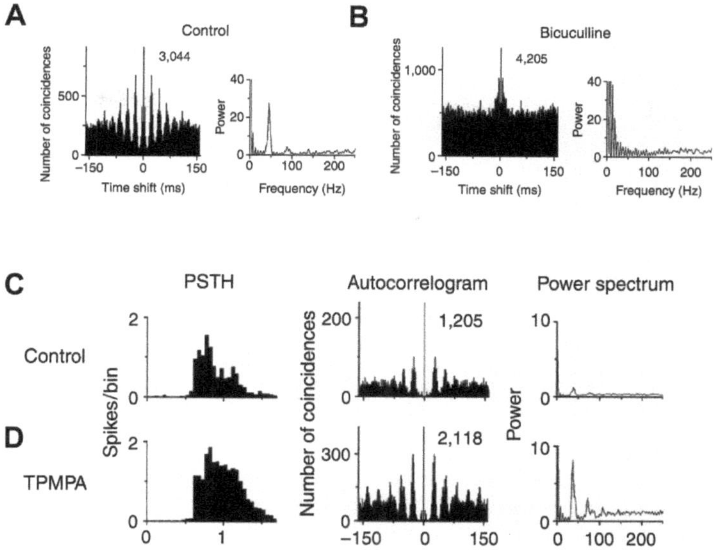

Fig. 14.4 Pharmacological manipulations of synchronous oscillations in dimming-detector RGCs. (**A**) *Left*, autocorrelation from the ensemble spike train of a population of dimming-detector RGCs responding to a looming dark spot. Number above plot indicates the total spike count. *Right*, power spectrum of the autocorrelation. (**B**) Same population as (**A**) in the presence of the $GABA_A$ receptor antagonist bicuculline. (**C**) *Left*, ensemble spike response over time during the presentation of the looming spot. *Middle*, right, autocorrelation, and power spectrum as in **B**. (**D**) Same as (**C**) but in the presence of the $GABA_C$ receptor antagonist TPMPA. *(Modified from Ishikane et al. (2005).)*

They went on to show that the oscillations were diminished by the $GABA_A$ receptor antagonist, bicuculline but enhanced by the $GABA_C$ receptor antagonist, TPMPA (Arai et al., 2004). Ishikane et al. (2005) used this pharmacological toolkit to link the synchronous oscillations of dimming detectors to frogs' jumping behavior to escape from looming stimuli (Fig. 14.4).

First, the authors showed that robust looming-evoked escape behaviors in the animals were eliminated by intraocular injection of bicuculline but unaffected by saline (Fig. 14.5A–E). Recordings of RGCs from the frogs confirmed that bicuculline suppressed synchronous oscillations in dimming detectors, but importantly, it increased their firing rate and had no significant effect on the firing rates of other RGC classes. TPMPA also increased the firing rates of dimming detectors without significantly altering those of other RGC classes, but it enhanced synchronous oscillations within the population of dimming-detector RGCs. Remarkably, intraocular injections of

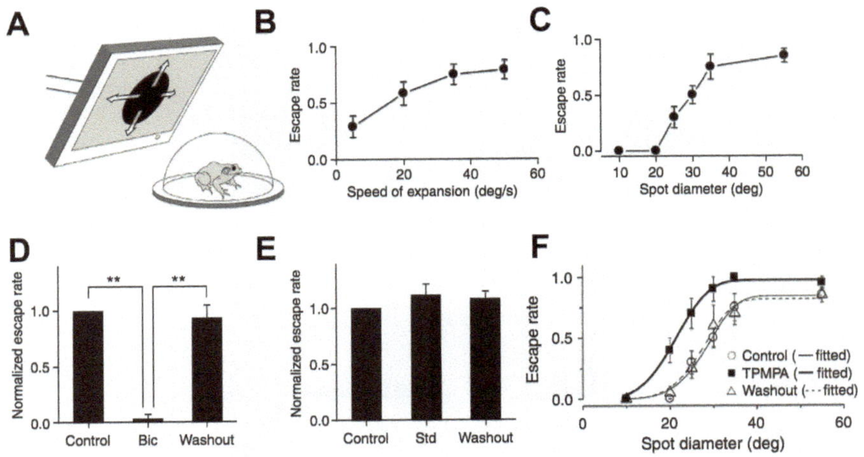

Fig. 14.5 Bidirectional modification of escape behavior by intraocular injection of GABA receptor antagonists. (**A**) Experimental paradigm for measuring escape responses in frogs. (**B** and **C**) Escape rate as a function of expansion speed (**B**) and spot diameter (**C**). (**D**) Normalized escape rate before drugs, upon intraocular injection of bicuculline, and after washout. (**E**) Same as (**D**) but with the injection of standard saline. (**F**) Escape rate as a function of spot diameter in control conditions, after intraocular injection of TPMPA, and after washout. *(Modified from Ishikane et al. (2005).)*

TPMPA enhanced the sensitivity of the frogs' escape response (Fig. 14.5F). A different visual behavior, called the optokinetic reflex, in which animals track a drifting grating with their eye position, was unaffected by either pharmacological intervention, suggesting specificity in threat-detection circuits.

Retinal input to brain circuits for threat detection

The optic tectum (OT) and its homolog in mammals, the superior colliculus (SC), are major input structures for visual information from RGCs; most studies on visual pathways for threat-detection involve OT or SC. This has been an especially active area of research over the last 5 years, producing a large amount of data as well as several discrepancies. One study in zebrafish identified looming-sensitive RGC axon terminals in OT (Temizer et al., 2015) while another study claimed that the computation takes place within OT, with the output neurons of the structure being the first to show robust selectivity for looming over dimming stimuli (Dunn et al., 2016). These two studies also differed in their report of the size threshold that elicited an escape response: 20 degrees of visual angle (Temizer et al., 2015) versus 70 degrees

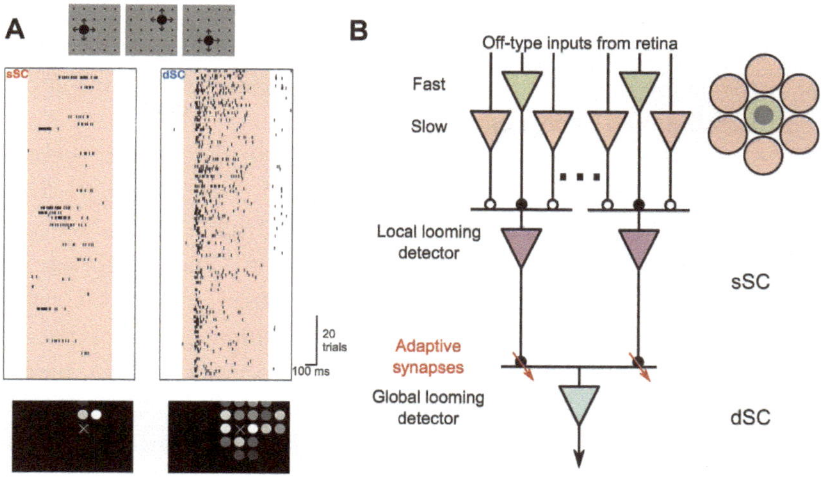

Fig. 14.6 Looming-sensitive responses in mouse superior colliculus. (**A**) *Top*, schematic of experiment. Expanding dark circles were presented at a grid of 25 different locations. *Middle*, scatter plots for a single neuron in superficial (s) SC *(left)* and deep (d) SC *(right)* simultaneously recorded. Rows are single trials at different stimulus positions. *Bottom*, data from *middle* presented as a heatmap where the brightness of the circle indicates the strength of response for a looming stimulus at the corresponding grid position. (**B**) Working model of how the looming computation derives from fast and slow RGCs synapsing onto local looming detectors in superficial SC, which then use adapting synapses to create global looming detectors downstream in deep SC. *(Modified from Lee et al. (2020).)*

(Dunn et al., 2016). In mice, location-invariant looming-sensitive responses have been recorded in SC (Fig. 14.6) (Lee et al., 2020), but which SC neurons they are and where they project remains controversial. Innate fear circuits in mice from looming stimuli through SC have been reported to pass through the ventral tegmental area (Zhou et al., 2019), the parabigeminal nucleus (Shang et al., 2015, 2018), and the lateral posterior thalamic nucleus (Shang et al., 2018; Wei et al., 2015). One study even reported a direct retinal projection to the dorsal raphe nucleus, the main serotonergic center in the brain, as being involved in fear and escape responses to looming stimuli (Huang et al., 2017).

Future directions

As a conserved behavior with obvious evolutionary significance, strong constraints on speed and accuracy, and a short neural pathway from sensation to behavior, detection of visual threats will surely continue to be a central paradigm in systems neuroscience in the coming years. While research in retinal processing in zebrafish continues to grow at a rapid pace (Bollmann, 2019), our knowledge of RGC types is still most advanced in mouse, among vertebrate species, so the next major steps in defining the retinal contribution to threat detection will likely come in this species.

The search for a link between a single RGC type and a specific behavior brings up some more general questions in retinal computation for which threat detection serves as an apt example. Can we really expect a 1:1 correspondence between RGC types and innate behaviors? The retinal code for a salient stimulus like a looming threat may be distributed among RGCs of different types. It is worth noting that the essential circuit mechanisms described in OFF transient alpha RGCs (nonlinear spatial integration and fast, postsynaptic crossover inhibition) may exist in other RGC types as well. Additionally, multiple visual features are surely represented in the spike trains of each RGC. In particular, the OFF transient alpha RGC has been implicated in signaling image recurrence (Krishnamoorthy et al., 2017) and it even becomes an ON-OFF cell in low background luminance (Tikidji-Hamburyan et al., 2015). As mentioned earlier, there is potentially a rich repertoire of visual features beyond looming that drive threat detection. It will undoubtedly be a major accomplishment if someone discovers a retinal circuit that recognizes a snake/cucumber.

Landmark paper: Münch, T.A., Da Silveira, R.A., Siegert, S., Viney, T.J., Awatramani, G.B., and Roska, B. (2009). Approach sensitivity in the retina processed by a multifunctional neural circuit. Nature Neuroscience *12*(10), 1308–1316.

Discussion questions:

1. What do the results in Fig. 2 tell us about the way approach sensitivity works in PV-5 RGCs? Do you agree with the authors' interpretation of this experiment? Can you think of additional experiments that might help distinguish approach sensitivity from more general properties of OFF RGCs?
2. How does the mechanism for approach sensitivity achieve invariance to motion speed, as shown in Fig. 1e? How is this different from other models of motion detection like the classic Reichardt detector (see Chapter 11)? Is speed invariance helpful for looming detection or could it pose a problem for the behavior?
3. Explain in your own words why the drug cocktail of CPP + NBQX isolates the AII amacrine cell as a source of inhibition. Are there alternative explanations?
4. What is the significance of panel 4i to the argument about the mechanism of approach sensitivity in PV-5 RGCs?
5. Fig. 5 shows that PV-6 RGCs, like PV-5 RGCs, have OFF excitation and ON inhibition. Why are they not also approach sensitive?

Landmark paper: Ishikane, H., Gangi, M., Honda, S., and Tachibana, M. (2005). Synchronized retinal oscillations encode essential information for escape behavior in frogs. Nature Neuroscience *8*(8), 1087–1095.

Discussion questions:

1. Have you ever tried to catch a frog by reaching around from the sides without casting a shadow over its head?
2. Explain why the autocorrelogram and the power spectrum are useful plots for examining synchrony in oscillations. What do the authors mean when they state that synchrony is not altered by these drugs (as in Fig. 4e), but synchronous oscillations are affected?
3. Why would a putative downstream looming detector care about the synchronous oscillations of RGCs even in the absence of a change in firing rate?
4. Given what you know about the distribution of GABA receptors throughout the retina, are you surprised by the specificity of bicuculline and TPMPA in this study? How might you reconcile the broad distribution of GABA receptors throughout the retina with this result?
5. What two factors make the TPMPA behavioral result (Fig. 6a) more remarkable (in my opinion at least) than the bicuculline result (Fig. 1)?
6. Is the result shown in Fig. 8 a satisfying control? Why or why not?

Programming exercise

To access the programming exercise, please visit https://www.elsevier.com/books-and-journals/book-companion/9780128198964.

This exercise explores how a population of OFF RGCs could distinguish a dark looming stimulus from other types of stimuli, like smooth motion of a dark object or linear dimming, that should still cause spikes but do not elicit a behavioral escape response.

The model creates a hexagonal mosaic of OFF RGCs with linear RFs plus gain control. The temporal filter and adaptation filter are in the corresponding .mat files. Open "loom_script.m" to see an example of how to generate the mosaic of cells, create stimuli, run the model, and plot the results.

1. Why does the response to the linear loom peak before the end of the stimulus (as the spot is still growing) while the response to the "real" loom (simulating an approaching object) peaks at the end?
2. Explain why the response to linear dimming peaks so early and why the total population response exceeds that for the linear loom, yet no cells exceed the activation threshold of 0.3.

3. Play with different values of the activation threshold (activation_thres on line 47) to see how this affects the fraction of active cells. Can you set a threshold that could result in a loom-detector? What about a threshold that distinguishes linear loom from an approaching object?
4. Open "makeStim" and read the documentation to see how to alter the stimuli. How does the response to the looming stimulus depend on the speed-to-size ratio (IoverV)? Is this consistent with how zebrafish respond to this parameter (see data in Temizer et al., 2015)?
5. How wide does a bar need to be before it elicits as strong a response as a loom? What default assumptions of this model affect this conclusion?
6. How does the population response amplitude depend on the speed of the bar? Does the response vs. speed function have a local maximum? Why?
7. What role does the spatial structure of the model play in its response properties? Alter the spacing between cells and the RF width to explore this relationship.
8. What role does gain control play in this model? Try different values of "beta_adap" from zero (no adaption) to 0.1 (10 times stronger than the original model).

References

Arai, I., Yamada, Y., Asaka, T., and Tachibana, M. (2004). Light-evoked oscillatory discharges in retinal ganglion cells are generated by rhythmic synaptic inputs. Journal of Neurophysiology *92*(2), 715–725.

Baden, T., Berens, P., Franke, K., Román Rosón, M., Bethge, M., and Euler, T. (2016). The functional diversity of retinal ganglion cells in the mouse. Nature *529*(7586), 345–350.

Ball, W., and Tronick, E. (1971). Infant responses to impending collision: optical and real. Science *171*(3973), 818–820.

Bollmann, J.H. (2019). The zebrafish visual system: from circuits to behavior. Annual Review of Vision Science *5*, 269–293.

Card, G.M. (2012). Escape behaviors in insects. Current Opinion in Neurobiology *22*(2), 180–186.

De Franceschi, G., Vivattanasarn, T., Saleem, A.B., and Solomon, S.G. (2016). Vision guides selection of freeze or flight defense strategies in mice. Current Biology *26*(16), 2150–2154.

De Vries, S.E.J., and Clandinin, T.R. (2012). Loom-sensitive neurons link computation to action in the Drosophila visual system. Current Biology *22*(5), 353–362.

Dunn, T.W., Gebhardt, C., Naumann, E.A., Riegler, C., Ahrens, M.B., Engert, F., and Del Bene, F. (2016). Neural circuits underlying visually evoked escapes in larval zebrafish. Neuron *89*(3), 613–628.

Fotowat, H., and Gabbiani, F. (2007). Relationship between the phases of sensory and motor activity during a looming-evoked multistage escape behavior. Journal of Neuroscience *27*(37), 10047–10059.

Gabbiani, F., Krapp, H.G., and Laurent, G. (1999). Computation of object approach by a wide-field, motion-sensitive neuron. Journal of Neuroscience *19*(3), 1122–1141.

Gabbiani, F., Cohen, I., and Laurent, G. (2005). Time-dependent activation of feed-forward inhibition in a looming-sensitive neuron. Journal of Neurophysiology *94*(3), 2150–2161.

Holmqvist, M.H. (1994). A visually elicited escape response in the fly that does not use the giant fiber pathway. Visual Neuroscience *11*(6), 1149–1161.

Homann, J., and Freed, M.A. (2017). A mammalian retinal ganglion cell implements a neuronal computation that maximizes the SNR of its postsynaptic currents. The Journal of Neuroscience *37*(6), 1468–1478.

Huang, L., Yuan, T., Tan, M., Xi, Y., Hu, Y., Tao, Q., Zhao, Z., Zheng, J., Han, Y., Xu, F., et al. (2017). A retinoraphe projection regulates serotonergic activity and looming-evoked defensive behaviour. Nature Communications *8*, 14908.

Isbell, L.A. (2006). Snakes as agents of evolutionary change in primate brains. Journal of Human Evolution *51*(1), 1–35.

Isbell, L.A. (2019). The Fruit, the Tree, and the Serpent. (Harvard University Press).

Ishikane, H., Kawana, A., and Tachibana, M. (1999). Short- and long-range synchronous activities in dimming detectors of the frog retina. Visual Neuroscience *16*(6), 1001–1014.

Ishikane, H., Gangi, M., Honda, S., and Tachibana, M. (2005). Synchronized retinal oscillations encode essential information for escape behavior in frogs. Nature Neuroscience *8*(8), 1087–1095.

Kim, T., Shen, N., Hsiang, J.C., Johnson, K.P., and Kerschensteiner, D. (2020). Dendritic and parallel processing of visual threats in the retina control defensive responses. Science Advances *6*. https://doi.org/10.1126/sciadv.abc9920.

King, S.M., and Cowey, A. (1992). Defensive responses to looming visual stimuli in monkeys with unilateral striate cortex ablation. Neuropsychologia *30*(11), 1017–1024.

King, S.M., Dykeman, C., Redgrave, P., and Dean, P. (1992). Use of a distracting task to obtain defensive head movements to looming visual stimuli by human adults in a laboratory setting. Perception *21*(2), 245–259.

King, J.G., Lettvin, J.Y., and Gruberg, E.R. (1999). Selective, unilateral, reversible loss of behavioral responses to looming stimuli after injection of tetrodotoxin or cadmium chloride into the frog optic nerve. Brain Research *841*(1–2), 20–26.

Krishnamoorthy, V., Weick, M., and Gollisch, T. (2017). Sensitivity to image recurrence across eye-movement-like image transitions through local serial inhibition in the retina. eLife *6*, e22431.

Lee, S., Zhang, Y., Chen, M., and Zhou, Z.J.J. (2016). Segregated glycine-glutamate co-transmission from vGluT3 amacrine cells to contrast-suppressed and contrast-enhanced retinal circuits. Neuron *90*, 27–34. https://doi.org/10.1016/j.neuron.2016.02.023.

Lee, K.H., Tran, A., Turan, Z., and Meister, M. (2020). The sifting of visual information in the superior colliculus. eLife *9*, e50678.

Lettvin, J., Maturana, H., McCulloch, W., and Pitts, W. (1959). What the frog's eye tells the frog's brain. Proceedings of the IRE *47*(11), 1940–1951.

Münch, T.A., Da Silveira, R.A., Siegert, S., Viney, T.J., Awatramani, G.B., and Roska, B. (2009). Approach sensitivity in the retina processed by a multifunctional neural circuit. Nature Neuroscience *12*(10), 1308–1316.

Murphy, G.J., and Rieke, F. (2011). Electrical synaptic input to ganglion cells underlies differences in the output and absolute sensitivity of parallel retinal circuits. The Journal of Neuroscience *31*(34), 12218–12228.

Nakagawa, H., and Hongjian, K. (2010). Collision-sensitive neurons in the optic tectum of the bullfrog, *Rana catesbeiana*. Journal of Neurophysiology *104*(5), 2487–2499.

Pang, J.-J., Gao, F., and Wu, S.M. (2003). Light-evoked excitatory and inhibitory synaptic inputs to ON and OFF alpha ganglion cells in the mouse retina. Journal of Neuroscience *23*(14), 6063–6073.

Preuss, T., Osei-Bonsu, P.E., Weiss, S.A., Wang, C., and Faber, D.S. (2006). Neural representation of object approach in a decision-making motor circuit. Journal of Neuroscience *26*(13), 3454–3464.

Schiff, W., Caviness, J.A., and Gibson, J.J. (1962). Persistent fear responses in rhesus monkeys to the optical stimulus of "looming" Science *136*(3520), 982–983.

Shang, C., Liu, Z., Chen, Z., Shi, Y., Wang, Q., Liu, S., Li, D., and Cao, P. (2015). A parvalbumin-positive excitatory visual pathway to trigger fear responses in mice. Science *348*, 1472–1477.

Shang, C., Chen, Z., Liu, A., Li, Y., Zhang, J., Qu, B., Yan, F., Zhang, Y., Liu, W., Liu, Z., et al. (2018). Divergent midbrain circuits orchestrate escape and freezing responses to looming stimuli in mice. Nature Communications *9*, 1232.

Soares, S.C., Lindström, B., Esteves, F., and Öhman, A. (2014). The hidden snake in the grass: superior detection of snakes in challenging attentional conditions. PLoS ONE *9*(12), e114724.

Sun, H., and Frost, B.J. (1998). Computation of different optical variables of looming objects in pigeon nucleus rotundus neurons. Nature Neuroscience *1*(4), 296–303.

Temizer, I., Donovan, J.C., Baier, H., and Semmelhack, J.L. (2015). A visual pathway for looming-evoked escape in larval zebrafish. Current Biology *25*(14), 1823–1834.

Tikidji-Hamburyan, A., Reinhard, K., Seitter, H., Hovhannisyan, A., Procyk, C.A., Allen, A.E., Schenk, M., Lucas, R.J., and Münch, T.A. (2015). Retinal output changes qualitatively with every change in ambient illuminance. Nature Neuroscience *18*(1), 66–74.

Turner, M.H., and Rieke, F. (2016). Synaptic rectification controls nonlinear spatial integration of natural visual inputs. Neuron *90*(6), 1257–1271.

Waldeck, R.F., and Gruberg, E.R. (1995). Studies on the optic chiasm of the leopard frog. I. Selective loss of visually elicited avoidance behavior after optic chiasm hemisection. Brain, Behavior and Evolution *46*(2), 84–94.

Wang, F., Li, E., De, L., Wu, Q., and Zhang, Y. (2021). OFF-transient alpha RGCs mediate looming triggered innate defensive response. Current Biology *31*, 2263–2273.e3. https://doi.org/10.1016/j.cub.2021.03.025.

Wei, P., Liu, N., Zhang, Z., Liu, X., Tang, Y., He, X., Wu, B., Zhou, Z., Liu, Y., Li, J., et al. (2015). Processing of visually evoked innate fear by a non-canonical thalamic pathway. Nature Communications *6*, 6756.

Yamamoto, K., Nakata, M., and Nakagawa, H. (2003). Input and output characteristics of collision avoidance behavior in the frog *Rana catesbeiana*. Brain, Behavior and Evolution *62*(4), 201–211.

Yang, X., Liu, Q., Zhong, J., Song, R., Zhang, L., and Wang, L. (2020). A simple threat-detection strategy in mice. BMC Biology *18*(1), 93.

Yilmaz, M., and Meister, M. (2013). Rapid innate defensive responses of mice to looming visual stimuli. Current Biology *23*(20), 2011–2015.

Zhou, Z., Liu, X., Chen, S., Zhang, Z., Liu, Y., Montardy, Q., Tang, Y., Wei, P., Liu, N., Li, L., et al. (2019). A VTA GABAergic neural circuit mediates visually evoked innate defensive responses. Neuron *103*(3), 473–488.e6.

PART 5

Other computations

The visual environment is full of periodicities. Many motor patterns, like the wing beating of this hummingbird, follow a periodic pattern. Retinal neurons entrain to periodic sequences of light and dark and then respond to their disruptions.

CHAPTER 15

Periodic sequence entrainment

Gregory William Schwartz[a,b,c]
[a]Departments of Ophthalmology and Physiology, Feinberg School of Medicine, Northwestern University, Chicago, IL, United States
[b]Department of Neurobiology, Weinberg College of Arts and Sciences, Northwestern University, Chicago, IL, United States
[c]Northwestern Interdepartmental Neuroscience Program (NUIN), Northwestern University, Chicago, IL, United States

Prediction is a major component of sensory systems, and a large conceptual framework for brain function is built around the notion that creating and updating models of future events and encoding model errors is a fundamental design principle of the nervous system (Bar, 2011). Examples of predictive coding in retinal computation include adaptation to luminance (Chapter 2) and contrast (Chapter 5) and motion anticipation (Chapter 13). This chapter describes how the retina encodes the prediction of another property of a visual stimulus: periodicity.

Temporal or spatiotemporal periodicity in the visual environment can be caused by the physical properties of inanimate objects, like the regular oscillations of luminance in shallow water from ripples on the surface (McFarland and Loew, 1983), or by the motor systems of other animals, like the periodic movements of the legs of a cricket or the wings of a fly or a bird. Analogous to adaptation to luminance or contrast—computations that allow the visual system to accentuate changes in these statistical properties of light—adaptation to periodicity enables the visual system to amplify changes to an ongoing pattern that might be behaviorally relevant. A disruption in the pattern of luminance underwater might indicate a predator striking the water above, and a change in the pattern of wing flaps of a nearby dragonfly might indicate a time to strike.

While periodicity might seem like a complex and high-level property of a visual stimulus, it is also well suited to be represented in the biophysical properties of neurons and synapses (e.g., in hair cells of the cochlea; Art and Fettiplace, 1987; Crawford and Fettiplace, 1981; Fettiplace, 1987), so it might be a natural computation to occur early in the visual system, perhaps even at the first synapse. I will first describe evidence that retinal neurons encode a prediction of periodic sequences, and then I will cover what is known about the circuit mechanisms of this computation.

Retinal Computation
https://doi.org/10.1016/B978-0-12-819896-4.00016-0

Retinal responses to "omitted" stimuli

My own journey in retinal computation began with the study of a relatively obscure phenomenon called the omitted stimulus response (OSR). In the 1990s, Theodore Bullock and colleagues had recorded local field potentials at various locations in the early visual system of fish and reptiles as they responded to periodic sequences of flashes (Bullock et al., 1990, 1993; Karamürsel and Bullock, 1994; Prechtl and Bullock, 1994). They described a phenomenon called the "omitted stimulus potential" (OSP) in which a strong response was recorded following the omission of a flash from a periodic sequence. A defining characteristic of the OSP was that it occurred at a fixed latency after the omitted flash across different temporal frequencies. In one of these papers, the authors recorded the OSP in the optic nerve and from the optic nerve head in the dissociated eye, demonstrating an origin within the retina (Fig. 15.1) (Bullock et al., 1990). With this background, my first project as a graduate student working with Michael Berry was to search for an omitted stimulus response (OSR) in the spikes of retinal ganglion cells (RGCs) of the salamander and mouse.

In both species, we found RGCs that responded strongly to omitted stimuli. Importantly, the spikes occurred at a fixed latency from the time of the omitted flash, not from the time of the last flash (Schwartz et al., 2007). Thus, the firing events embodied a prediction of the frequency of

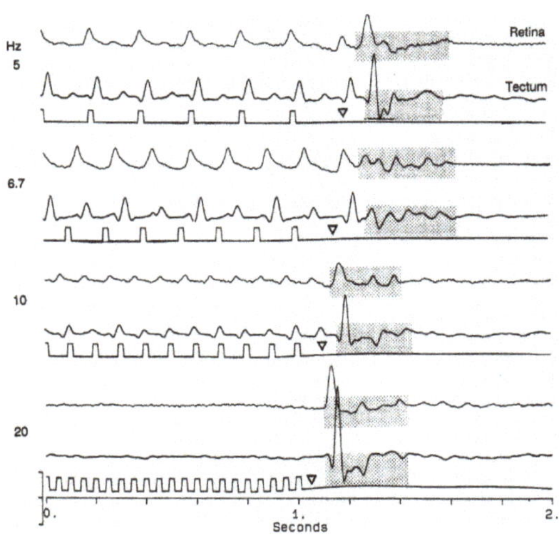

Fig. 15.1 Omitted stimulus potentials in the retina and tectum of fish. Recordings of local field potentials from the retina *(top traces)* and optic tectum *(bottom traces)* of fish. Flash trains of four different frequencies are shown and aligned to the time of the last flash. *Triangle* indicates time of the next expected flash (omitted flash time). *(Modified from Bullock et al. (1990).)*

the preceding sequence (Fig. 15.2) (Berry and Schwartz, 2011). We went on to demonstrate that the OSR occurred even when the flash sequence retained the same mean luminance as the background and that it was robust to other variables, like the flash duration, whether stimuli were square or sine waves, and whether the patterns were purely temporal or spatiotemporal.

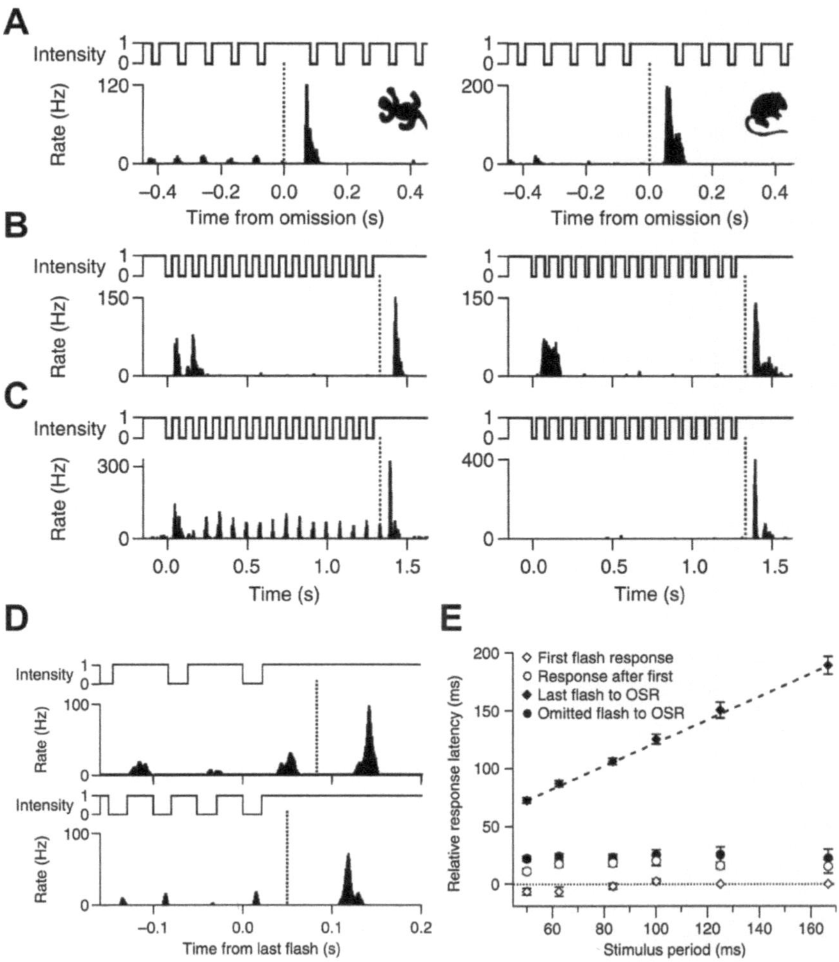

Fig. 15.2 Omitted stimulus responses in the spikes of salamander and mouse RGCs. (**A**) Example RGCs from salamander *(left)* and mouse *(right)* responding to the omission of a dark flash in a periodic sequence. *Dashed line* indicates time of omitted flash in (**A–D**). (**B** and **C**) Examples of different types of responses over the course of the flash train. (**D**) Responses of the same salamander RGC to flash trains at two different frequencies aligned to the time of the last flash. (**E**) Response latency for different response features as indicated. *Dashed line* is unity slope. *(Modified from Schwartz et al. (2007).)*

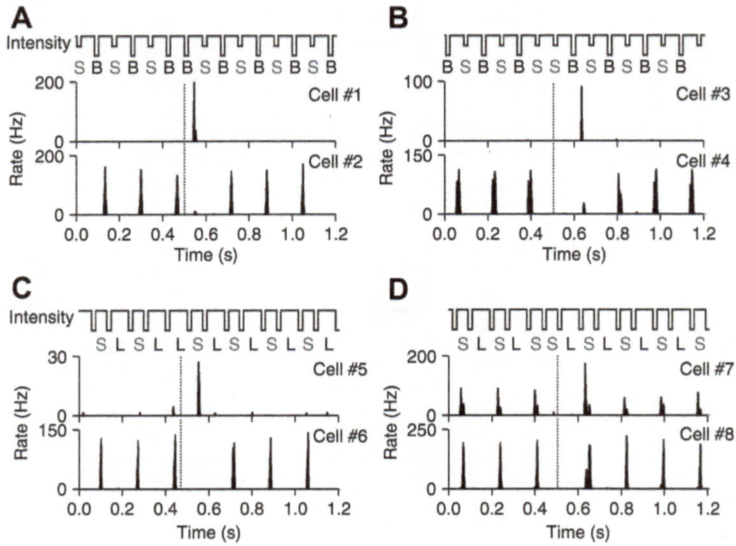

Fig. 15.3 OSRs for violations of more complex patterns. (**A**) Two RGCs that respond to violations of a pattern of alternating small and big flashes. (**B**) Same as (**A**) but following the opposite sequence violation: repeated big flashes. (**C** and **D**) Same as (**A** and **B**) but for short and long flash intervals. *(Modified from Schwartz et al. (2007).)*

We also showed OSRs for violations of more complex patterns, like alternating big and small flashes or long and short intervals (Fig. 15.3) (Schwartz et al., 2007). Subsequent work revealed an OSR in the electroretinogram (ERG) in humans (Fig. 15.4) (Gowrisankaran et al., 2013; McAnany and Alexander, 2009) and its absence in a patient with a mutation associated with signal transduction in ON bipolar cells (BCs) that caused congenital stationary night blindness (McAnany et al., 2013).

Circuit location and mechanism

Pharmacological experiments provided insights into the retinal mechanism of the OSR. Block of inhibitory GABA and glycine receptors failed to eliminate the OSR, while block of ON pathways with L-AP4 eliminated the OSR in salamander RGCs (Fig. 15.5) (Schwartz and Berry, 2008; Werner et al., 2008). These results, along with the absence of the OSR in a human patient with ON BC dysfunction (McAnany et al., 2013), suggested that the response originates somewhere in the ON BC pathway between

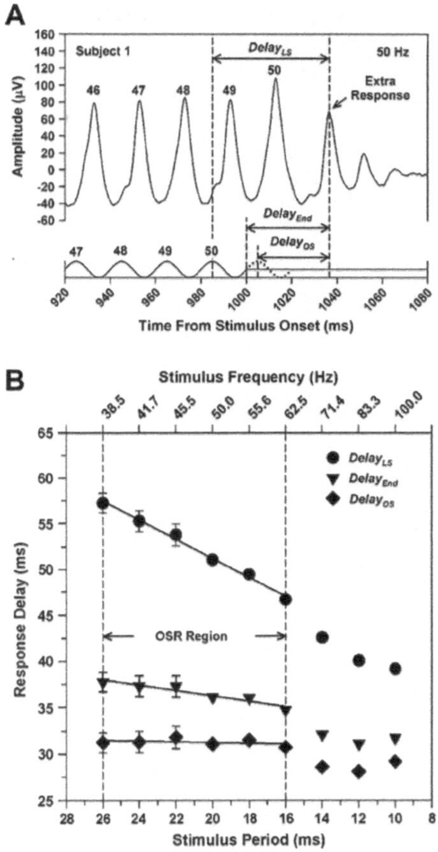

Fig. 15.4 OSR in human ERG. (A) Electroretinogram (ERG) measured in a human subject during the presentation of a sinusoidally modulated stimulus. Extra response following the omitted stimulus period is indicated. (**B**) Delays for different response features as indicated by different symbols versus stimulus period. Line through circles indicates a slope of −1 in the "OSR Region" for stimulus periods between 26 and 16 ms. *(Modified from McAnany and Alexander (2009).)*

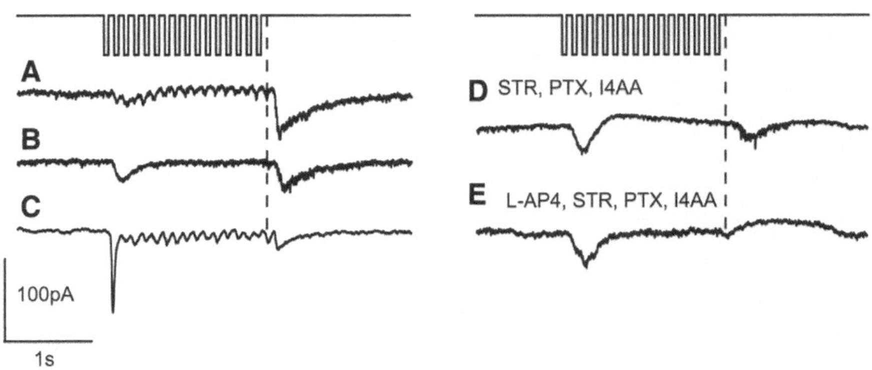

Fig. 15.5 Pharmacology of the OSR in excitatory input currents to salamander RGCs. (**A–C**) Three examples of excitatory input currents to salamander RGCs that show an OSR. (**D**) A cocktail of inhibitory blockers does not block the OSR. (**E**) The addition of L-AP4 to block ON pathways eliminates the OSR. *(From Werner et al. (2008).)*

photoreceptors and ON BC output synapses. Werner et al. (2008) showed an OSR in the excitatory inputs to salamander RGCs and used a dual ON-OFF linear-nonlinear (LN) model to capture some aspects of the phenomenon, but not others, like its robustness to stimuli with no mean luminance change or its sensitivity to complex patterns.

Several aspects of the OSR suggest a mechanism based on electrical resonance. First, the response develops over only a few flashes, which is inconsistent with slower changes in synaptic weight. Second, we observed cells with multiple OSRs that "ring" after the end of the stimulus train or with complex harmonic responses during the train (Fig. 15.6) (Schwartz and Berry, 2008). There is also precedent for oscillations in the membrane potential of BCs (Burrone and Lagnado, 1997; Protti et al., 2000). There have been no further experimental publications on the retinal mechanism of the OSR, so many questions remain, most importantly, how a mechanism based on oscillations in BCs can dynamically adjust the oscillation frequency to match the stimulus over the frequency range of the OSR (~20 Hz). A detailed model suggested that the effect of Ca^{2+} and Ca^{2+}-dependent

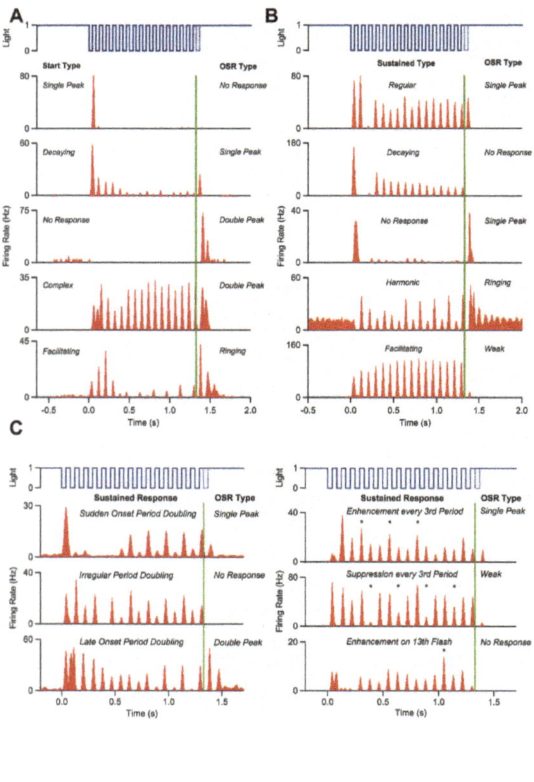

Fig. 15.6 Evidence for oscillations in the spike responses of salamander RGCs to periodic flash sequences. (**A** and **B**) Examples of the variety of response types to periodic flash sequences found among salamander RGC grouped either by OSR type (**A**) or by sustained type (**B**). (**C**) Examples showing oscillations at different frequencies than that of the flash train. *(Modified from Schwartz and Berry (2008).)*

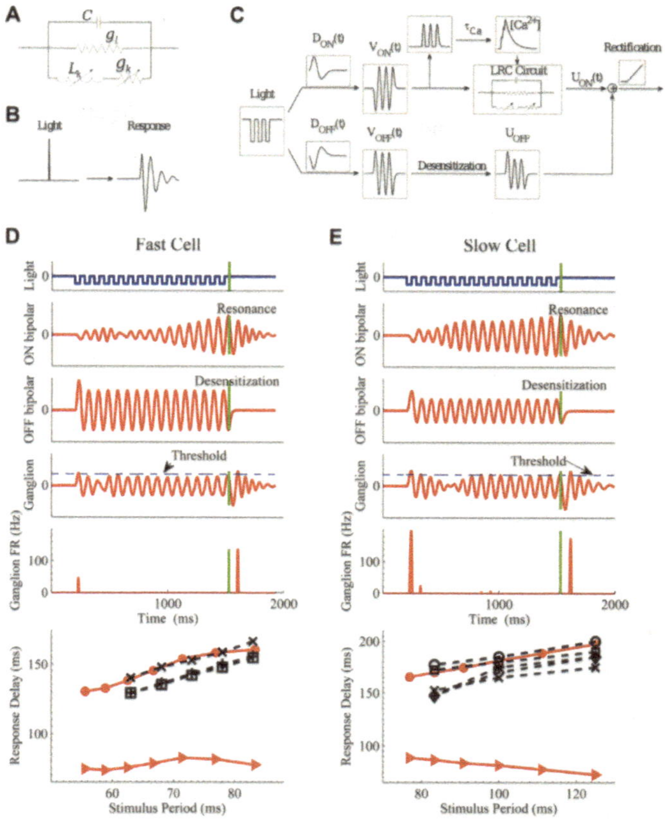

Fig. 15.7 A dynamic resonance model for the OSR. (A) Schematic of the LRC circuit (consisting of an inductor, two resistors, and a capacitor) that forms the core of the model. (**B**) The impulse response of an LRC circuit is a damped oscillation. (**C**) Full schematic of the two-channel (ON-OFF) OSR model. (**D**) *Top*, flash train and corresponding responses of different elements of the model. *Bottom*, response latency from last flash *(circles)* and from omitted flash *(triangles)* as a function of stimulus period. (**E**) Same as (**D**) but for a different parameter range of the LRC circuit. *(Modified from Gao et al. (2009).)*

K^+ channels in BCs could create a dynamic oscillator (Fig. 15.7) (Gao et al., 2009). Another possibility is that different BCs have electrical resonance at different frequencies and that those matching the stimulus either get amplified or pull the others with them through gap-junction connectivity (Menzler and Zeck, 2011; Schwartz and Berry, 2008), though such a model failed to capture all the features of the OSR (Gao et al., 2009). In preliminary data from my thesis, we recorded the responses of salamander BCs to flash sequences and observed depolarizations after the end of the train that appeared to be OSRs (Fig. 15.8) (Schwartz, 2008).

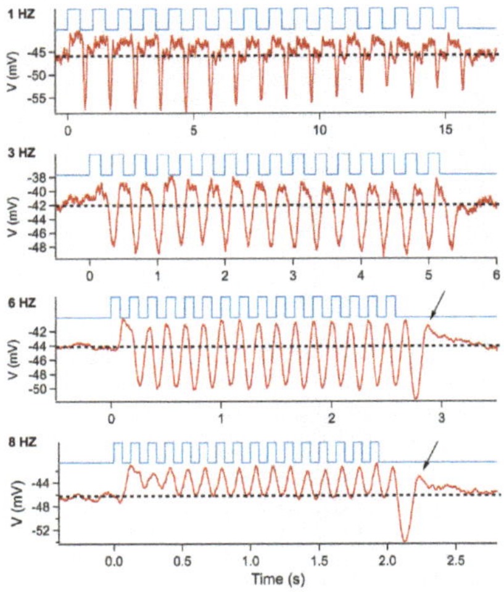

Fig. 15.8 Bipolar cell responses to periodic flash trains. Current-clamp recordings from an ON bipolar cell in salamander responding to bright flash sequences at four different frequencies. OSRs are indicated by arrows. *(Modified from Schwartz (2008).)*

Future directions

Despite being found in a wide range of species, from fish to mice to humans, the OSR remains an obscure research area in retinal computation. Because it appears to be present in BCs and does not rely on amacrine cells, a detailed understanding of the mechanism seems well within reach. Behavioral experiments may also offer insights into how the OSR could be used in vision. Nearly 20 years after I began my career in retinal neuroscience by exploring this fascinating computation, perhaps it is time for another student to take up the effort.

Landmark paper: Schwartz, G., and Berry, M.J. (2008). Sophisticated temporal pattern recognition in retinal ganglion cells. Journal of Neurophysiology *99*(4), 1787–1798.

Discussion questions:

1. Which data from Figs. 3 and 4 suggest resonance in RGCs or their upstream circuits?
2. How do the authors conclude that the timing of the OSR can be updated in a single flash? What insights does this observation provide about mechanism?
3. The paper contains the statement "... the retina seems to maintain simultaneous representations of both the last time interval between flashes and the average time interval in the recent past." What

evidence supports this? Can you explain the behavior in each of the cells in Fig. 7 and annotate the firing events at the end of the sequence for each one?
4. L-AP4 (APB) and CPPG have opposite effects on the mGluR6 receptors on ON BCs. How do you explain their similar effect on the OSR in Fig. 9?

Landmark paper: Gao, J., Schwartz, G., Berry, M.J., and Holmes, P. (2009). An oscillatory circuit underlying the detection of disruptions in temporally-periodic patterns. Network: Computation in Neural Systems *20*(2), 106–135.

Discussion questions:

1. What are "L," "R," and "C" in the LRC circuit? Make a table of the biological element in the model that corresponds to each electrical element in this circuit.
2. Equation 8 computes the resonant frequency of BCs in the model. The authors point to four important terms in the equation. Explain in your own words how each affects the resonant frequency. How does the calcium concentration "self-tune" the model?
3. The authors assume a linear relationship between ON BC voltage and glutamate release. In reality, this relationship has quite a steep nonlinearity, with a Hill coefficient of 4–6. How might this effect the conclusions?
4. What aspects of the OSR behavior in salamander RGCs are not tested in this paper? Do you think some of them would not be well predicted? Why or why not?
5. What specific measurements would be most useful in testing the model in this paper?
6. Why did Werner et al. (2008) conclude that a dual-LN model was able to capture the OSR but the authors of this paper disagree? What are the key arguments on each side of this controversy?

Programming exercise

To access the programming exercise, please visit https://www.elsevier.com/books-and-journals/book-companion/9780128198964.

This exercise is based on the OSR model in Gao et al. (2009). I implemented the core parts of the model in the attached files. The script "OSR_model.m" allows you to run the model up to the summation of ON and OFF BC channels and plot the results.

1. From the summed response, compute the latency from the beginning of each flash to the response peak and then compute the OSR latency, both with respect to the last flash and with respect to the omitted flash. Change the stimulus frequency, collect these values, and plot them as in Fig. 15.2E. Over what range of frequencies does the OSR predict the time of the omitted flash?
2. Implement the final part of the model from Gao et al. (2009) that converts the summed BC responses into a firing rate. Does this affect the OSR performance? Try a different form for the spike nonlinearity, like a sigmoidal function.
3. Now add gain control to the spike rate (see Chapters 2 and 13) as described by Berry et al. (1999). How does this affect the timing of the OSR?
4. The stimulus flashes are set to 25% duty cycle by default. Change this to 10% duty cycle and 50% duty cycle. How does this affect the OSR amplitude and timing? Now try a fixed pulse duration. Is the OSR robust to these changes? Why or why not?
5. Make a stimulus with dark flashes but in which the mean luminance remains fixed (see Supplemental Information in Schwartz et al., 2007). Does this model still give you an OSR? Why or why not? Did you have to change other parameters to make it work?
6. Try to produce as many of the response patterns in Fig. 15.6 as you can by altering the "ON_scale" parameter in the model. What have you learned from this?

Advanced:

7. Gao et al. (2009) state that the parameters "d" and "b" in their model (which you will find in "ON_LRC_model.m") are primarily responsible for tuning the resonance by the calcium concentration. Perform a search of this 2D parameter space to optimize the range over which the OSR tracks the missing flash time predictively. Explain the metric you decided to use to guide your search function.

References

Art, J.J., and Fettiplace, R. (1987). Variation of membrane properties in hair cells isolated from the turtle cochlea. The Journal of Physiology *385*(1), 207–242.

Bar, M. (2011). Predictions in the Brain: Using Our Past to Generate a Future (Oxford University Press).

Berry, M.J., and Schwartz, G. (2011). The retina as embodying predictions about the visual world. In Predictions in the Brain: Using Our Past to Generate a Future, M. Bar, ed. (Oxford University Press).

Berry, M.J., Brivanlou, I.H., Jordan, T.A., and Meister, M. (1999). Anticipation of moving stimuli by the retina. Nature *398*(6725), 334–338.

Bullock, T.H., Hofmann, M.H., Nahm, F.K., New, J.G., and Prechtl, J.C. (1990). Event-related potentials in the retina and optic tectum of fish. Journal of Neurophysiology *64*(3), 903–914.

Bullock, T.H., Karamürsel, S., and Hofmann, M.H. (1993). Interval-specific event related potentials to omitted stimuli in the electrosensory pathway in elasmobranchs: an elementary form of expectation. Journal of Comparative Physiology A *172*(4), 501–510.

Burrone, J., and Lagnado, L. (1997). Electrical resonance and Ca2+ influx in the synaptic terminal of depolarizing bipolar cells from the goldfish retina. Journal of Physiology *505*(3), 571–584.

Crawford, A.C., and Fettiplace, R. (1981). An electrical tuning mechanism in turtle cochlear hair cells. The Journal of Physiology *312*(1), 377–412.

Fettiplace, R. (1987). Electrical tuning of hair cells in the inner ear. Trends in Neurosciences *10*(10), 421–425.

Gao, J., Schwartz, G., Berry, M.J., and Holmes, P. (2009). An oscillatory circuit underlying the detection of disruptions in temporally-periodic patterns. Network: Computation in Neural Systems *20*(2), 106–135.

Gowrisankaran, S., McAnany, J.J., and Alexander, K.R. (2013). Poststimulus response characteristics of the human cone flicker electroretinogram. Visual Neuroscience *30*(4), 147–152.

Karamürsel, S., and Bullock, T.H. (1994). Dynamics of event-related potentials to trains of light and dark flashes: responses to missing and extra stimuli in elasmobranch fish. Electroencephalography and Clinical Neurophysiology *90*(6), 461–471.

McAnany, J.J., and Alexander, K.R. (2009). Is there an omitted stimulus response in the human cone flicker electroretinogram? Visual Neuroscience *26*(2), 189–194.

McAnany, J.J., Alexander, K.R., Kumar, N.M., Ying, H., Anastasakis, A., and Fishman, G.A. (2013). Electroretinographic findings in a patient with congenital stationary night blindness due to a novel NYX mutation. Ophthalmic Genetics *34*(3), 167–173.

McFarland, W.N., and Loew, E.R. (1983). Wave produced changes in underwater light and their relations to vision. Environmental Biology of Fishes *8*(3–4), 173–184.

Menzler, J., and Zeck, G. (2011). Network oscillations in rod-degenerated mouse retinas. Journal of Neuroscience *31*(6), 2280–2291.

Prechtl, J.C., and Bullock, T.H. (1994). Event-related potentials to omitted visual stimuli in a reptile. Electroencephalography and Clinical Neurophysiology *91*(1), 54–66.

Protti, D.A., Flores-Herr, N., and Von Gersdorff, H. (2000). Light evokes Ca2+ spikes in the axon terminal of a retinal bipolar cell. Neuron *25*(1), 215–227.

Schwartz, G. (2008). Computation and Coding in the Retina (Doctoral dissertation, Prinecton University). https://search.proquest.com/openview/b4b459009c5a1721586fe265e20c43ba/1?pq-origsite=gscholar&cbl=18750&diss=y.

Schwartz, G., and Berry, M.J. (2008). Sophisticated temporal pattern recognition in retinal ganglion cells. Journal of Neurophysiology *99*(4), 1787–1798.

Schwartz, G., Harris, R., Shrom, D., and Berry, M.J. (2007). Detection and prediction of periodic patterns by the retina. Nature Neuroscience *10*(5), 552–554.

Werner, B., Cook, P.B., and Passaglia, C.L. (2008). Complex temporal response patterns with a simple retinal circuit. Journal of Neurophysiology *100*(2), 1087–1097.

Color vision is central to the human sensory experience, and it forms the basis of many aspects of art. But the way that we see color represents only a fraction of how it can be perceived. Color processing is incredibly variable across the animal kingdom. Eyes from different species contain different numbers of color channels tuned to different wavelengths of light. The mouse eye is tuned for green in the lower visual field and ultraviolet in the upper visual field. This artistic rendering of the mouse-eye view gives you a sense of the ethological purpose of this arrangement.

CHAPTER 16

Color processing

Gregory William Schwartz[a,b,c]
[a]Departments of Ophthalmology and Physiology, Feinberg School of Medicine, Northwestern University, Chicago, IL, United States
[b]Department of Neurobiology, Weinberg College of Arts and Sciences, Northwestern University, Chicago, IL, United States
[c]Northwestern Interdepartmental Neuroscience Program (NUIN), Northwestern University, Chicago, IL, United States

Color plays a crucial role in visual processing for many animals, including humans. As wavelength is an elementary property of light, color processing begins at the first stage of vision with the relative spectral sensitivities of different photoreceptor types. Subsequent retinal processing creates the "opponent" channels that drive color perception and behavior.

Color vision is an enormous field with scholarship dating back to the time of Issac Newton (Mollon, 2003). Several entire volumes have been written on the subject (Backhaus et al., 2011; Gegenfurtner and Sharpe, 1999; Gouras, 2009; Kremers et al., 2016; Mollon, 2003; Shevell, 2003), as well as an excellent recent review (Baden and Osorio, 2019). Investigations of color vision have also included an impressive diversity of species across all taxa. Thus, this chapter will be shallow by necessity in covering fascinating topics such as the evolution of opsins and photoreceptors, specializations in photoreceptor biophysics, optical specializations for color in different eyes, spectral measurements of visual environments, and color-selective behaviors across species. However, unlike many of the other chapters, a mouse-centered focus is inappropriate given that color processing is so inextricably linked to visual ecology and circuits differ substantially between species (Baden, 2020). In addition to mice, I will highlight examples of color-processing circuits in zebrafish, ground squirrels, rabbits, macaques, and humans while also comparing elements of the circuits in these species to those that are known in amphibians, reptiles, and birds.

Photoreceptors form the foundation for color vision

Color vision begins with opsin molecules that have different spectral sensitivity functions. The probability that an opsin molecule absorbs a photon depends on that photon's wavelength. This sensitivity is determined by both

Retinal Computation
https://doi.org/10.1016/B978-0-12-819896-4.00017-2

the opsin's molecular structure and its bound chromophore that undergoes a photon-initiated confirmational change. Fortunately for vision scientists, it takes only a single parameter, λ_{max} (the wavelength with peak sensitivity), to describe the spectral sensitivity function of known opsin molecules (Govardovskii et al., 2000). While most opsin molecules bind only one type of chromophore, there are opsin types that can switch between two chromophores, called A1 and A2, to adjust their spectral sensitivity dynamically, for example, seasonally (Enright et al., 2015; Loew and Govardovskii, 2001). Individual photoreceptor types typically express a single opsin gene (but not always; Dalton et al., 2014), but their spectral sensitivity in vivo is also affected by the optical filtering of the rest of the eye. The lens and cornea of different animals are adapted for different spectral filtering, and birds and reptiles use colored oil droplets at the base of their cones to narrow their spectral bandwidth (Bowmaker et al., 1997; Goldsmith and Butler, 2003; Hailman, 1976; Hart and Vorobyev, 2005; Loew et al., 2002; Wilby and Roberts, 2017).

In the human-centric view of color vision in popular culture, our trichromatic (blue, green, red) system is considered superior to the many other dichromatic mammals, like "color-blind" dogs, that lack a long-wavelength-sensitive (L) cone. In fact, our trichromatic vision is less complex than that of the majority of vertebrate species. The jawless fish of ~500 million years ago that were ancestors to all modern vertebrates had four different cone types with peaks in the UV, blue, green, and red ranges (Collin et al., 2009). Tetrachromacy has been retained in many species of fish, and additional photoreceptor types have been added in birds and reptiles, both of which have a fifth cone type (double-cones), and some amphibians, which have the main four cone types plus two spectrally distinct rod types (Korenyak and Govardovskii, 2013; Yovanovich et al., 2017). The addition of the L-opsin in mammals was an evolutionarily recent event confined to Old World primates, including humans (Dulai et al., 1999; Mollon et al., 1984; Nathans et al., 1986). As we will see in the following sections, this "new" L-opsin pathway is distinct in its retinal circuitry from the ancient L-cone pathways in nonmammals. While the retinal circuits for color vision in primates are more completely understood than in other vertebrate species, they may also be considerably simpler.

While the complement of photoreceptors provides the substrate for color vision, an individual photoreceptor type cannot convey spectral information since it has no way to disambiguate the intensity of light from its wavelength. Thus, color processing necessarily involves comparisons between photoreceptors of different spectral types. Theoretical considerations about the nature of these comparisons are discussed in the next section, followed by physiological studies of retinal circuits for color processing in different species. The chapter concludes with a discussion of the regionalization of color-processing circuits in the retina and a summary of some of the major unresolved questions in the field.

Theories of efficient and behaviorally relevant color processing

Theories of color processing largely focus on how opsin comparisons evolved to be represented in retinal circuits. Consider a retinal neuron downstream of the photoreceptors. The input from each of *n* photoreceptor types can be negative (sign conserving), zero (no connectivity), or positive (sign reversing). Since vertebrate photoreceptors are OFF cells (they reduce glutamate release for increases in light), by convention, sign conserving synapses are assigned a negative polarity to denote OFF responses. Therefore, the total number of possible connectivity patterns is $3^n - 1$ (excluding the pattern with zero connectivity to any photoreceptor). For example, in a trichromatic retina with short (S), medium (M), and long (L) wavelength-sensitive cones, downstream neurons could be S^+ML^-, L^+, S^+M^-, SL^+M^-, etc. in 26 possible configurations. Tetrachromatic retinas have 80 such possible color channels (Baden and Osorio, 2019). But is there really a retinal circuit for each possible color channel? If not, what principles dictate which color channels survived natural selection and which were discarded?

Buchsbaum and Gottschalk (1983) explored this question from the perspective of information theory and argued that a broad class of light spectra could be efficiently represented by three opponent channels in any trichromatic system. These channels were ordered by the size of their contribution in representing the stimulus spectrum. The first (most informative) channel was achromatic (SML^+ vs. SML^-), representing total luminance. The next channel was the color-opponent red versus green (M^+L^- vs. M^-L^+) with essentially no contribution from S cones. The final channel was the color-opponent blue-yellow channel ($S^+M^-L^+$ vs. $S^-M^+L^-$). The opposite signs of the M and L cones in their model created a triphasic channel, but the authors noted that both M and L contributions were much smaller than the dominant S component, so a biphasic channel (S^+ML^- vs. S^-ML^+) was also consistent with their calculations (Fig. 16.1A). Ruderman et al. (1998) validated this model using spectra of natural images (for humans) captured with a hyperspectral camera. These authors showed that natural images have a stereotyped pattern of correlation in the sensitivity spectra

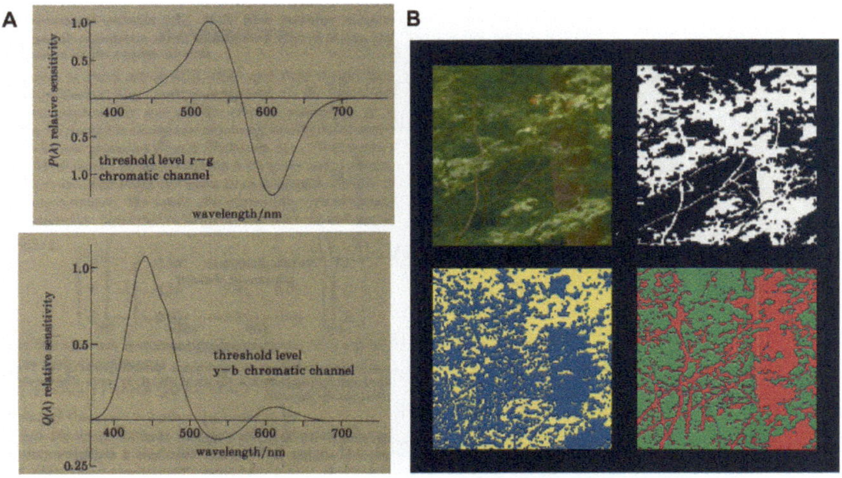

Fig. 16.1 Optimal trichromatic color channels based on information theory. (**A**) Weights derived for an optimal L vs. M channel *(top)* and S vs. LM channel *(bottom)* based on the spectral distribution of natural scenes. (**B**) A natural image *(top left)* and its projection onto the first three (most informative) spectral axes: achromatic *(top right)*, *blue-yellow (bottom left)*, and *red-green (bottom right)*. *(From (A) Buchsbaum and Gottschalk (1983) and (B) Ruderman et al. (1998).)*

of S, M, and L cones and that three channels—achromatic, red-green, and blue-yellow—were optimal for decorrelating the spectra for efficient representation (Fig. 16.1B). Thus, efficient coding theory posits that color-processing circuits evolved to encode the spectral distribution of natural scenes most efficiently.

Of course, the visual system is not a hyperspectral camera in the sense that its goals are more complex than the faithful representation of the spectral content of light. Thus, many researchers have taken a different perspective that incorporates the coevolution of color-vision circuits with the reflectance spectra of behaviorally relevant objects in the environment, like food sources, predators, and prey, or the coloration of conspecifics for mating and other visual communication (Lind et al., 2017; Osorio and Vorobyev, 1996; Vorobyev et al., 1998). One instructive example of this approach is work on the selective pressure that supported the gene duplication event leading to separate M- and L-opsins in Old World primates (Dulai et al., 1999; Mollon et al., 1984). Unlike the relatively equal spectral spacing of the four ancestral opsins (Baden and Osorio, 2019; Collin et al., 2009), primate M- and L-opsins are closely spaced at 533 nm and 565 nm, respectively, and highly overlapping. Some researchers have hypothesized that primate L-opsin evolved to support fine discriminations in this range among shades of green, yellow, and red that contain important information about types of leaves and/or the ripeness of fruit (Dominy and Lucas, 2001; Lucas et al., 2003; Osorio and Vorobyev, 1996). Others have suggested that red-green discrimination aids in attention (Gegenfurtner and Rieger, 2000; Wichmann et al., 2002), predator detection (Pessoa et al., 2014), mate selection (Waitt et al., 2003), and social dynamics (Changizi et al., 2006).

The existence of separate red-green and blue-yellow opponent pathways has strong support in human psychophysics (Gegenfurtner and Kiper, 2003; Mollon, 1982; Stockman and Brainard, 2010) and, as discussed later, clear physiological correlates in the primate retina. As we will see, such a correspondence between efficient color-coding theory and retinal circuits is less clear in other species. It also remains unclear whether retinal circuits must represent both polarities of a color-opponent channel. For example, is an S^+ML^- circuit sufficient to represent the range of blue-yellow contrasts, or is a separate S^-ML^+ circuit needed?

Chromatic circuits in the outer retina

Horizontal cells

Horizontal cells (HCs) are poised to create color-opponent circuits at the first synapse in the retina by receiving excitatory inputs selectively from certain photoreceptor types and making inhibitory outputs selectively onto others. HC circuits represent a prime example where species differences are critical in studying color vision. The mouse retina has a single HC type that has been implicated in rod-cone opponency (described later), but no cone-type-specific HC connectivity has been found in the outer retina. In contrast, adult zebrafish have five HC types, four of them color-opponent, all with selective connectivity among the four cone types (Fig. 16.2A) (Li et al., 2009). Larval zebrafish have at least three distinct HC types with cone-type-specific connectivity already present at 7–8 days postfertilization (Yoshimatsu et al., 2020a) (Fig. 16.2B). Chickens also have four HC types, though their physiology and connectivity have not been as extensively characterized as in zebrafish (Seifert et al., 2020). A comprehensive review by Twig et al. (2003) covers the great variety of color-opponent HC types found in more than a dozen species of fish, amphibians, and reptiles.

Mammalian retinas, in general, lack color-opponent HCs even when they have more than one HC type. Primates, including humans, have

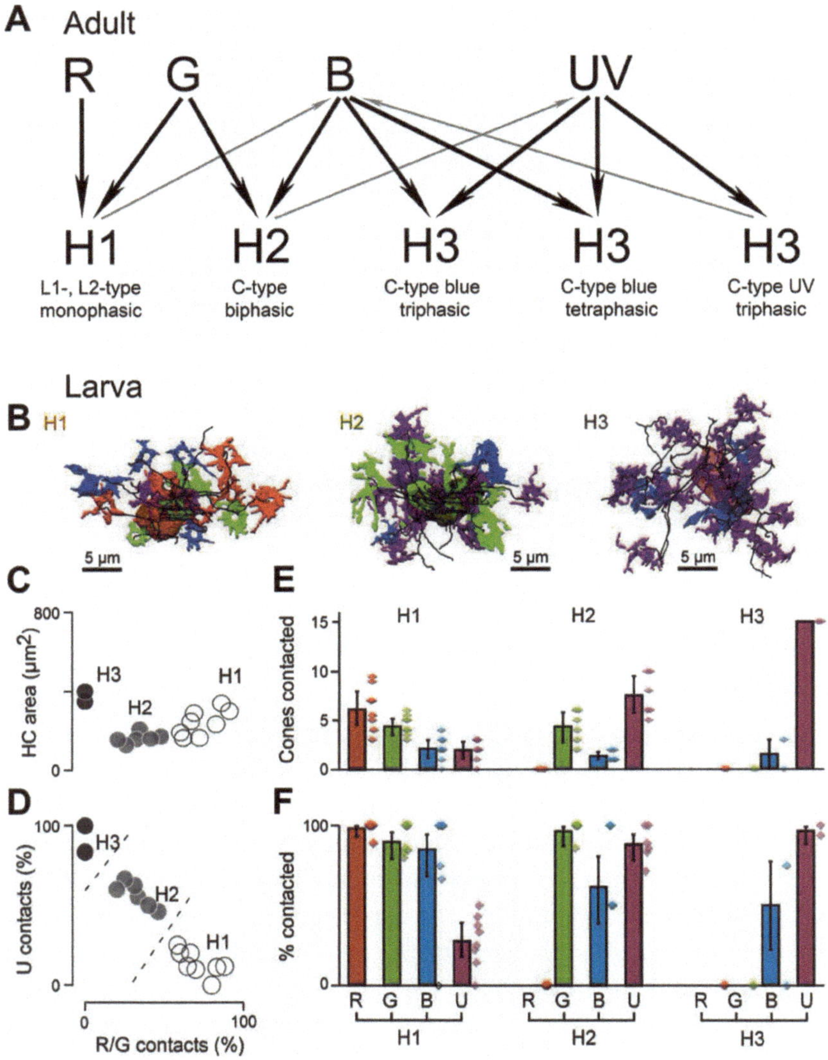

Fig. 16.2 Connections between cones and horizontal cells in the zebrafish retina. (**A**) Summary of cone-HC connectivity in adult zebrafish retina. *Black arrows* indicate synapses from cones onto HCs. *Gray arrows* indicate feedback synapses from HCs to cones. (**B**) Electron microscopic reconstructions of each of the three morphological types of HCs in larval zebrafish along with the set of cones they contact. Cones are colored by their spectral type: UV *(purple)*, S *(blue)*, M *(green)*, L *(red)*. (**C**) Dendritic area of each reconstructed HC (y-axis) versus the percentage of contacts onto L or M cones within its dendritic field (x-axis). (**D**) Percentage of UV cones contacted versus L or M cones. (**E**) Number of cones of each spectral type contacted by H1, H2, and H3 HCs. (**F**) Same as (**E**) but converted to a percentage of cones within each HC's dendritic field. *((A) From Li et al. (2009). (B–F) Modified from Yoshimatsu et al. (2020a).)*

two HC types (Ahnelt and Kolb, 2000; Dacey et al., 1996). H1 HCs receive input from M and L cones, avoiding S cones, and they provide feedback to M and L cones. Thus, they carry an ML$^-$ signal, since they hyperpolarize to increases in light intensity. H2 HCs receive input from all three cone types (SML$^-$) and provide feedback to all three types with a predominance of contacts to S cones (Dacey et al., 1996; Packer et al., 2010). Because S cones are only 8%–10% of the cones of the primate retina, as in most mammalian retinas (Calkins, 2001; Curcio et al., 1991; Kryger et al., 1998; Marc, 1982; Nadal-Nicolás et al., 2020), H2 HCs are dominated by the M- and L-cone components. Thus, despite being nonopponent themselves, they provide a yellow (Fig. 16.3) (ML$^-$) surround to S cones to support blue-yellow color opponency (Packer et al., 2010). We will see in a following section how H1 HCs make an important contribution to the red-green color-opponent circuit in primates (Dacey et al., 1996).

Bipolar cells

Selective connectivity between photoreceptors and bipolar cells (BCs) can bring color information from the outer retina to the inner retina, where it can influence retinal ganglion cells (RGCs). Thus, a key question is whether all cone bipolar cell (CBC) types contact cones indiscriminately within their dendritic fields or whether some CBC types receive input from a subset of cones based on their chromatic type. This question has been addressed most completely in zebrafish and several mammalian species (mouse, ground

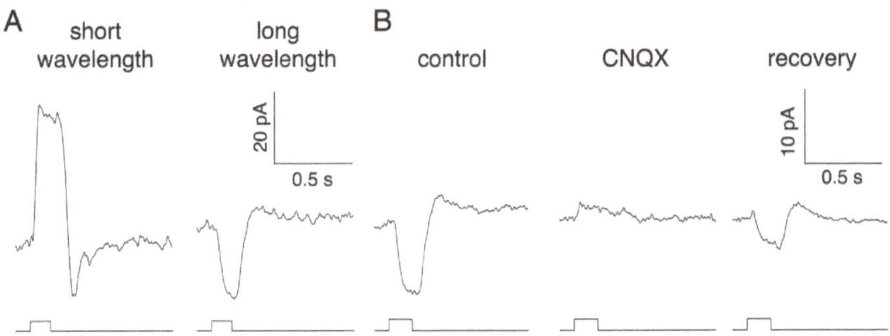

Fig. 16.3 Horizontal cell feedback is required for color opponency in primate S cones. (**A**) Voltage-clamp recordings from a primate S cone showing outward current form S-activating *(blue)* light and inward current for LM-activating *(yellow)* light. (**B**) Recording from another S cone responding to yellow light in control conditions and in the presence of the AMPA receptor antagonist CNQX to block HC feedback. *(From Packer et al. (2010).)*

squirrel, rabbit, and macaque), using various techniques including electrophysiology, functional calcium imaging, and anatomy.

Cone to BC connectivity is complex in fish

Bipolar cells of the retina of most teleost fish are diverse and complex (Connaughton et al., 2004; Sherry and Yazulla, 1993). Many types have multiple, functionally distinct terminals in different lamina of the inner plexiform layer (IPL) (Baden et al., 2014), and some types contact both rods and cones (Ishida et al., 1980). Zebrafish retina, where their morphology has been most extensively characterized, contains ~20 different BC types (Connaughton et al., 2004). Physiologically, color opponency has been measured both at the soma of goldfish BCs (Fig. 16.4) (Shimbo et al., 2000) and at the axon terminals of zebrafish BCs (Fig. 16.5) (Zimmermann et al., 2018), though in both cases, color-opponent cells were somewhat rare (~10%–15% of BCs). Importantly, however, a BC need not be color-opponent to carry chromatic information. For example, selective connectivity with UV cones could create a pure UV^+ BC, but the opponent (e.g., SML^-) signal might arrive at the RGC through a different BC or an amacrine cell (AC). Thus, we must make the distinction between color-

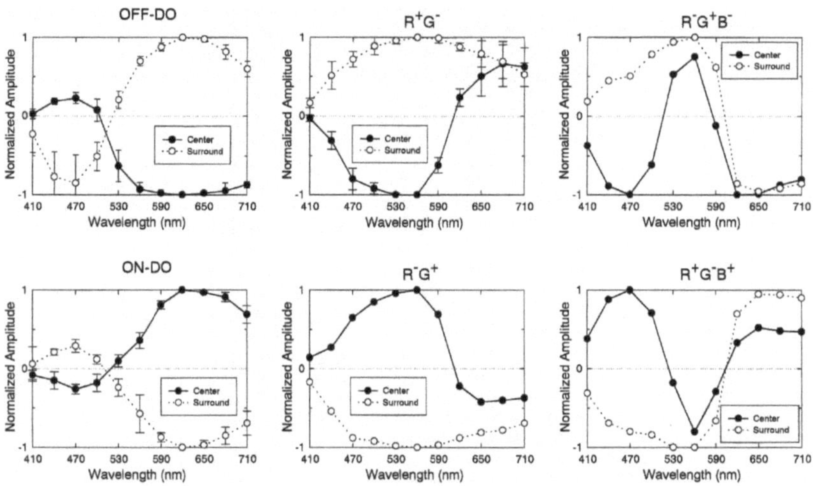

Fig. 16.4 Color-opponent responses in goldfish bipolar cells. Examples of different types of spectral tuning curves found in goldfish bipolar cells for RF center *(filled symbols)* and surround *(open symbols)* stimulation. Each cell is assigned to a response type based on its RF center response. Cells in the left column are "double-opponent" (DO) because they show opposite opponency for RF center vs. surround stimulation. *(From Shimbo et al. (2000).)*

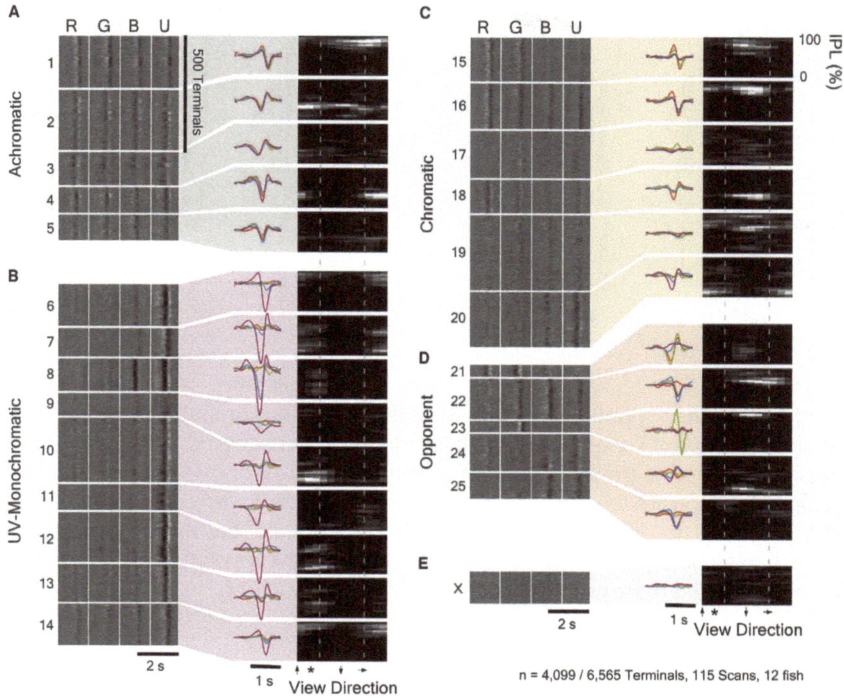

Fig. 16.5 Clusters of chromatic responses from zebrafish BC terminals. (**A**) *Left*, temporal filters for each BC terminal shown as a line in the heatmap beneath each of the four LED colors (R, G, B, U). Terminals are grouped into five achromatic clusters. *Middle*, mean temporal filter for each wavelength (colors) for each of the five clusters. *Right*, spatial distribution of terminals from each group. *Arrows* at the bottom indicate viewing angle. *Asterisk* indicates the strike zone. (**B–E**) Same as (**A**) for UV-Monochromatic, Chromatic, Opponent, and Unresponsive groups of BC terminals. *(Modified from Zimmermann et al. (2018).)*

opponent BCs and spectral cone-type-selective BCs. The latter category is defined physiologically by chromatic weights that differ substantially and systematically from the underlying cone mosaic or anatomically by a strong bias in cone connectivity within the dendritic field.

A large-scale survey of over 4000 BC terminals in larval zebrafish clustered their chromatic responses into 26 groups, including 5 that were color-opponent and 15 that were spectrally selective (Fig. 16.5) (Zimmermann et al., 2018). The asymmetric distribution of these response clusters as a function of retinal position is discussed in the following section on regionalization. An equally impressive anatomical survey of the cone contacts of 321 BCs in adult zebrafish identified up to 18 cone connectivity patterns (with 9

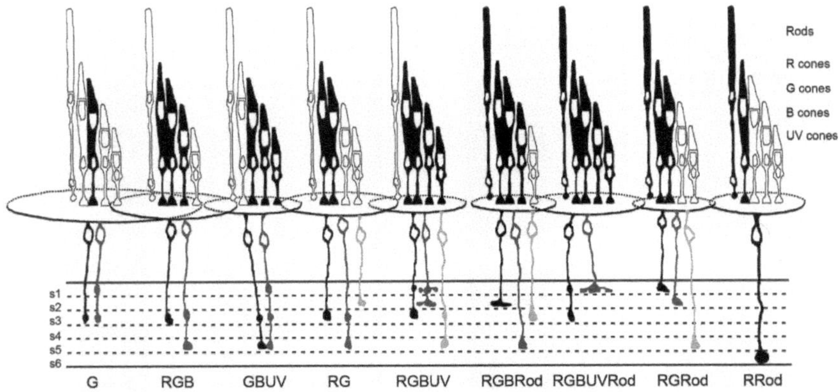

Fig. 16.6 Patterns of photoreceptor to BC connectivity in zebrafish. Schematic of the nine major groups of cone-to-BC connectivity found in the adult zebrafish retina. Cone types are indicated to the right. Discs show the relative sizes of the BC dendritic fields, and drawings indicate the stratification patterns observed in each group. Cone types providing input to this BC group are colored *black*. In each group, the most prevalent morphological BC pattern is drawn in *black*, the second most prevalent in *gray*, and the third most prevalent in *silver*. *(From Li et al. (2012).)*

dominant ones) (Li et al., 2012) (Fig. 16.6). While the number of cone-to-BC connectivity patterns is likely smaller than the 80 possible combinations in this tetrachromatic retina, it is nonetheless clearly a complex and sophisticated set of color-processing circuits that we are only beginning to understand.

Mammals have a conserved S-ON CBC but may differ in additional chromatic BCs

Cone contacts among BC types have been studied extensively in several mammalian species. A common motif in all of these species is the presence of a single S-cone selective ON BC type. S-cone selective ON BCs have been identified in Old World (Kouyama and Marshak, 1992; Mariani, 1984) and New World primates (Ghosh and Grünert, 1999), rabbits (Mills et al., 2014), ground squirrels (Li and DeVries, 2006), and mice (Breuninger et al., 2011; Haverkamp et al., 2005; Nadal-Nicolás et al., 2020), and they share strikingly similar morphology across these species (Fig. 16.7).

Cone-type selectivity for M and L cones appears to be more divergent across mammalian species and is more controversial. M- and L-cone connectivity to midget and diffuse BCs is mostly random, though there is a small preference for groupings of cones of the same type in the periphery (Field et al., 2010). There is a single report of a possible M- or L-cone

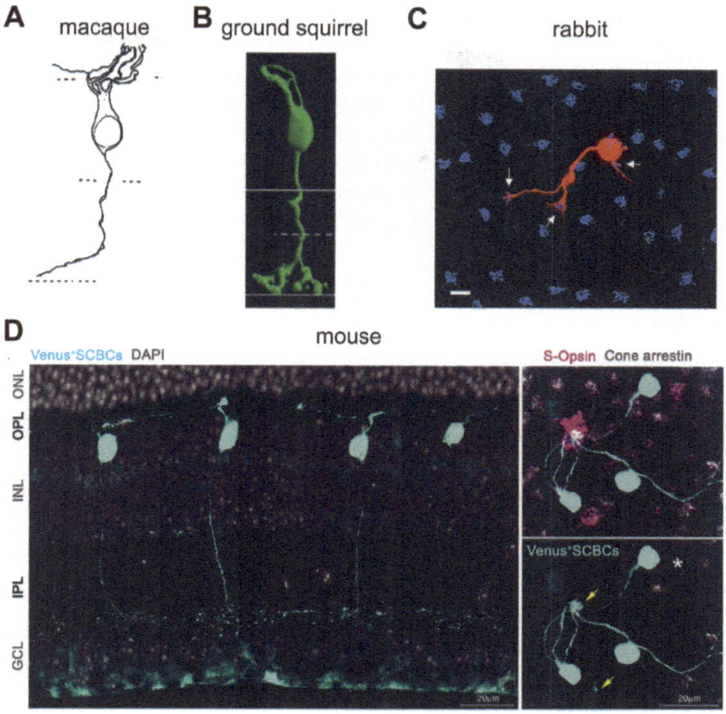

Fig. 16.7 S-cone ON BCs in different mammals. (**A**) Drawing of an S-cone BC from macaque. (**B**) Reconstruction of an imaged S-cone BC in ground squirrel. (**C**) *Top view* of an S-cone BC in rabbit retina *(red)* with arrows indicating contacts with S cones *(blue)*. (**D**) Transgenically labeled S-cone BCs in mouse in side view *(left)* and in top view *(right)* where they can be seen making specific contacts with S cones *(magenta)*. *(From (A) Kouyama and Marshak (1992), (B) Li and DeVries (2006), (C) Mills et al. (2014), and (D) Nadal-Nicolás et al. (2020).)*

selective giant ON bipolar cell in the macaque (Joo et al., 2011), but to my knowledge, no other reports of a selective M-ON BC in a mammalian retina. Ground squirrel retina contains two OFF BCs that receive only M-cone input (Fig. 16.8) (Li and DeVries, 2006), though their role, if any, in color processing, is unclear. Evidence for a specific M-OFF BC in mice (possibly type 1 BCs) is mixed (Behrens et al., 2016; Breuninger et al., 2011; Szatko et al., 2020).

S-OFF BC pathways also show heterogeneity among mammalian species. There is no dedicated S-OFF BC in primates, though S cones sometimes make contacts with OFF (but not ON) midget BCs (Field et al., 2010; Klug et al., 2003). In the central retina of primates, midget BCs collect input from single cones, effectively creating a sparse set of S-OFF BCs (Klug et al.,

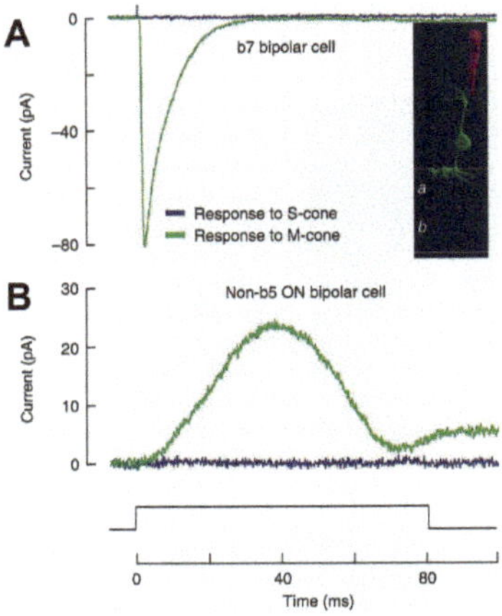

Fig. 16.8 M-OFF BC in ground squirrel. (**A**) Evoked currents in an M-OFF (b7) BC in ground squirrel from paired recordings with a presynaptic M cone *(green)* and S cone *(blue)*. Inset shows the recorded M cone and postsynaptic BC. Inset width is 27 μm. (**B**) Same as (**A**) for an ON BC (not type 5). *(From Li and DeVries (2006).)*

2003; Patterson et al., 2019) (though not in the fovea where S cones are absent; Calkins, 2001). While an S-OFF BC seems to be lacking in mice (Behrens et al., 2016), both physiological and anatomical data suggest that there is an S-OFF BC in the rabbit retina (Mills et al., 2014).

Chromatic circuits in the inner retina

Due in part to the large number of cell types (e.g., ~45 RGC types and 60+ AC types in mice as opposed to 15 BC types; Bae et al., 2018; Shekhar et al., 2016; Yan et al., 2020), the field has a less complete understanding of how chromatic signals are processed in the inner retina than in the outer retina. However, several circuit motifs recurring in multiple mammalian species are described in the following sections. Inner retinal color-processing circuits in tetrachromatic vertebrates are almost entirely unknown, but a hint of their complexity comes from RGC recordings in turtle that have demonstrated 12 separate color-opponent channels (Rocha et al., 2008).

Amacrine cells invert S-ON signals into S-OFF

As mentioned earlier, the S-ON CBC is conserved in mammals while an S-OFF BC has been found only in the rabbit. Nonetheless, color-opponent S-OFF/M(L)-ON RGCs have been found not only in rabbit (Mills et al., 2014) but also in ground squirrel (Sher and Devries, 2012), guinea pig (Yin et al., 2009), and macaque (Dacey et al., 2003). Since these species all have S-ON BCs, but many lack S-OFF BCs, there had been speculation that an AC could invert the S-ON signal into an S-OFF signal in these circuits. Chen and Li (2012) were able to record from such an AC in the ground squirrel and prove that it indeed receives input from S-ON CBCs (Fig. 16.9) and likely inhibits downstream S-OFF/M-ON RGCs (Sher

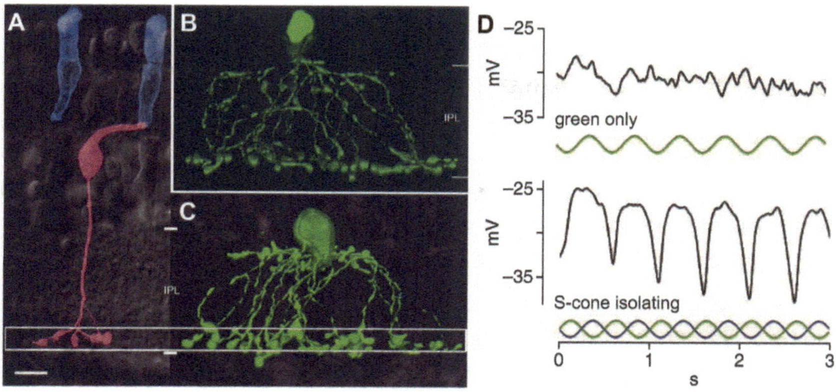

Fig. 16.9 An S-cone selective amacrine cell in the ground squirrel retina. (A–C) Images of S cones *(blue)*, an S-cone BC *(magenta)*, and the S-cone AC *(green)*. Scale bar = 10 μm. (**D**) Current-clamp recordings from an S-cone AC in response to a sinusoidally modulated green light *(top)* and an S-cone isolating stimulus *(bottom)*. *(Modified from Chen and Li (2012).)*

and Devries, 2012). Pharmacological experiments in rabbit suggest that, despite the presence of an S-OFF BC, a similar sign-inverting AC pathway provides the S-OFF signal (S-ON inhibition) to at least one color-opponent RGC type (Mills et al., 2014), which has also been found in guinea pig (Yin et al., 2009). Critically, this pathway from the S-ON CBC through an AC is blocked by the type III metabotropic glutamate receptor agonist, L-AP4, because it depends on an ON CBC that uses these receptors (Slaughter and Miller, 1981). Direct pathways through an S-OFF BC are insensitive to L-AP4 since OFF CBCs use ionotropic glutamate receptors. A morphologically similar S-OFF/ML-ON RGC type in macaque may use a homologous circuit (Dacey et al., 2003), but voltage-clamp and pharmacology experiments for circuit dissection have not been performed on these RGCs. So far, the ground squirrel is the only species in which the identity of the S-cone inverting AC has been determined, but improving genetic access to ACs in mice (Yan et al., 2020), as well as electron-microscopic reconstructions of retinal circuits in mice (Helmstaedter et al., 2013), rabbits (Marc et al., 2013), and primates (Patterson et al., 2019) promise to reveal new insights into these and potentially other AC types involved in color-processing in the mammalian retina.

Color-opponent RGCs in mammals

While blue-ON/yellow-OFF RGCs in primates were recorded as early as 1960 by Hubel and Wiesel (1960), Dacey and Lee (1994) are credited with discovering a morphological S-ON/ML-OFF RGC type in primate retina called the small bistratified cell. These RGCs, which are the fifth most numerous RGC type in the macaque and human retina (after ON and OFF midget and ON and OFF parasol) (Dacey, 1993), have dendrites in the inner portion of the IPL, where they receive S^+ input from S-ON CBCs, and in the outer portion of the IPL, where they receive ML^- input from OFF CBCs (Dacey, 2000). While the OFF CBCs synapsing onto small bistratified RGCs do not appear to be ML-selective, the much larger fraction of M and L cones than S cones in the primate retina makes all nonselective CBCs ML-dominant (Dacey, 2000). Dacey et al. (2003) later recorded from populations of sparse RGC types in the macaque after retrograde tracing from dorsal lateral geniculate nucleus and found two additional color-opponent RGC types: a large bistratified S-ON/ML-OFF cell and an S-OFF/ML-ON cell that was monostratified in the ON portion of the IPL (Fig. 16.10).

Color opponency without selective wiring: The red-green system in Old World primates

Since the gene duplication event that created L cones in Old World primates was so recent (probably less than 30 million years ago; Dulai et al., 1999), perhaps red-green color opponency in these animals, including humans, "piggybacked" onto existing retinal circuits rather than evolving its own dedicated circuits. With a few caveats discussed next, this is essentially the

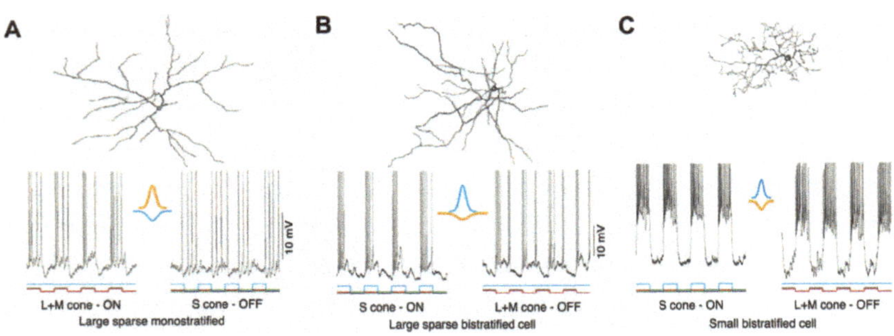

Fig. 16.10 ***Blue-yellow*** **opponent RGCs of the primate retina.** (**A**) Traced image of a large sparse monostratified RGC *(top)* along with its responses to square-wave modulation of L + M stimulating light *(bottom left)* and S-stimulating light *(bottom right)*. Inset shows the spatial structure of the RF with similarly sized LM-ON *(yellow)* and S-OFF *(blue)* regions. (**B** and **C**) Same as (**A**) for two additional color-opponent RGC types in the primate retina. *(Modified from Dacey and Packer (2003).)*

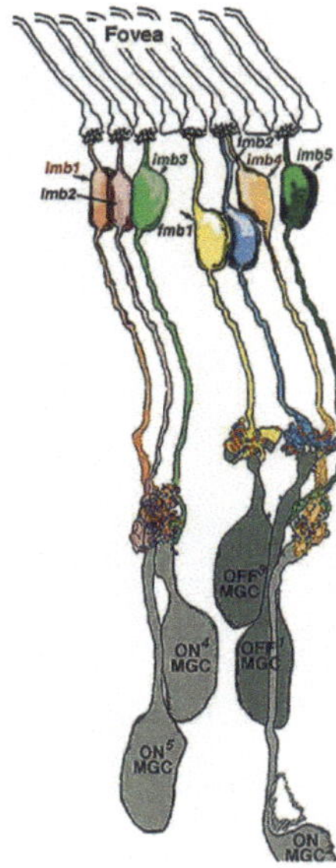

Fig. 16.11 The 1:1:1 pathway of the primate fovea. Drawing of the midget pathway in the central primate fovea from an electron microscopic reconstruction. Midget BCs are individually colored and labeled as invaginating (imb, ON) or flat (fmb, OFF) based on the morphology of their cone contacts. *(From Kolb and Marshak (2003).)*

accepted view (Dacey and Packer, 2003; Lennie, 2000). Red-green color-opponency in primates does not appear to require selective wiring in the inner or the outer retina. Instead, it is largely a consequence of limited (or zero) divergence in the midget pathway. Specialized circuits in the primate fovea connect single cones to single midget BCs to single midget RGCs (Fig. 16.11) (Kolb and Marshak, 2003; Paulus and Kröger-Paulus, 1983). Thus, even in the absence of selective wiring, the RF center of a midget RGC is spectrally pure since it receives input from a single M or L cone (there are no S cones in the fovea; Hofer et al., 2005). The RF surround of midget RGCs is provided by H1 horizontal cells (Packer and Dacey, 2002). Because horizontal cells collect from multiple cones, even in the fovea where their electrical coupling is weak or absent (Packer and Dacey, 2002), the RF surround is spectrally mixed (LM^-).

But what happens outside the fovea, where multiple cones converge onto midget BCs and multiple midget BCs converge onto midget RGCs? Indeed, red-green opponency degrades in humans toward the periphery, but it does not go to zero (Mullen and Kingdom, 1996), and strong red-green opponency has been measured in RGCs in peripheral primate retina (Martin et al., 2001). While it is far from spectrally pure, nonrandom functional clustering of inputs from cones of the same type has been observed in peripheral midget RGCs in macaque (Fig. 16.12) (Field et al., 2010).

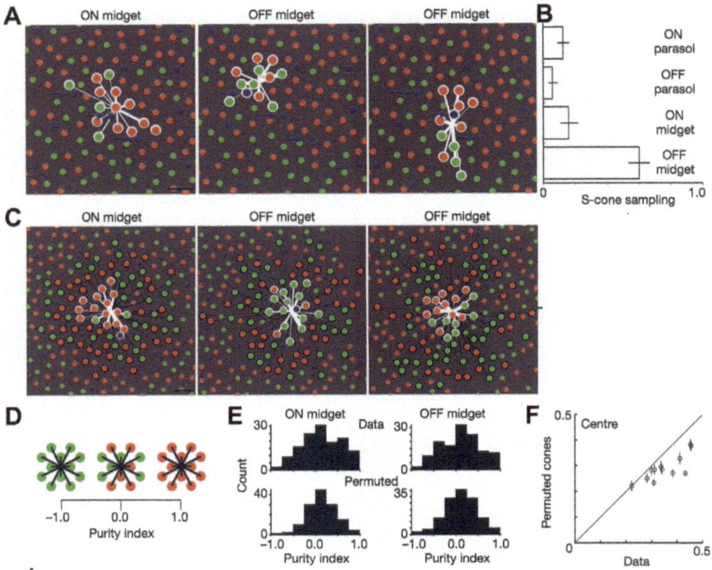

Fig. 16.12 Nonrandom sampling of M and L cones by parasol and midget RGCs in peripheral primate retina. (**A**) Computed weights of each cone in the mosaic from peripheral macaque retina to a particular RGC. Cones are drawn at their respective locations and colored according to spectral type. Circled cones provide input to the recorded RGC, and the width of the white line from the cone to the RGC RF center indicates the strength of the connection. The *leftmost* data shows an ON midget RGC that avoids S cones within its dendritic field *(blue arrows)* while the *middle* and *right* data show OFF midget RGCs that do receive S-cone input. (**B**) Fraction of S cones within the dendritic field sampled by each RGC type. (**C**) Similar plots as in (**A**) for three additional midget RGCs. The left and middle RGCs receive relatively pure L- or M-cone input. The cell on the right receives mixed input. (**D**) Schematic of purity index. (**E**) Histogram of purity index for ON and OFF midget RGCs with measured cone identities *(top)* and with randomly permuted cone identities *(bottom)*. (**F**) Mean purity index for all midget RGCs in each recording (each point) for measured (x-axis) or permuted cone identities (y-axis). *(Modified from Field et al. (2010).)*

Not only is the L-M cone mosaic in primates random with (nearly) random wiring to BCs, but the overall ratio of L:M cones varies enormously between different people with normal color vision. Remarkably, a person with an 80:20 L:M ratio performs the same on color identification and discrimination tasks as a person with a 50:50 ratio or a 20:80 ratio (Hofer et al., 2005). Thus, the prevailing view is that the perception of color in the red-green system develops through synaptic plasticity in the brain. In support of this view, dichromatic New World primates (Mancuso et al., 2009) and even mice (Jacobs et al., 2007) can learn to perform trichromatic color discrimination after the random viral expression of human L-opsin in a subset of cones. The plasticity of brain circuits for color vision is even more noteworthy considering that these experiments did not create pure L cones but instead added random amounts of L-opsin to cones already expressing their own opsins. Based on these results, viral strategies to introduce missing opsins are being investigated as potential treatments for color blindness in humans (Neitz and Neitz, 2014).

Color circuits with rods

Rods are specialized for sparse photon detection (see Chapter 1), and our typical perception of rod-based vision is that it lacks color. As mentioned earlier, some amphibians have two rod types with different spectral sensitivities, and behavioral experiments have demonstrated color discrimination in extremely dim luminance in frogs and toads (Korenyak and Govardovskii, 2013; Yovanovich et al., 2017). Most species, however, including all mammals, have a single rod type with a spectrum typically peaking at a slightly shorter wavelength than that of M cones. Thus, rods and cones would need to be coactive to support color vision in dim conditions.

Indeed there are luminance ranges, typically referred to as "mesopic," under which rods and cones both contribute to vision. In humans, the mesopic range is somewhat narrow, but color vision has been reported in people lacking both M and L cones, and its spectral properties are consistent with a system using S cones and rods (Reitner et al., 1991). In mice, the mesopic range is much larger (Nathan et al., 2006) and, in fact, rods and cones may be coactive even in very bright conditions because rods escape saturation (Rhim et al., 2020; Tikidji-Hamburyan et al., 2017). Joesch and Meister (2016) described a circuit in an RGC type in mice (the J-RGC) in which an S-OFF center is opposed by a rod-ON surround to create a UV^-/$green^+$ (mouse S cones peak in the UV at 365 nm). As there is no S-OFF BC in mice, how did these RGCs achieve their UV-dominated RF center? As discussed next, opsin expression is regionalized in mice such that S-opsin dominates in the ventral retina. Thus, RGCs in this region could achieve rod-cone color opponency without a dedicated S-cone BC. Recent evidence from large-scale calcium and glutamate imaging suggests that J-RGCs are not unique in their rod-cone opponency in ventral retina (Fig. 16.13) (Szatko et al., 2020). Through different types of experiments—pharmacology (Joesch and Meister, 2016) and glutamate imaging from cones (Szatko et al., 2020)—both studies pointed to horizontal cells as key players in the rod-cone opponent circuits in mice. If horizontal cells receive input from rods (even nonselectively since they are numerically dominant in mice) and provide feedback selectively onto cones, they could support such a color-opponent circuit.

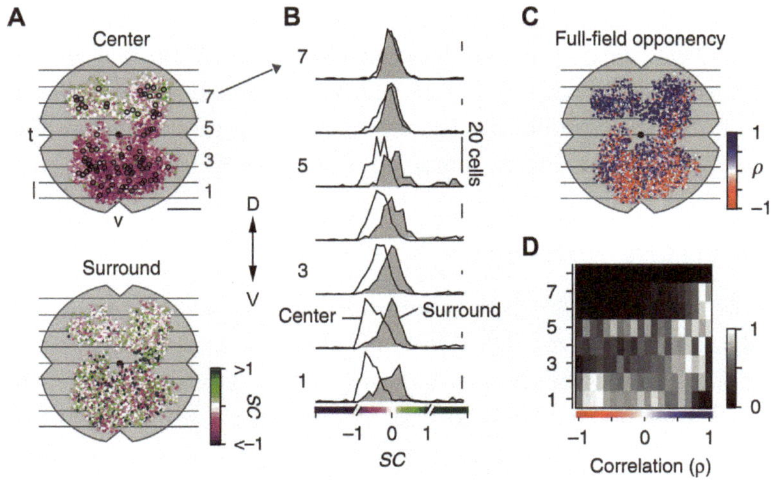

Fig. 16.13 Color opponency in mouse RGCs across the retina. (**A**) RGCs recorded with two-photon calcium imaging shown at their positions in the retina and colored by the spectral contrast of their responses to RF center stimulation *(top)* and RF surround stimulation *(bottom)*. Temporal (t) and ventral (v) directions indicated. Scale bar = 1 mm. (**B**) Data from (**A**) plotted as histograms at seven locations along the dorsoventral axis. (**C**) Full-field color opponency for each RGC defined as the minimum correlation for green vs. UV flashes (at either light onset of offset). (**D**) Histograms of full-field opponency vs. dorsoventral position as in (**B**). *(Modified from Szatko et al. (2020).)*

Regional specialization

Mice are indeed a prime example of regional specialization in color circuits. Consistent with the rod-cone opponent RGCs found predominantly in the ventral retina, color-opponent responses in the dorsal lateral geniculate nucleus and behavior appear to be restricted to the upper visual field (Denman et al., 2017, 2018) (the part of visual space projected onto the ventral retina) (but see Mouland et al., 2021). The dorsal-ventral gradient in S-opsin expression is not unique to mice. It is found in many mammals, from hamsters to hyenas (Peichl, 2005). The evolutionary basis for this regional specialization is thought to involve detecting dark objects (often predators) against the UV-bright sky. This theory is supported by contrast-sensitivity measurements from cones showing a bias toward OFF-sensitivity in the ventral retina that is well aligned to the color statistics of natural scenes as viewed through the mouse eye (Baden et al., 2013). Primates lack a dorsal-ventral opsin gradient, but as discussed earlier, the specialization of the primate fovea creates red-green opponency based on single cones. Regional specialization of retinal anatomy, including cell types and densities, has been measured in hundreds of species (Baden et al., 2020), but relatively less is known about how function, and specifically color processing, vary in the retinas of different species with position in the eye.

Recent work in Tom Baden's lab on zebrafish has established this species as a strong model for studying regional specialization in color processing. In large-scale in vivo imaging studies, they measured the color sensitivity distributions of cones, BCs (Yoshimatsu et al., 2020a,b; Zimmermann et al., 2018) and RGCs (Zhou et al., 2020) across retinal position (Fig. 16.14). They found

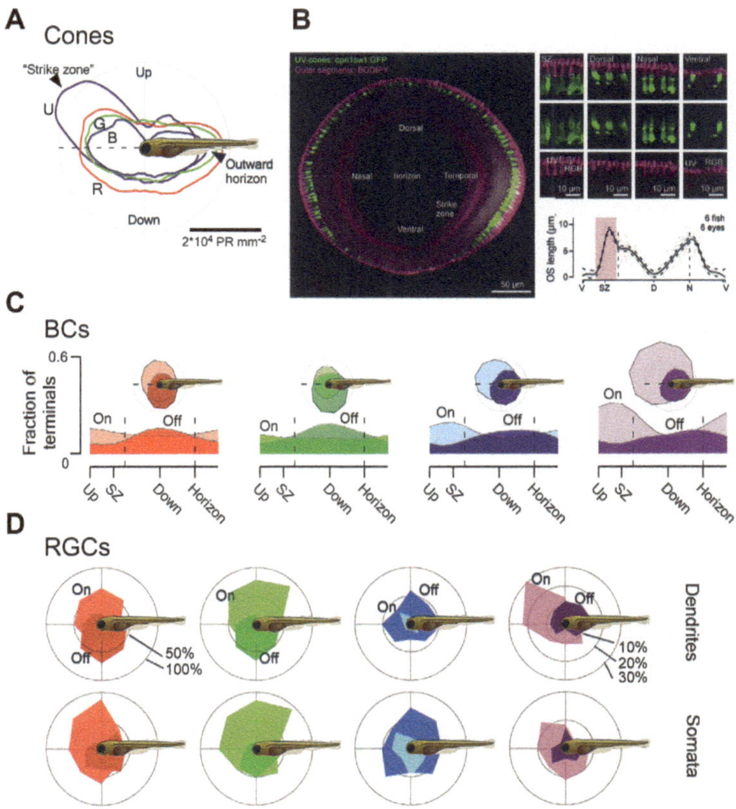

Fig. 16.14 Regionalization of color channels in the zebrafish retina. (**A**) Distribution of each spectral cone type in the larval zebrafish as a function of position in visual space. (**B**) *Left*, image of a larval zebrafish retina in which UV cones are green and the outer segments of all cones are magenta. *Right top*, magnified images of regions from the strike zone (SZ) and from dorsal, nasal, and ventral retina. *Right bottom*, outer segment length of UV cones as a function of retinal position. (**C**) Distributions of BC terminals as separated by spectral sensitivity with position in the retina. *From left* to *right*, *red*, *green*, *blue*, and UV-sensitive terminals each separated into ON and OFF groups as indicated. (**D**) Similar to (**C**) but for the calcium responses of RGC dendrites *(top)* and somas *(bottom)*. *(From (A, C) Zimmermann et al. (2018), (B) Yoshimatsu et al. (2020b), and (D) Zhou et al. (2020).)*

a wide variety of asymmetric distributions of color-opponent and color-selective cell types. While the regionalization of color channels had some close associations with measurements of the spectral content of zebrafish natural scenes (Zimmermann et al., 2018), there were also some notable differences between BCs and RGCs, like the predominance of blue-OFF RGCs despite similar numbers of blue-ON and blue-OFF BCs (Zhou et al., 2020). One "striking" example of regionalization in the zebrafish retina is the existence of a "strike zone" in the ventral-temporal retina where anatomical, molecular, and functional specializations in cones and downstream circuits support the detection of small, UV-bright prey in a region in front of and above the animal. Indeed, this is the region of the visual field in which zebrafish attack paramecia, which reflect UV light, and the fish rely on UV illumination and UV cones to hunt effectively (Yoshimatsu et al., 2020b).

Summary

Color vision is a field with a long and storied history, but unlike another established retinal computation—Photon Detection—where this book began, our understanding of color processing in the retina is not nearly as advanced. The diversity and complexity of retinal circuits for color vision, particularly in non-mammals, are both daunting and exciting. Color processing is also the area of retinal computation that has been most closely tied to visual ecology, with incredible comparative studies from birds to fish to mammals and thousands more in the many invertebrate species that have amazing specializations for sensing color in their own evolutionary niches. The review cited at the beginning of this chapter pointed to some specific questions for future work (Baden and Osorio, 2019), and I will not repeat all of them here. However, one particular question raised by these authors is worth consideration as it relates to previous chapters. How do the retinal circuits for color support adaptation and the percept of color constancy in different illumination conditions?

Landmark paper: Mills, S.L., Tian, L.M., Hoshi, H., Whitaker, C.M., and Massey, S.C. (2014). Three distinct blue-green color pathways in a mammalian retina. Journal of Neuroscience *34*(5), 1760–1768.

Discussion questions:

1. How does the comparison of in-phase vs. out-of-phase modulation of blue and green LEDs in Fig. 1 demonstrate color opponency? Is it problematic that both LEDs activate both S and M cones to some extent? Why or why not?

2. What are cone-isolating stimuli? How are they constructed from LEDs that are not perfectly aligned to the peak wavelength sensitivity of each opsin?
3. Explain the pharmacology results for the S-ON RGC in Fig. 2. Why does L-AP4 eliminate all responses? What does HEPES do to the response pattern and why?
4. Contrast the pharmacology results in the S-ON RGC with those in the bistratified S-OFF RGC. What do the authors conclude from these differences about the distinct circuits involved in color opponency in these RGCs?
5. The authors compare the inverted S-OFF RGC to an RGC type found in the ground squirrel and for which the inverting amacrine cell was found (see Fig. 16.9). What aspect of their pharmacology results in Fig. 2 is not consistent with a rabbit homolog of this amacrine cell being the sole source of the S-OFF signal in this RGC? How do the authors address this discrepancy?
6. Which drug inverts the M-ON response in bistratified S-OFF RGCs (Fig. 3)? Why?
7. Are you convinced by the anatomical evidence for S-OFF bipolar cells shown in Fig. 5? Why or why not? What additional experiments would you like to see to prove their existence?
8. What do the authors say about the purpose of the outer IPL stratification of the bistratified S-OFF RGCs? Which RGC types mentioned in other chapters share this anatomical feature?

Landmark paper: Zimmermann, M.J.Y., Nevala, N.E., Yoshimatsu, T., Osorio, D., Nilsson, D.E., Berens, P., and Baden, T. (2018). Zebrafish differentially process color across visual space to match natural scenes. Current Biology *28*(13), 2018–2032.e5.

Discussion questions:

1. What is Snell's window? Why is color opponency unlikely to be used in the part of the retina that typically looks at Snell's window?
2. Compare and contrast the results in Fig. 1 with those in the papers described in the section "Theories of 'optimal' color processing" and Fig. 16.1 for trichromatic systems. This work introduces the additional parameter of spatial position (elevation in this case) in analyzing the spectrum of natural images. What is the main finding about how color varies with elevation in the natural environment of the zebrafish? How do you think this plays out in terrestrial animals? What about birds?
3. What is the most dramatic asymmetry the authors discovered in the photoreceptor distribution? Compare and contrast this to the primate fovea.

4. What are the key regional asymmetries the authors found in BC color responses? How precisely is color processing divided into different layers of the IPL?
5. Is the presence of the "strike zone" supported by efficient coding theory? Why or why not? In general, what external information beyond the statistics of natural scenes is important for thinking about the optimization of color-processing circuits?

Programming exercise

To access the programming exercise, please visit https://www.elsevier.com/books-and-journals/book-companion/9780128198964.

This exercise explores color coding theory from the perspective of discriminating potential food items. Load the file "sampleSpectra_lowNoise.mat." This file contains the following variables:

1. wavelength: (1 × 501) wavelength in nm from 300 nm to 800 nm.
2. food_spectra (500 × 501):500 examples of spectra from edible, but not desirable fruit for our made-up organism. Each row is a different example. The columns are reflectance (%) for each for the 501 wavelengths.
3. good_food_spectra (500 × 501): same as "food_spectra," but for a highly desirable fruit.
4. poison_spectra (500 × 501): same format as the previous ones, but this time for toxic fruit.

Problems

1. Plot the mean spectrum for each of the fruit types. Add error bars. Which wavelength ranges do you think will be most useful for distinguishing between them? Why?
2. Use the function "opsin_template.m" to create sensitivity spectra for four different kinds of cones. This function uses the equations in Govardovskii et al. (2000) to estimate the sensitivity curve from the single parameter λ_{max}, the peak sensitivity wavelength. Use λ_{max} values from the four opsins in goldfish (UV = 356 nm, S = 447, M = 537, L = 623). Plot the curves in different colors on the same graph. Which color channels (e.g., UVS^+/ML^-) do you think will be most useful for the fruit discrimination task?
3. Compute the activation of each of the four cones by each of the 1500 fruit examples. Hint: if your cone spectra are sampled the same as the fruit spectra, the activation can be computed in a single line as a dot product (MATLAB * operation).
4. Assemble all the results together into a 1500 × 4 matrix (fruits by cone activation) and run principal components analysis (MATLAB "pca"). What fraction of the variance in the data is explained by each of

these four principal components (PCs)? Plot the coefficients for the first PC. These can be thought of as weights on the four cone types. Does the PCA result fit with your intuition from the mean fruit spectra in part 1? Why or why not?

5. Now make your matrix of activation values into a MATLAB table and assign each column variable a name ("UV," "S," "M," "L") matching its cone type. Add a fifth (string) column called "fruit_type" and assign values "poison," "edible," and "tasty," each to the matching 500 examples. Open MATLAB's "Classification Learner" app and load your table. Make sure the "fruit_type" column is set to be the "response" and cross-validation (5-fold is fine) is on. Your predictors should be the four cone activation columns.
6. Explore the scatter plots of your data. What do they tell you about which cones are best for which discriminations? Train a Linear Support Vector Machine (SVM). What accuracy did it achieve? Look at the confusion matrix. What kind of errors were made? Were any of the errors potentially deadly?
7. Export your trained model to the MATLAB workspace. This (rather complex) structure contains the weights the model used to make each of the binary classifications. To get them out, do the following: If your trained model structure is called "trainedModel," first run trainedModel.ClassificationSVM.ClassNames. This will give you the names of the classes in order. For example, {"edible," "tasty," "poison"}. Then run trainedModel.ClassificationSVM.BinaryLearners {1}.Beta. This returns the four cone weights for the first classification (edible vs. tasty in my example) ordered by their column order in the table.trainedModel.ClassificationSVM.BinaryLearners{2}. Beta returns the weights for the second classification (edible vs. poison) and trainedModel.ClassificationSVM.BinaryLearners{3}. Beta returns the weights for the last classification (tasty vs. poison). Compare the color channels the model fit for each of the three classifications.
8. We can weight the model to avoid deadly mistakes. Click on "Misclassification Costs" and change the values where the true class is "poison" but the predicted class was "edible" or "tasty" to 100. Retrain the model. What is the accuracy now? How does the confusion matrix change? How do the cone weights change?
9. Repeat the exercise using "sampleSpectra_highNoise." This data was generated from the same underlying spectra, but has considerably more noise. Does the noise look uniform across wavelength? How does this change the resulting model cone weights?
10. Compare the model cone weights to the results from PCA. In general, what has this exercise taught you about the role of behavioral constraints in shaping color channels and what you might miss by only analyzing the statistics of natural scenes?

References

Ahnelt, P.K., and Kolb, H. (2000). The mammalian photoreceptor mosaic-adaptive design. Progress in Retinal and Eye Research *19*(6), 711–777.

Backhaus, W.G.K., Kliegl, R., Werner, J.S., eds. 2011. Color Vision: Perspectives from Different Disciplines (De Gruyter).

Baden, T., and Osorio, D. (2019). The retinal basis of vertebrate color vision. Annual Review of Vision Science *5*, 177–200.

Baden, T. (2020). Vertebrate vision: lessons from non-model species. Seminars in Cell and Developmental Biology *106*, 1–4.

Baden, T., Euler, T., and Berens, P. (2020). Understanding the retinal basis of vision across species. Nature Reviews Neuroscience *21*(1), 5–20.

Baden, T., Nikolaev, A., Esposti, F., Dreosti, E., Odermatt, B., and Lagnado, L. (2014). A synaptic mechanism for temporal filtering of visual signals. PLoS Biology *12*(10), e1001972.

Baden, T., Schubert, T., Chang, L., Wei, T., Zaichuk, M., Wissinger, B., and Euler, T. (2013). A tale of two retinal domains: near-optimal sampling of achromatic contrasts in natural scenes through asymmetric photoreceptor distribution. Neuron *80*(5), 1206–1217.

Bae, J.A., Mu, S., Kim, J.S., Turner, N.L., Tartavull, I., Kemnitz, N., Jordan, C.S., Norton, A.D., Silversmith, W.M., Prentki, R., et al. (2018). Digital museum of retinal ganglion cells with dense anatomy and physiology. Cell *173*(5), 1293–1306.e19.

Behrens, C., Schubert, T., Haverkamp, S., Euler, T., and Berens, P. (2016). Connectivity map of bipolar cells and photoreceptors in the mouse retina. eLife *5*, e20041.

Bowmaker, J.K., Heath, L.A., Wilkie, S.E., and Hunt, D.M. (1997). Visual pigments and oil droplets from six classes of photoreceptor in the retinas of birds. Vision Research *37*(16), 2183–2194.

Breuninger, T., Puller, C., Haverkamp, S., and Euler, T. (2011). Chromatic bipolar cell pathways in the mouse retina. Journal of Neuroscience *31*(17), 6504–6517.

Buchsbaum, G., and Gottschalk, A. (1983). Trichromacy, opponent colours coding and optimum colour information transmission in the retina. Proceedings of the Royal Society of London B: Biological Sciences *220*(1218), 89–113.

Calkins, D.J. (2001). Seeing with S cones. Progress in Retinal and Eye Research *20*(3), 255–287.

Changizi, M.A., Zhang, Q., and Shimojo, S. (2006). Bare skin, blood and the evolution of primate colour vision. Biology Letters *2*(2), 217–221.

Chen, S., and Li, W. (2012). A color-coding amacrine cell may provide a blue-Off signal in a mammalian retina. Nature Neuroscience *15*(7), 954–956.

Collin, S.P., Davies, W.L., Hart, N.S., and Hunt, D.M. (2009). The evolution of early vertebrate photoreceptors. Philosophical Transactions of the Royal Society B: Biological Sciences *364*(1531), 2925–2940.

Connaughton, V.P., Graham, D., and Nelson, R. (2004). Identification and morphological classification of horizontal, bipolar, and amacrine cells within the zebrafish retina. Journal of Comparative Neurology *477*(4), 371–385.

Curcio, C.A., Allen, K.A., Sloan, K.R., Lerea, C.L., Hurley, J.B., Klock, I.B., and Milam, A.H. (1991). Distribution and morphology of human cone photoreceptors stained with anti-blue opsin. Journal of Comparative Neurology *312*(4), 610–624.

Dacey, D.M. (1993). Morphology of a small-field bistratified ganglion cell type in the macaque and human retina. Visual Neuroscience *10*(6), 1081–1098.

Dacey, D.M. (2000). Parallel pathways for spectral coding in primate retina. Annual Review of Neuroscience *23*, 743–775.

Dacey, D.M., and Lee, B.B. (1994). The "blue-on" opponent pathway in primate retina originates from a distinct bistratified ganglion cell type. Nature *367*(6465), 731–735.
Dacey, D.M., Lee, B.B., Stafford, D.K., Pokomy, J., and Smith, V.C. (1996). Horizontal cells of the primate retina: cone specificity without spectral opponency. Science *271*(5249), 656–659.
Dacey, D.M., and Packer, O.S. (2003). Colour coding in the primate retina: diverse cell types and cone-specific circuitry. Current Opinion in Neurobiology *13*(4), 421–427.
Dacey, D.M., Peterson, B.B., Robinson, F.R., and Gamlin, P.D. (2003). Fireworks in the primate retina: in vitro photodynamics reveals diverse LGN-projecting ganglion cell types. Neuron *37*, 15–27.
Dalton, B.E., Loew, E.R., Cronin, T.W., and Carleton, K.L. (2014). Spectral tuning by opsin coexpression in retinal regions that view different parts of the visual field. Proceedings of the Royal Society B: Biological Sciences *281*(1797), 1–9.
Denman, D.J., Luviano, J.A., Ollerenshaw, D.R., Cross, S., Williams, D., Buice, M.A., Olsen, S.R., and Reid, R.C. (2018). Mouse color and wavelength-specific luminance contrast sensitivity are non-uniform across visual space. eLife 7, e31209.
Denman, D.J., Siegle, J.H., Koch, C., Reid, R.C., and Blanche, T.J. (2017). Spatial organization of chromatic pathways in the mouse dorsal lateral geniculate nucleus. The Journal of Neuroscience *37*(5), 1102–1116.
Dominy, N.J., and Lucas, P.W. (2001). Ecological importance of trichromatic vision to primates. Nature *410*(6826), 363–366.
Dulai, K.S., Von Dornum, M., Mollon, J.D., and Hunt, D.M. (1999). The evolution of trichromatic color vision by opsin gene duplication in new world and old world primates. Genome Research *9*(7), 629–638.
Enright, J.M., Toomey, M.B., Sato, S.Y., Temple, S.E., Allen, J.R., Fujiwara, R., Kramlinger, V.M., Nagy, L.D., Johnson, K.M., Xiao, Y., et al. (2015). Cyp27c1 red-shifts the spectral sensitivity of photoreceptors by converting Vitamin A1 into A2. Current Biology *25*(23), 3048–3057.
Field, G.D., Gauthier, J.L., Sher, A., Greschner, M., MacHado, T.A., Jepson, L.H., Shlens, J., Gunning, D.E., Mathieson, K., Dabrowski, W., et al. (2010). Functional connectivity in the retina at the resolution of photoreceptors. Nature *467*(7316), 673–677.
Gegenfurtner, K.R., and Kiper, D.C. (2003). Color vision. Annual Review of Neuroscience *26*, 181–206.
Gegenfurtner, K.R., and Rieger, J. (2000). Sensory and cognitive contributions of color to the recognition of natural scenes. Current Biology *10*(13), 805–808.
Gegenfurtner, K.R., Sharpe, L.T., eds. 1999. Color Vision: From Genes to Perception (Cambridge University Press).
Ghosh, K.K., and Grünert, U. (1999). Synaptic input to small bistratified (blue-ON) ganglion cells in the retina of a new world monkey, the marmoset *Callithrix jacchus*. Journal of Comparative Neurology *413*(3), 417–428.
Goldsmith, T.H., and Butler, B.K. (2003). The roles of receptor noise and cone oil droplets in the photopic spectral sensitivity of the budgerigar, *Melopsittacus undulatus*. Journal of Comparative Physiology A: Neuroethology, Sensory, Neural, and Behavioral Physiology *189*(2), 135–142.
Gouras, P. (2009). Color Vision. In Encyclopedia of Neuroscience (Academic Press).
Govardovskii, V.I., Fyhrquist, N., Reuter, T., Kuzmin, D.G., and Donner, K. (2000). In search of the visual pigment template. Visual Neuroscience *17*(4), 509–528.
Hailman, J.P. (1976). Oildroplets in the eyes of adult anuran amphibians: a comparative survey. Journal of Morphology *148*(4), 453–468.

Hart, N.S., and Vorobyev, M. (2005). Modelling oil droplet absorption spectra and spectral sensitivities of bird cone photoreceptors. Journal of Comparative Physiology A: Neuroethology, Sensory, Neural, and Behavioral Physiology *191*(4), 381–392.

Haverkamp, S., Wässle, H., Duebel, J., Kuner, T., Augustine, G.J., Feng, G., and Euler, T. (2005). The primordial, blue-cone color system of the mouse retina. Journal of Neuroscience *25*(22), 5438–5445.

Helmstaedter, M., Briggman, K.L., Turaga, S.C., Jain, V., Seung, H.S., and Denk, W. (2013). Connectomic reconstruction of the inner plexiform layer in the mouse retina. Nature *500*(7461), 168–174.

Hofer, H., Carroll, J., Neitz, J., Neitz, M., and Williams, D.R. (2005). Organization of the human trichromatic cone mosaic. The Journal of Neuroscience *25*(42), 9669–9679.

Hubel, D.H., and Wiesel, T.N. (1960). Receptive fields of optic nerve fibres in the spider monkey. The Journal of Physiology *154*(3), 572–580.

Ishida, A.T., Stell, W.K., and Lightfoot, D.O. (1980). Rod and cone inputs to bipolar cells in goldfish retina. Journal of Comparative Neurology *191*(3), 315–335.

Jacobs, G.H., Williams, G.A., Cahill, H., and Nathans, J. (2007). Emergence of novel color vision in mice engineered to express a human cone photopigment. Science *315*(5819), 1723–1725.

Joesch, M., and Meister, M. (2016). A neuronal circuit for colour vision based on rod-cone opponency. Nature *532*(7598), 236–239.

Joo, H.R., Peterson, B.B., Haun, T.J., and Dacey, D.M. (2011). Characterization of a novel large-field cone bipolar cell type in the primate retina: evidence for selective cone connections. Visual Neuroscience *28*(1), 29–37.

Klug, K., Herr, S., Ngo, I.T., Sterling, P., and Schein, S. (2003). Macaque retina contains an S-cone OFF midget pathway. Journal of Neuroscience *23*(30), 9881–9887.

Kolb, H., and Marshak, D. (2003). The midget pathways of the primate retina. Documenta Ophthalmologica *106*(1), 67–81.

Korenyak, D.A., and Govardovskii, V.I. (2013). Photoreceptors and visual pigments in three species of newts. Journal of Evolutionary Biochemistry and Physiology *49*(4), 399–407.

Kouyama, N., and Marshak, D.W. (1992). Bipolar cells specific for blue cones in the macaque retina. Journal of Neuroscience *12*(4), 1233–1252.

Kremers, J., Baraas, R.C., Marshall, N.J., eds. 2016. Human Color Vision. (Springer International Publishing).

Kryger, Z., Galli-Resta, L., Jacobs, G.H., and Reese, B.E. (1998). The topography of rod and cone photoreceptors in the retina of the ground squirrel. Visual Neuroscience *15*(4), 685–691.

Lennie, P. (2000). Color vision: putting it together. Current Biology *10*(16), 589–591.

Li, W., and DeVries, S.H. (2006). Bipolar cell pathways for color and luminance vision in a dichromatic mammalian retina. Nature Neuroscience *9*(5), 669–675.

Li, Y.N., Matsui, J.I., and Dowling, J.E. (2009). Specificity of the horizontal cell-photoreceptor connections in the zebrafish (*Danio rerio*) retina. Journal of Comparative Neurology *516*(5), 442–453.

Li, Y.N., Tsujimura, T., Kawamura, S., and Dowling, J.E. (2012). Bipolar cell-photoreceptor connectivity in the zebrafish (*Danio rerio*) retina. Journal of Comparative Neurology *520*(16), 3786–3802.

Lind, O., Henze, M.J., Kelber, A., and Osorio, D. (2017). Coevolution of coloration and colour vision? Philosophical Transactions of the Royal Society B: Biological Sciences. *372*(1724), 20160338.

Loew, E.R., and Govardovskii, V.I. (2001). Photoreceptors and visual pigments in the red-eared turtle, *Trachemys scripta elegans*. Visual Neuroscience *18*(5), 753–757.

Loew, E.R., Fleishman, L.J., Foster, R.G., and Provencio, I. (2002). Visual pigments and oil droplets in diurnal lizards: a comparative study of Caribbean anoles. Journal of Experimental Biology *205*(7), 927–938.

Lucas, P.W., Dominy, N.J., Riba-Hernandez, P., Stoner, K.E., Yamashita, N., Loría-Calderón, E., Petersen-Pereira, W., Rojas-Durán, Y., Salas-Pena, R., Solis-Madrigal, S., et al. (2003). Evolution and function of routine trichromatic vision in primates. Evolution *57*(11), 2636–2643.

Mancuso, K., Hauswirth, W.W., Li, Q., Connor, T.B., Kuchenbecker, J.A., Mauck, M.C., Neitz, J., and Neitz, M. (2009). Gene therapy for red-green colour blindness in adult primates. Nature *461*(7265), 784–787.

Marc, R.E. (1982). Chromatic Organization of the Retina. In McDevitt, D.S., ed. Cell Biology of the Eye. (Academic Press, New York), pp. 435–473.

Marc, R.E., Jones, B.W., Watt, C.B., Anderson, J.R., Sigulinsky, C., and Lauritzen, S. (2013). Retinal connectomics: towards complete, accurate networks. Progress in Retinal and Eye Research *37*, 141–162.

Mariani, A.P. (1984). Bipolar cells in monkey retina selective for the cones likely to be blue-sensitive. Nature *308*(5955), 184–186.

Martin, P.R., Lee, B.B., White, A.J.R., Solomon, S.G., and Rüttiger, L. (2001). Chromatic sensitivity of ganglion cells in the peripheral primate retina. Nature *410*(6831), 933–936.

Mills, S.L., Tian, L.M., Hoshi, H., Whitaker, C.M., and Massey, S.C. (2014). Three distinct blue-green color pathways in a mammalian retina. Journal of Neuroscience *34*(5), 1760–1768.

Mollon, J.D. (1982). Color vision. Annual Review of Psychology *33*, 41–85.

Mollon, J.D. (2003). The Origins of Modern Color Science. In Shevell, S.K., ed. (2nd ed.) The Science of Color (Elsevier), pp. 1–39.

Mollon, J.D., Bowmaker, J.K., and Jacobs, G.H. (1984). Variations of colour vision in a New World primate can be explained by polymorphism of retinal photopigments. Proceedings of the Royal Society of London B: Biological Sciences *222*(1228), 373–399.

Mouland, J.W., Pienaar, A., Williams, C., Watson, A.J., Lucas, R.J., and Brown, T.M. (2021). Extensive cone-dependent spectral opponency within a discrete zone of the lateral geniculate nucleus supporting mouse color vision. Current Biology. https://doi.org/10.1016/j.cub.2021.05.024.

Mullen, K.T., and Kingdom, F.A.A. (1996). Losses in peripheral colour sensitivity predicted from "hit and miss" post-receptoral cone connections. Vision Research *36*(13), 1995–2000.

Nadal-Nicolás, F.M., Kunze, V.P., Ball, J.M., Peng, B.T., Krisnan, A., Zhou, G., Dong, L., and Li, W. (2020). True S-cones are concentrated in the ventral mouse retina and wired for color detection in the upper visual field. eLife. *9*, e56840.

Nathan, J., Reh, R., Ankoudinova, I., Ankoudinova, G., Chang, B., Heckenlively, J., and Hurley, J.B. (2006). Scotopic and photopic visual thresholds and spatial and temporal discrimination evaluated by behavior of mice in a water maze. Photochemistry and Photobiology *82*(6), 1489.

Nathans, J., Thomas, D., and Hogness, D.S. (1986). Molecular genetics of human color vision: the genes encoding blue, green, and red pigments. Science *232*(4747), 196–202.

Neitz, M., and Neitz, J. (2014). Curing color blindness—mice and nonhuman primates. Cold Spring Harbor Perspectives in Medicine *4*(11), a017418.

Osorio, D., and Vorobyev, M. (1996). Colour vision as an adaptation to frugivory in primates. Proceedings of the Royal Society B: Biological Sciences *263*(1370), 593–599.

Packer, O.S., and Dacey, D.M. (2002). Receptive field structure of H1 horizontal cells in macaque monkey retina. Journal of Vision *2*(4), 272–292.

Packer, O.S., Verweij, J., Li, P.H., Schnapf, J.L., and Dacey, D.M. (2010). Blue-yellow opponency in primate S cone photoreceptors. Journal of Neuroscience *30*(2), 568–572.

Patterson, S.S., Kuchenbecker, J.A., Anderson, J.R., Bordt, A.S., Marshak, D.W., Neitz, M., and Neitz, J. (2019). An S-cone circuit for edge detection in the primate retina. Scientific Reports *9*(1), 11913.

Paulus, W., and KrÖger-Paulus, A. (1983). A new concept of retinal colour coding. Vision Research *23*(5), 529–540.

Peichl, L. (2005). Diversity of mammalian photoreceptor properties: adaptations to habitat and lifestyle? The Anatomical Record Part A: Discoveries in Molecular, Cellular, and Evolutionary Biology *287*(1), 1001–1012.

Pessoa, D.M.A., Maia, R., de Albuquerque Ajuz, R.C., De Moraes, P.Z.P.M.R., Spyrides, M.H.C., and Pessoa, V.F. (2014). The adaptive value of primate color vision for predator detection. American Journal of Primatology *76*(8), 721–729.

Reitner, A., Sharpe, L.T., and Zrenner, E. (1991). Is colour vision possible with only rods and blue-sensitive cones? Nature *352*(6338), 798–800.

Rhim, I., Coello-Reyes, G., and Nauhaus, I. (2020). Variations in photoreceptor throughput to mouse visual cortex and the unique effects on tuning. BioRxiv. https://doi.org/10.1101/2020.11.03.366682.

Rocha, F.A.F., Saito, C.A., Silveira, L.C.L., De Souza, J.M., and Ventura, D.F. (2008). Twelve chromatically opponent ganglion cell types in turtle retina. Visual Neuroscience *25*(3), 307–315.

Ruderman, D.L., Cronin, T.W., and Chiao, C.-C. (1998). Statistics of cone responses to natural images: implications for visual coding. Journal of the Optical Society of America A *15*(8), 2036.

Seifert, M., Baden, T., and Osorio, D. (2020). The retinal basis of vision in chicken. Seminars in Cell and Developmental Biology *106*, 106–115.

Shekhar, K., Lapan, S.W., Whitney, I.E., Tran, N.M., Macosko, E.Z., Kowalczyk, M., Adiconis, X., Levin, J.Z., Nemesh, J., Goldman, M., et al. (2016). Comprehensive classification of retinal bipolar neurons by single-cell transcriptomics. Cell *166*(5), 1308–1323.e30.

Sher, A., and Devries, S.H. (2012). A non-canonical pathway for mammalian blue-green color vision. Nature Neuroscience *15*(7), 952–953.

Sherry, D.M., and Yazulla, S. (1993). Goldfish bipolar cells and axon terminal patterns: a Golgi study. Journal of Comparative Neurology *329*(2), 188–200.

Shevell, S.K., ed. 2003. The Science of Color (2nd ed.) (Elsevier).

Shimbo, K., Toyoda, J.I., Kondo, H., and Kujiraoka, T. (2000). Color-opponent responses of small and giant bipolar cells in the carp retina. Visual Neuroscience *17* (4), 609–621.

Slaughter, M.M., and Miller, R.F. (1981). 2-Amino-4-phosphonobutyric acid: a new pharmacological tool for retina research. Science *211*(4478), 182–185.

Stockman, A., and Brainard, D.H. (2010). Color Vision Mechanisms. In Bass, M., ed. (3rd ed.) Handbook of Optics Volume III: Vision and Vision Optics. (The McGraw-Hill Companies, Inc), pp. 11.1–11.104.

Szatko, K.P., Korympidou, M.M., Ran, Y., Berens, P., Dalkara, D., Schubert, T., Euler, T., and Franke, K. (2020). Neural circuits in the mouse retina support color vision in the upper visual field. Nature Communications *11*(1), 3481.

Tikidji-Hamburyan, A., Reinhard, K., Storchi, R., Dietter, J., Seitter, H., Davis, K.E., Idrees, S., Mutter, M., Walmsley, L., Bedford, R.A., et al. (2017). Rods progressively escape saturation to drive visual responses in daylight conditions. Nature Communications *8*(1), 1813.

Twig, G., Levy, H., and Perlman, I. (2003). Color opponency in horizontal cells of the vertebrate retina. Progress in Retinal and Eye Research *22*(1), 31–68.

Vorobyev, M., Osorio, D., Bennett, A.T.D., Marshall, N.J., and Cuthill, I.C. (1998). Tetrachromacy, oil droplets and bird plumage colours. Journal of Comparative Physiology A *183*(5), 621–633.

Waitt, C., Little, A.C., Wolfensohn, S., Honess, P., Brown, A.P., Buchanan-Smith, H.M., and Perrett, D.I. (2003). Evidence from rhesus macaques suggests that male coloration plays a role in female primate mate choice. Proceedings of the Royal Society B: Biological Sciences *270*(Suppl 2), S144–S146.

Wichmann, F.A., Sharpe, L.T., and Gegenfurtner, K.R. (2002). The contributions of color to recognition memory for natural scenes. Journal of Experimental Psychology: Learning, Memory, and Cognition *28*(3), 509–520.

Wilby, D., and Roberts, N.W. (2017). Optical influence of oil droplets on cone photoreceptor sensitivity. Journal of Experimental Biology *220*(11), 1997–2004.

Yan, W., Laboulaye, M.A., Tran, N.M., Whitney, I.E., Benhar, I., and Sanes, J.R. (2020). Mouse retinal cell atlas: molecular identification of over sixty amacrine cell types. Journal of Neuroscience *40*(27), 5177–5195.

Yin, L., Smith, R.G., Sterling, P., and Brainard, D.H. (2009). Physiology and morphology of color-opponent ganglion cells in a retina expressing a dual gradient of S and M opsins. Journal of Neuroscience *29*(9), 2706–2724.

Yoshimatsu, T., Bartel, P., Schröder, C., Janiak, F.K., St-Pierre, F., Berens, P., and Baden, T. (2020a). Near-optimal rotation of colour space by zebrafish cones in vivo. BioRxiv. 2020.10.26.356089.

Yoshimatsu, T., Schröder, C., Nevala, N.E., Berens, P., and Baden, T. (2020b). Fovea-like photoreceptor specializations underlie single UV cone driven prey-capture behavior in zebrafish. Neuron *107*(2), 320–337.e6.

Yovanovich, C.A.M., Koskela, S.M., Nevala, N., Kondrashev, S.L., Kelber, A., and Donner, K. (2017). The dual rod system of amphibians supports colour discrimination at the absolute visual threshold. Philosophical Transactions of the Royal Society B: Biological Sciences *372*(1717), 20160066.

Zhou, M., Bear, J., Roberts, P.A., Janiak, F.K., Semmelhack, J., Yoshimatsu, T., and Baden, T. (2020). Zebrafish retinal ganglion cells asymmetrically encode spectral and temporal information across visual space. Current Biology *30*(15), 2927–2942.e7.

Zimmermann, M.J.Y., Nevala, N.E., Yoshimatsu, T., Osorio, D., Nilsson, D.E., Berens, P., and Baden, T. (2018). Zebrafish differentially process color across visual space to match natural scenes. Current Biology *28*(13), 2018–2032.e5.

Index

Note: Page numbers followed by *f* indicate figures, *t* indicate tables, and *b* indicate boxes.

Kolibri 360 GmbH
Ferdinand-Jühlke-Straße 7
99095 Erfurt, Deutschland
produktsicherheit@kolibri360.de